Frequency-Agile Antennas for Wireless Communications

For a complete listing of titles in the
Artech House Antennas and Propagation Series,
turn to the back of this book.

Frequency-Agile Antennas for Wireless Communications

Aldo Petosa

ARTECH HOUSE
BOSTON | LONDON
artechhouse.com

Library of Congress Cataloging-in-Publication Data
A catalog record for this book is available from the U.S. Library of Congress.

British Library Cataloguing in Publication Data
A catalogue record for this book is available from the British Library.

Cover design by Vicki Kane

ISBN 13: 978-1-60807-768-7

685 Canton Street
Norwood, MA 02062

10 9 8 7 6 5 4 3 2 1

Contents

Preface

As the result of the ever-increasing demand for wireless services due to the proliferation of such portable devices as smart phones and tablets, a large increase in wireless capacity will be required within the next few years. Many of the proposed solutions to deal with these demands will have a significant impact on antenna designs. Antennas with frequency agility are currently being considered as a promising technology to help implement these new solutions.

The intent of this book is to provide the reader with a sense of the capabilities of frequency-agile antennas (FAAs), the widely diverse methods for achieving tunability, the current achievable performance, and the challenges still facing FAA designs. The book explores the many aspects of FAAs, including the metrics used to evaluate their performance, the most commonly used antenna elements, the wide variety of mechanisms for achieving tunability, and diverse examples of FAA designs. The focus is on FAAs for wireless mobile communications with applications including handsets, laptops, and wireless machine-to-machine communications, as well as larger fixed designs, such as for cellular base station antennas.

Acknowledgments

This book is the by-product of research activities on FAAs that were carried out over the last several years at the Communications Research Centre Canada (CRC). As such, I would like to acknowledge several past and current colleagues and collaborators who, in various ways, have helped make the writing of this book possible. First of all, I owe a debt of gratitude to Michel Cuhaci, the recently retired research manager of the Advanced Antenna Technology Lab at CRC. As manager, Michel allowed his staff a wide latitude to select and carry out research on a broad range of topics, and he was supportive of the initiatives and of the activities undertaken on FAAs. Much of the information gathered during the background investigation carried out throughout the course of the research activities on FAAs has found its way into the book.

I would further like to recognize my former colleague Soulideth Thirakoune and collaborators Jason Desjardins and Dr. Derek McNamara for their important contributions to the research and development of frequency-tunable dielectric resonator antennas, one of the latest types of FAAs to be investigated.

I would also like to acknowledge my colleagues of the former Advanced Antenna Technology Lab and of the recently formed RF Technologies group at CRC, for their support, advice, and friendship over the years.

Finally, I wish to thank the staff at Artech House for their enthusiastic support and assistance throughout the various stages of development of this book.

Chapter 1

Introduction

It is expected that mobile data subscriptions will more than double and that mobile wireless traffic will increase by more than tenfold over the next few years. The proliferation of smart phones, tablets, laptops, and other portable devices is placing greater demands for services such as Web browsing, global positioning, video streaming, and video telephony, necessitating an expected increase in wireless capacity by a factor of 1,000 within the next few years. Methods for increasing the needed capacity include densification (the use of an increased number of smaller cells), improved efficiency of spectrum usage [by using techniques such as white space exploitation, incorporation of cognitive radio, interference mitigation techniques, or multiple-input multiple-output (MIMO) systems], and the use of new spectrum bands. Many of these proposed solutions will have a significant impact on antenna design.

With different frequency allocations for different services (see Table 1.1), mobile wireless devices are required to operate in two or more bands simultaneously, necessitating the need for wideband, multiband, or multiple antennas. Such antennas can be designed on low-cost printed circuit boards and can provide an inexpensive solution, which is of prime consideration for the consumer market. However, the small size of handsets and many portable devices places constraints on the dimensions and form factors of these antennas. For many portable wireless communication applications, the antennas need to be electrically small in order to be properly integrated within the devices. These electrically small antennas are governed by fundamental gain-bandwidth limitations that prevent them from simultaneously having both high radiation efficiency and wide bandwidth. Wideband antennas will also require filtering in order to remove unwanted signals in adjacent channels and to improve signal-to-noise ratio, which introduces losses, requires additional real estate on the circuit board, and adds to the overall device cost.

By designing antennas with frequency agility, small antennas with narrow instantaneous bandwidths can be tuned to operate over a larger range of frequencies, thus effectively enhancing bandwidth performance. These FAAs would not require the same level of filtering circuitry in the radio frequency (RF) front end as wideband antennas and would potentially have better efficiency. FAAs could also be used to optimize dynamically any detuning due to proximity effects of handsets by hand or head. They could also be used with adaptive MIMO systems, resulting in potentially higher performance improvement by joint optimization of the dynamically tunable antenna with adaptive space-time modulation techniques [1]. There is thus an emerging need for small, compact antennas with frequency agility. FAAs do, however, come with challenges, including increased complexity, higher power consumption, nonlinearities introduced by the tuning devices, RF power-handling limitations, and increased cost over passive wideband or multiband antennas.

Table 1.1

List of Selected Wireless Services

Service	Acronym	Frequency (MHz)
Advanced wireless service	AWS-1	1,710-1,755/2,110-2,155
Bluetooth	-	2,400-2,483.5
Digital communications systems	DCS1800	1,710-1,785/1,805-1,880
Digital enhanced cordless telecommunications	DECT	1,800-1,897
Digital video broadcasting-Handheld	DVB-H	170-230/470-862/1,452-1,492
FCC unlicensed band	-	57,000-64,000
Global System for Mobile Communications	GSM850	824-894
	GSM900	880-960
Industrial scientific and medical band	ISM	2,000-3,000
Long-term evolution	LTE	698-806
Personal communication services	PCS1900	1,850-1,910/1,930-1,990
Public safety bands	-	150/400/700/800/900/4,900
Terrestrial digital multimedia broadcasting	T-DMB	174-216
Universal mobile telecommunications system	UMTS	1,920-2,200
Wideband code division multiple access	WCMDA	1,920-2,180
Wireless broadband internet	WiBro	2,300-2,400
Worldwide interoperability for microwave access	WiMax	3,400-3,600
Wireless local area network	WLAN	2,400-2,484 (802.11b&g) 5,150-5,350 (802.11a indoor) 5,725-5,825 (802.11a outdoor)
Zigbee	-	688-686.8 (Europe) 902-928 (United States) 2,400-2,834 (world)

1.1 DEFINITION OF FAAs

Before proceeding further, the term FAA should be clarified in the context of this book. Referring to Figure 1.1, an FAA can be considered as belonging to the larger class of antennas known as active integrated antennas (AIAs). The latter are antennas that have one or more active devices integrated within or connected to the antenna. Active antennas made their appearance in the 1960s and 1970s with the advent of high-frequency microwave transistors. This continued into the 1980s and 1990s where the integration of such active devices as transistors, oscillators, mixers, and multipliers was used for applications including spatial power combining, grid oscillators, frequency converters, transceivers, and beam-scanning arrays [2]. In 1999, the U.S. Defense Advanced Research Projects Agency (DARPA) initiated the Reconfigurable Aperture Program (RECAP), funding research on antennas whose beam patterns, polarization, or frequency could be dynamically reconfigured by use of some form of active device integrated into the antenna. These reconfigurable antennas are different from phased arrays, where the active devices (amplifiers and phase shifters) are packaged in transceivers connected to the antennas or are integrated in external power divider networks, and not in the antenna elements themselves.

FAAs can be considered the subset of reconfigurable antennas that are designed to have a dynamically tunable frequency. For the purposes of this text, FAAs are distinguished from an antenna system comprising a tunable matching circuit, as shown in Figure 1.2(a). This text will focus on antennas of the form of Figure 1.2(b), where the active devices are integrated within the antenna element itself (or directly connected nearby, but not connected to the matching network). The two configurations shown in Figure 1.2 both lead to frequency agility, and there are advantages and disadvantages to each approach. Placing the active devices in the matching network ensures that these devices do not significantly interfere with the radiation pattern. There is often more room, and it is usually less complicated to integrate devices within the matching network rather than within the antenna element. However, embedding devices within the antenna may result in lower insertion losses and offer a more compact design, which is important in many handheld and mobile wireless devices. Since the principles of tunable matching networks are fairly well established, this book is dedicated to the tunable antennas of the form of Figure 1.2(b), which are designated as FAAs.

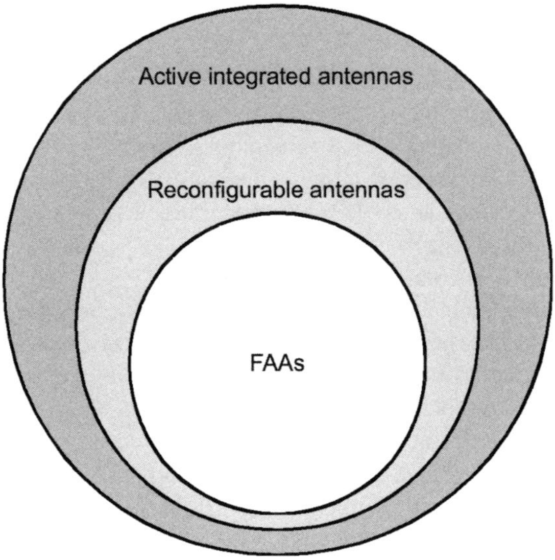

Figure 1.1 Active integrated antennas.

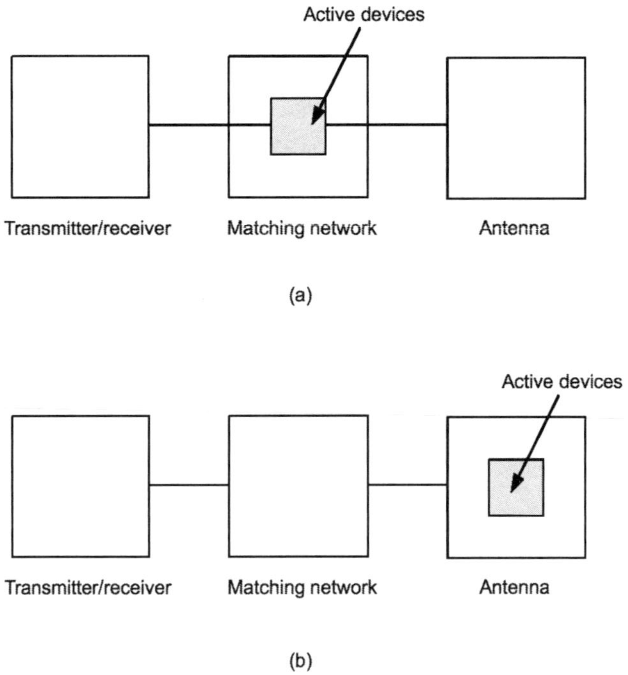

Figure 1.2 Difference between (a) a tunable matching network and (b) an FAA.

1.2 OUTLINE

The intent of this book is to provide the reader with a sense of the capabilities of FAAs, the widely diverse methods for achieving tunability, the current achievable performance, and the challenges still facing FAA designs. The book explores the many aspects of FAAs, including the metrics used to evaluate their performance, the most commonly used antenna elements, the wide variety of mechanisms for achieving tunability, and a comprehensive survey of diverse examples of FAA designs. The focus is on FAAs for wireless mobile communications with applications including handsets, tablets, laptops, and wireless machine-to-machine communications (such as computer mice, printers, and other peripherals), as well as larger, fixed designs such as for cellular base station antennas. Many of the FAAs surveyed in this book were designed to operate in one or more of the services listed in Table 1.1.

The book is divided into the following chapters. Chapter 2 defines the various parameters used to quantify the performance of FAAs. Included are traditional quantities applicable to all antennas (directivity, gain, radiation efficiency, polarization, and bandwidth) along with parameters exclusive to FAAs (tuning mode, tuning range, total spectrum, tunable bandwidth, and tuning efficiency). Metrics for characterizing the performance of a particular tuning technique, such as DC power consumption, maximum bias level, tuning speed, maximum RF power handling, and bias circuit complexity, are considered. Chapter 2 also introduces a classification system for FAAs that is then adopted throughout the book.

Chapter 3 presents an overview of the most common antenna elements suitable for FAAs for wireless communication applications. These are typically low-gain antennas including dipoles, monopoles, printed antennas, slot antennas, planar inverted-F antennas (PIFAs), and dielectric resonator antennas (DRAs). Chapter 3 also contains a review of the fundamentals of operation, along with a basic analysis and the design equations for each antenna type.

Chapter 4 examines the various tuning methods being considered for FAAs. These methods are grouped into three categories: mechanical actuation, tunable substrates, and integrated electronic devices. An FAA in the mechanical actuation class undergoes a physical change in one or more of its dimensions, resulting in a shift in the resonant frequency. Various types of actuators are highlighted such as electromechanical, piezoelectric, hydraulic, and pneumatic systems, and the key metrics for evaluating their performance are reviewed. FAAs in the second category incorporate materials such as ferrites, ferroelectrics, or semiconductors whose permeability, permittivity, or conductivity can be altered by a bias field, which, in turn, causes a change in the resonant frequency. A brief review of the physics of how these materials behave under applied bias fields is included. The third class of FAAs makes use of discrete electronic devices, such as PIN diodes, varactors, and micromechanical systems (MEMS) switches, that are used to load the antenna and cause a shift in resonance. The modes of operation of these electronic devices are described, and simple circuit models are presented to help predict performance behavior.

Chapters 5–8 contain numerous examples of a wide variety of FAA designs and are organized as follows. Chapter 5 examines FAAs based on mechanical actuation. Antenna types examined include microstrip antennas, PIFAs, helical antennas, slot dipoles, and DRAs. The types of actuators used include electrostatic, piezoelectric, magnetostatic, hydraulic, and pneumatic.

Chapter 6 looks at FAAs incorporating tunable substrates. Ferrites, ferroelectrics, liquid crystals, and semiconductor substrates combined with antenna elements such as microstrip patches, cavities, DRAs, and coplanar slot antennas are reviewed.

Chapter 7 covers the use of electronic devices for achieving FAA designs with continuous frequency tuning. The electronic devices are divided into four categories: diodes, transistors, MEMS, and varactors. These devices are integrated into various antenna elements including slot antennas, PIFAs, coplanar waveguide (CPW) antennas, and DRAs, as well as several types of printed antennas including loops, dipoles, patches, rings, monopoles, and helices.

Chapter 8 reviews FAAs designed for discrete frequency tuning. Electronic devices used include diodes, transistors, MEMS, and optical switches. These devices have been incorporated into antenna elements such as slot antennas, PIFAs, CPW antennas, DRAs, and a variety of printed antennas. Chapter 8 concludes with a few examples of hybrid FAA designs, combining both discrete and continuous tuning modes.

References

[1] Cetiner, B. A., et al., "A MIMO System with Multifunctional Reconfigurable Antennas," *IEEE Antennas and Wireless Propagation Letters,* Vol. 5, December 2006, pp. 463-466.

[2] Lin, J., and T. Itoh, "Active Integrated Antennas," *IEEE Transactions on Microwave Theory and Techniques,* Vol. 42, No. 12, December 1994, pp. 2186-2194.

Chapter 2

Characterizing FAAs

Antennas, in general, can be classified by a variety of attributes. The typical classification is based either on an antenna's physical traits (such as wire, printed, or aperture), on its electromagnetic behavior (resonant or nonresonant), or on its radiation characteristics (whether it is omnidirectional, has low or high gain, and is linearly or circularly polarized). All antennas have a set of common parameters that can be used to measure or characterize their performance. FAAs require an additional set of metrics to quantify the added tuning capability both in terms of electromagnetic behavior of the antenna and on the performance of the tuning mechanism itself. This chapter examines the various quantities used to characterize FAAs. Section 2.1 reviews the major parameters that quantify the performance of all antennas. Section 2.2 then presents the metrics used specifically for FAAs. Finally, Section 2.3 introduces a classification of FAAs based on the type of tuning mechanism incorporated. This classification is used throughout the book to categorize the various FAAs.

2.1 GENERAL ANTENNA PARAMETERS

This section reviews the most commonly used parameters for characterizing the electromagnetic performance of an antenna. The radiation performance of an antenna is measured by such parameters as directivity, gain, half-power beamwidth, sidelobe levels (SLLs), polarization, and radiation efficiency. The circuit performance of an antenna is characterized by the input impedance, the reflection coefficient, the voltage standing wave ratio (VSWR), the return loss, and the mismatch loss. The bandwidth of an antenna is also an important parameter, and can be defined either in terms of its circuit performance or its radiation performance.

2.1.1 Radiation Pattern

The radiation pattern of an antenna is a three-dimensional (3D) graphical representation of the antenna's radiation intensity (radiated power per unit solid angle), measured in the far field of the antenna. The minimum far-field distance, R, is usually given by

$$R = \frac{2d^2}{\lambda_o} \tag{2.1}$$

where d is the largest dimension of the antenna aperture, and λ_o is the operating wavelength in free space. At and beyond the far-field distance, the radiated fields from the antenna essentially do not have any radial component, with the electric and magnetic fields being orthogonal to each other and to the direction of propagation (i.e., the electromagnetic fields exhibit plane-wave behavior), and they can be expressed solely as a function of the spherical coordinates (θ,ϕ), as shown in Figure 2.1.

In theory, the shapes of the radiation patterns that various antennas can produce are almost limitless. In practice, however, there are a finite number of useful pattern shapes that most practical antennas will produce. Electrically small antennas (ones whose dimensions can be enclosed in a sphere of radius $r = \lambda_o/2\pi$) produce nearly isotropic (spherical) patterns. Linear antennas (such as dipoles) and some wire loop antennas produce omnidirectional (donut-shaped) patterns. Low-gain antennas (such as microstrip patches and dielectric resonator antennas) produce quasi-hemispherical patterns. Other antennas (such as helices, horns, and reflectors) produce directive (pencil-beam-shaped) patterns, similar to the one shown in Figure 2.1. Chapter 3 will examine the typical radiation patterns of specific antennas in more detail.

Although radiation patterns are 3D, it is often more convenient to represent them in two-dimensional (2D) slices, termed pattern cuts. The pattern cut is usually generated by keeping one of the two spherical coordinates (θ,ϕ) constant, while varying the other coordinate over 360°. Figure 2.2 is a polar plot of a normalized directive radiation pattern (with the peak value set to 0 dB) showing pattern cuts in two orthogonal planes, defined in this example by $\phi = 0°$ and $\phi = 90°$. If any two such orthogonal planes also each pass through the peak of the antenna radiation pattern, these planes are referred to as principal planes. In many cases, plotting the 2D patterns in two principal planes is sufficient to obtain most of the information required to characterize a radiation pattern.

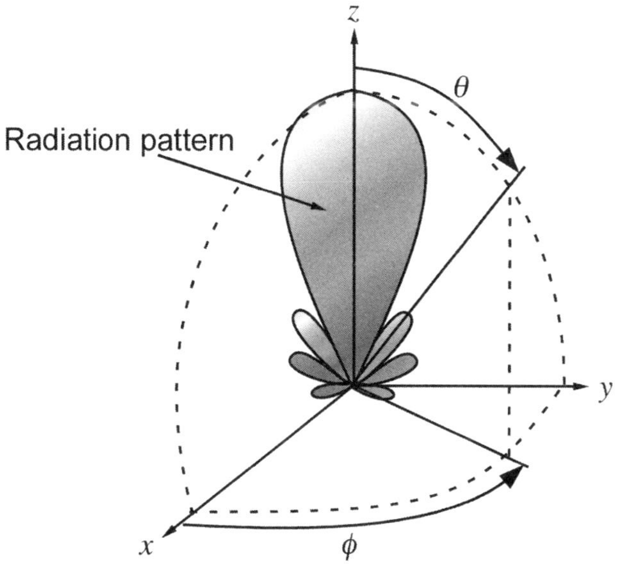

Figure 2.1 Spherical coordinate system for describing radiation patterns.

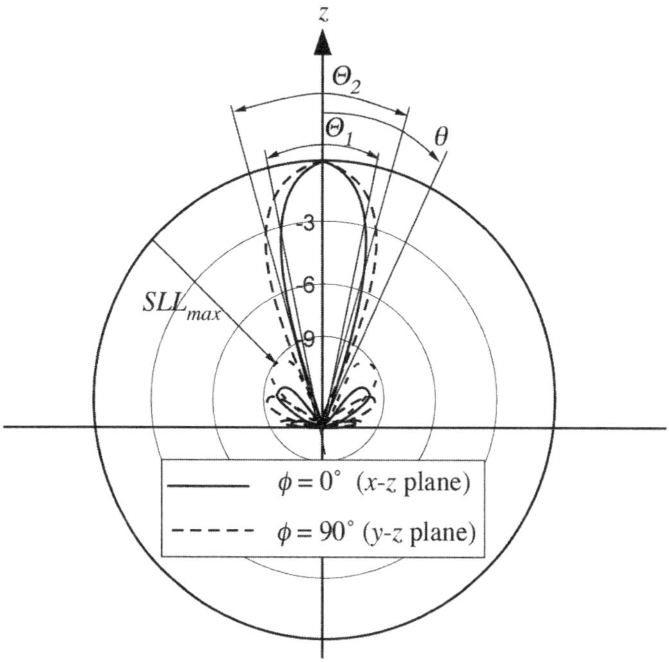

Figure 2.2 Two-dimensional polar plot of a directive radiation pattern.

The main beam of a directive pattern can be characterized by its principal-plane half-power beamwidths (labeled Θ_1 and Θ_2 in Figure 2.2), which represent the angular range over which the radiation intensity is equal to or greater than half (or -3 dB) of the peak value. Another important parameter is the maximum SLL (SLL_{max}), which indicates the maximum value of the minor lobes of the pattern.

2.1.2 Directivity

The directivity of an antenna is a measure of how well its radiation pattern is concentrated. The more concentrated, or the narrower the radiation beam, the higher the directivity. The directivity of an antenna can be expressed as the ratio of its radiation intensity to that of an isotropic antenna (a fictitious ideal antenna that radiates a uniform spherical-shaped pattern). The radiation intensity U_o of the isotropic antenna is given by

$$U_o = \frac{P_{rad}}{4\pi} \tag{2.2}$$

where P_{rad} is the total power radiated by the isotropic antenna, and U_o has units of watts per solid angle. The directivity $D(\theta,\phi)$ of an antenna can be defined as

$$D(\theta,\phi) = \frac{U(\theta,\phi)}{U_o} = \frac{4\pi\, U(\theta,\phi)}{P_{rad}} \tag{2.3}$$

where $U(\theta,\phi)$ is the radiation intensity of the antenna. Although directivity is a function of the angular spherical coordinates (θ,ϕ), it is common to characterize the antenna by a single value equal to the maximum value of (2.3). This value is commonly expressed in decibels. The total radiated power P_{rad} of the antenna is obtained by integration of its radiation intensity over an enclosed surface, usually taken to be a sphere. In spherical coordinates

$$P_{rad} = \int_0^{2\pi}\int_0^{\pi} U(\theta,\phi)\sin\theta\, d\theta\, d\phi \tag{2.4}$$

Substituting (2.4) into (2.3), the directivity can be determined by solving

$$D(\theta,\phi) = \frac{4\pi\, U(\theta,\phi)}{\int_0^{2\pi}\int_0^{\pi} U(\theta,\phi)\sin\theta\, d\theta\, d\phi} \tag{2.5}$$

In practice, the directivity is often not easy to measure accurately since the 3D radiation pattern is often not easily obtained. It is now common for designers to make use of one of a number of commercially available full-wave electromagnetic solvers to predict the directivity of an antenna. Also, a frequently used simple approximation of directivity for antennas that produce directive (pencil-beam-shaped) radiation patterns is

$$D = 10\log\left(\frac{32{,}400}{\Theta_1\ \Theta_2}\right)\ (\text{dB}) \tag{2.6}$$

where Θ_1 and Θ_2 represent the half-power beamwidths (expressed in degrees) in two principal-plane cuts of the radiation pattern. Figure 2.3 plots the approximate directivity versus the half-power beamwidths, based on (2.6) for the case where $\Theta_1 = \Theta_2$.

2.1.3 Gain and Radiation Efficiency

The gain of an antenna is closely linked to its directivity. It accounts for any losses within the antenna and is defined as

$$G(\theta,\phi) = \frac{4\pi U(\theta,\phi)}{P_{in}} \tag{2.7}$$

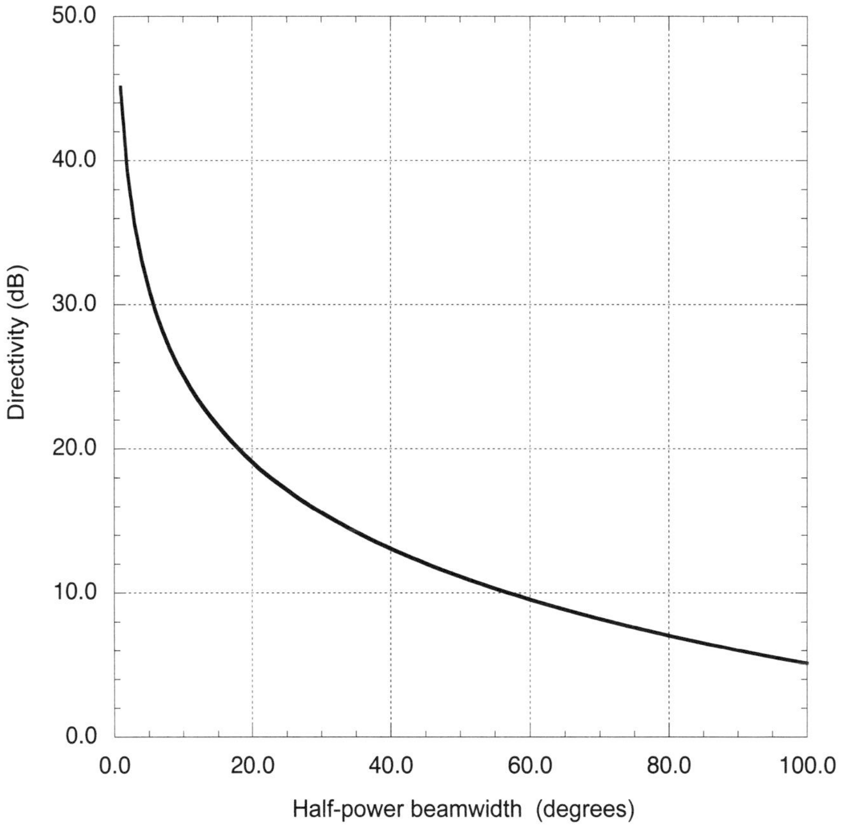

Figure 2.3 Directivity versus half-power beamwidth for $\Theta_1 = \Theta_2$.

where P_{in} is the power at the input of the antenna terminals. In general, an antenna will have some losses (such as ohmic or dielectric losses) so that for the case of a transmitting antenna, the total power radiated (P_{rad}) from the antenna is less than the power entering its terminals (P_{in}). The relationship between the input power and radiated power can be expressed as

$$P_{rad} = e_{rad} P_{in} \tag{2.8}$$

where e_{rad} is defined as the radiation efficiency of the antenna and has a value that ranges between 0 and 1. Combining (2.3), (2.7), and (2.8), the gain of an antenna can then be related to its directivity by

$$G(\theta,\phi) = e_{rad} D(\theta,\phi) \tag{2.9}$$

As was the case for directivity, an antenna's gain is often quoted by a single value, which is understood to mean its maximum gain. It should be noted that the gain does not account for any mismatch losses between the antenna and the transmission line connected to the input terminals of the antenna.

The gain of an antenna is easier to measure than its directivity. A common method of determining gain is by a measurement technique known as the gain substitution method. In this method, the received power (P_{ra}) of the antenna with unknown gain (G_a) is compared with the received power (P_{rs}) of a gain standard antenna (one whose gain has been previously established). The unknown gain can be determined using

$$G_a = G_s + (P_{ra} - P_{rs}) \tag{2.10}$$

where G_s is the gain of the gain standard. In (2.10) the gains are expressed in decibels and the powers are in decibel watts (dBW) or decibel milliwatts (dBm) (where 1 W = 0 dBW = 30 dBm) and it is valid for the case when the polarizations of the two antennas are the same. Gain standards are readily commercially available, and the gain substitution method is a simple technique for determining the gain of an antenna.

2.1.4 Polarization

The polarization of an antenna refers to the time-varying behavior of the electric field of the electromagnetic wave produced by the antenna. At any fixed coordinate (θ_o, ϕ_o), the electric field will vary in magnitude and phase in such a way that it traces out an elliptical curve over one period, as shown in Figure 2.4. This ellipse is quantified by three parameters. The axial ratio (AR) of the ellipse is defined as

$$\text{AR} = \frac{A}{B} \tag{2.11}$$

where A and B are the lengths of the major and minor axes of the ellipse, respectively. The polarization tilt angle (τ) is the angle formed by the major axis with the E_θ-axis. The third parameter is the sense of rotation of the magnitude of the electric field

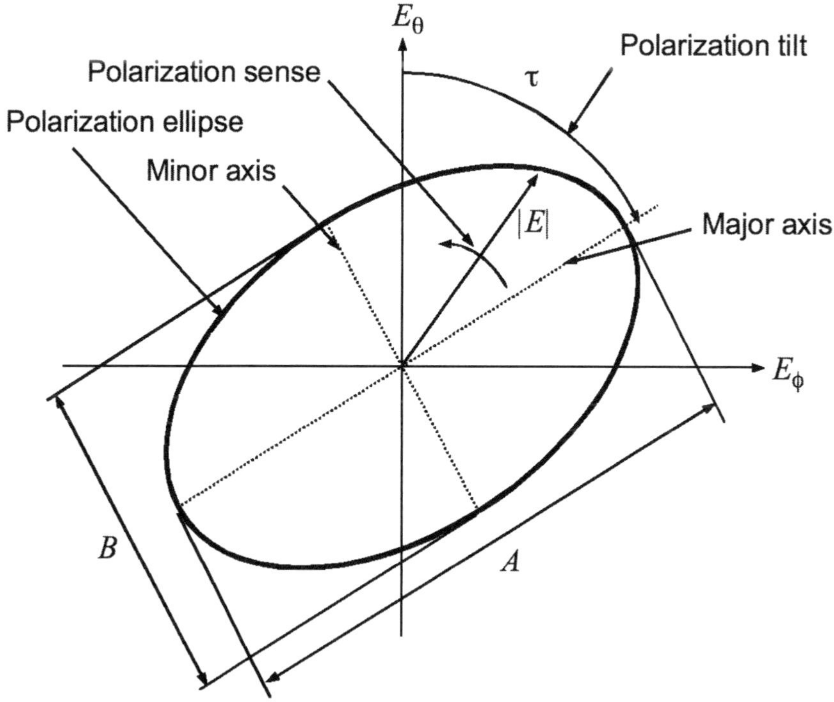

Figure 2.4 Polarization ellipse.

(clockwise or counterclockwise, as seen by an observer looking in the direction of propagation).

Two special cases of polarization are linear polarization (AR = ∞) and circular polarization (AR = 1). Many antennas produce predominantly linear polarized radiation patterns, while others are designed to achieve nearly circular polarization. Linearly polarized antennas are typically used for terrestrial wireless applications, while circularly polarized antennas are commonly used for satellite communication applications. In order to maximize the power transferred between two antennas, they require identical polarizations.

2.1.5 Bandwidth

The bandwidth of an antenna is the range of frequencies over which the antenna performance is considered acceptable. The impedance bandwidth of an antenna is often specified by the maximum acceptable magnitude of the complex reflection coefficient ($|\Gamma|$) or of the VSWR generated when the antenna is connected to a transmission line. The relationship between reflection coefficient and VSWR is given by

$$VSWR = \frac{1+|\Gamma|}{1-|\Gamma|} \tag{2.12}$$

The reflection coefficient arising from the mismatch between an antenna with an input impedance Z_{ant} connected to a transmission line with characteristic impedance Z_o is determined by

$$\Gamma = \frac{Z_{ant} - Z_o}{Z_{ant} + Z_o} \tag{2.13}$$

In general, the antenna input impedance (Z_{ant}) is complex and varies with frequency; thus Γ is also complex and a function of frequency. For many applications, antennas with VSWR ≤ 2 ($|\Gamma| \leq 0.33$) are considered acceptable, and the frequency range over which this condition is met defines the antenna's input impedance bandwidth.

The reflection coefficient is often expressed as a return loss (RL) defined as

$$\text{RL} = -20\log|\Gamma| = -20\log|S_{11}| \text{ (dB)} \tag{2.14}$$

where S_{11} is the scattering parameter equal to Γ. Values of RL $\geq$ 10 dB are often quoted as being acceptable.

The magnitude of the reflection coefficient is also used to determine the mismatch loss (ML), where

$$\text{ML} = -10\log\left(1-|\Gamma|^2\right) \text{ (dB)} \tag{2.15}$$

The ML is a measure of the fraction of the incident power that is transferred from the transmission line to the input terminal of the antenna.

Other measures of an antenna's bandwidth are based on minimum gain requirements or on specific attributes of the radiation pattern (such as beamwidth, sidelobe levels, or cross-polarization levels). Certain applications will require specific antenna performance with respect to some or all of the various antenna parameters. For such cases, the antenna bandwidth would then be defined as the range of frequencies over which all of the required specifications are met.

2.2 FAA-SPECIFIC PARAMETERS

To classify and assess the performance of antennas with frequency agility, additional metrics are necessary. Parameters are required not only to characterize the electromagnetic behavior of the FAA, but also to measure the performance of the associated tuning mechanism. This section begins with a definition of the tuning mode of an FAA, then introduces two sets of parameters to deal with both the electromagnetic behavior (Sections 2.2.2 and 2.2.3) and tuning mechanism performance (Section 2.2.4).

2.2.1 Tuning Mode

The nature of the tuning of an FAA can be considered to be either discrete or continuous. Discrete frequency tuning techniques can only alter the frequency to a fixed number of distinct (and often predetermined) values, while in continuous tuning, the frequency can take on any value within the tuning range. The discrete or continuous nature of the frequency tuning shall be referred to as the *tuning mode*.

Discrete frequency tuning can be used to cover a wide frequency range, since a common approach is to switch between antenna elements that can be designed with widely different resonant frequencies. Electronic switches are typically used to implement discrete tuning. FAAs designed with continuous tuning usually have a narrower tuning range, since this technique typically relies on altering the properties of a single radiating element. Techniques for continuous tuning include the use of variable capacitors (varactors) or of tunable substrates (such as ferrites or ferroeletrics) whose material properties can be altered through the appropriate bias.

2.2.2 Parameters Related to Tuning Frequency

Several parameters can be defined that quantify the range of frequencies over which the FAA operates. The *tuning range* (TR) for a resonant antenna is usually defined as

$$\mathrm{TR} = \frac{2\left(f_{oh} - f_{ol}\right)}{\left(f_{oh} + f_{ol}\right)} \times 100\% \tag{2.16}$$

where f_{oh} and f_{ol} correspond to the upper and lower resonant frequencies, respectively, at which the antenna can operate. These frequencies are often identified by the minima in the $|S_{11}|$ plots, as shown in Figure 2.5.

Another parameter, closely related to the tuning range, is the *tuning ratio* (TX), defined as

$$\mathrm{TX} = \frac{f_{oh}}{f_{ol}} \times 100\% \tag{2.17}$$

The TR is not commonly used to characterize FAAs, but is more often used to characterize the capacitance range of varactor diodes. The TR and TX are directly related and each parameter can be expressed in terms of the other:

$$\mathrm{TR} = 2\frac{\mathrm{TX} - 1}{\mathrm{TX} + 1} \tag{2.18}$$

and

$$\mathrm{TX} = \frac{2 + \mathrm{TR}}{2 - \mathrm{TR}} \tag{2.19}$$

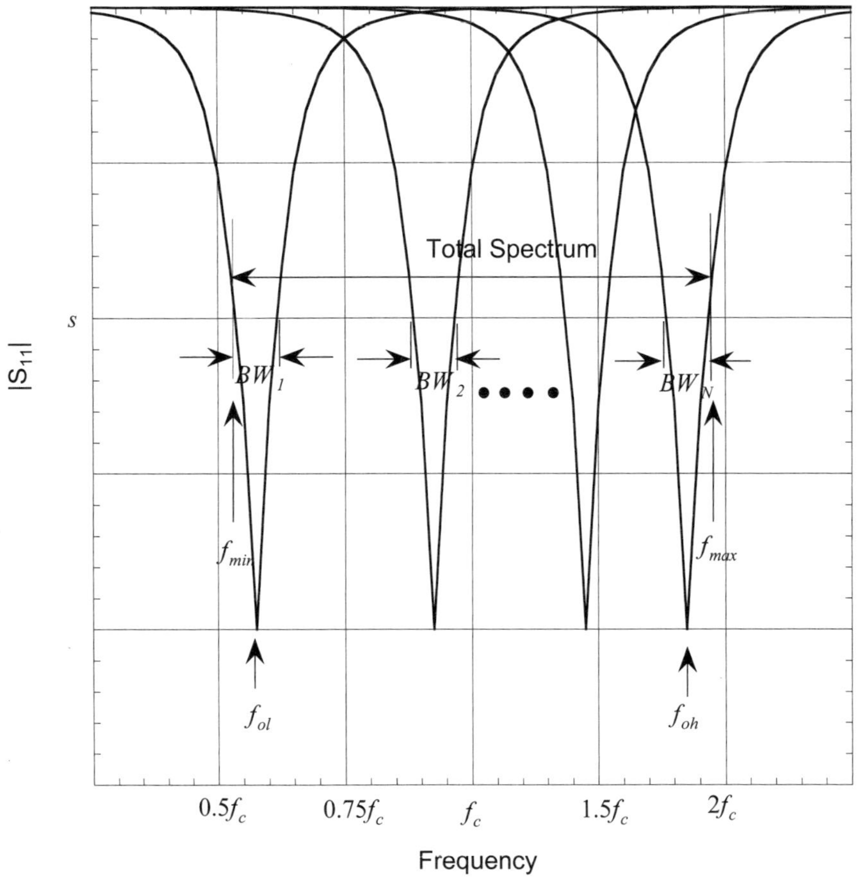

Figure 2.5 Reflection coefficient responses of an FAA tuned to its highest and lowest frequency values.

Some publications define tuning range based on the minimum (f_{min}) and maximum (f_{max}) usable frequencies, which could be based on the maximum acceptable input reflection coefficient (as shown in Figure 2.5, where $|S_{11}| \leq s$), or on the minimum acceptable antenna gain, or some other suitable metric. In this text, the term *total spectrum* (TS) is used to refer to this broader tuning range, defined as

$$\text{TS} = \frac{2\left(f_{max} - f_{min}\right)}{\left(f_{max} + f_{min}\right)} \times 100\% \qquad (2.20)$$

The definition for TS would be more suitable for nonresonant antennas, or for ones that do not demonstrate sharp resonances.

The TR, TX, and TS quantify the frequency range over which the antenna can be adjusted, but they do not provide any information as to what percentage of this band the antenna exhibits acceptable performance. A parameter termed *tunable bandwidth* (BW_T) can be defined to account for antenna performance. Since the acceptable performance of an antenna is application dependent, there is no unique way of defining BW_T. For FAAs, one useful definition for tunable bandwidth would be in terms of the range of frequencies where a minimum value (s) of the VSWR is achieved. Denoting this as the *tunable impedance bandwidth* (BW_{TZ}), and referring to Figure 2.5, then

$$BW_{TZ} = \frac{2\sum_{i=1}^{N} BW_i}{f_{min} + f_{max}} \times 100\% \tag{2.21}$$

where BW_i represent the instantaneous nonoverlapping subbands within the total spectrum where the minimum acceptable VSWR is met, and N denotes the number of states of the FAA. For FAAs operating in the continuous tuning mode, the value of N would be determined by considering only the bias states where there would be no overlap between the frequencies of adjacent instantaneous bands. For continuous tuning FAAs, one would expect BW_{TZ} to be high, whereas for FAAs with discrete tuning, the BW_{TZ} could be low, especially if the TR is wide and the number of discrete frequencies is small. A similar definition could be applied to the *tunable gain bandwidth* (BW_{TG}) for a minimum acceptable antenna gain. Alternatively, the BW_{TZ} of the FAA could be reported along with the peak gain and maximum gain variation over the spectrum coverage.

As an example, Figure 2.6 shows the $|S_{11}|$ frequency response of two hypothetical FAAs, both with discrete tuning modes. The first FAA has only two states (the two solid curves in Figure 2.6). With f_{ol} = 4.3 GHz and f_{oh} = 7.7 GHz, the TR for this antenna is determined from (2.16) to be TR = 56.7%. Defining the minimum acceptable $|S_{11}|$ to be s = -10 dB, then f_{min} = 4.1 GHz and f_{max} = 7.85 GHz, and the total spectrum using (2.20) is TS = 62.8%. The instantaneous bandwidth for this antenna (based on $|S_{11}| = s$) is BW_i = 300 MHz for both states (i = 1, 2), so that the tunable impedance bandwidth from (2.21) is BW_{TZ} = 10%. The second FAA has six states (the two solid curves plus the four dashed curves of Figure 2.6). It would thus have the same TR and TS as the first antenna. However, with four more states, its has BW_{TZ} = 30%.

2.2.3 Tuning Efficiency

Tuning efficiency (η_T) is a measure of the insertion loss introduced by the mechanism used to achieve frequency agility. Depending on the tuning technique adopted, it may be difficult to evaluate the tuning efficiency directly. One approach is to compare the gain of the FAA with its non-frequency-agile counterpart. The tuning efficiency would then be calculated as

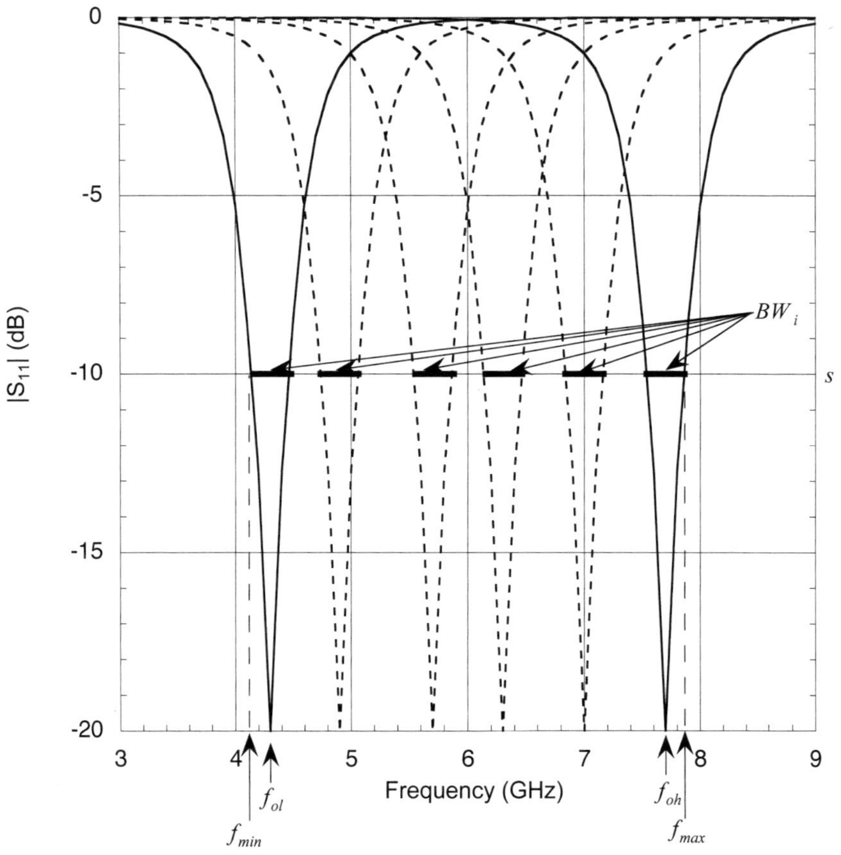

Figure 2.6 Example of the tuning response of two hypothetical FAAs.

$$\eta_T = 10^{(G_a - G_r)/10} \times 100\% \qquad (2.22)$$

where G_a and G_r are the gains of the FAA and the reference antenna, respectively, both expressed in decibels. In some cases, an equivalent non-FAA might not exist, and some other method would need to be implemented for evaluating the tuning efficiency.

2.2.4 Characterizing the Tuning Method

As will be seen in later chapters, there are numerous and widely diverse techniques available for achieving frequency agility. There are, however, a set of metrics that can be applied to nearly all these tuning techniques to help quantify their performance. These common parameters are considered in this section.

All tuning methods, whether they involve mechanical actuators, tunable substrates, or electronic devices, ultimately require a drive voltage and/or bias current. For the majority of tuning techniques, these voltages (currents) are direct current (DC), but there are occasional exceptions (such as certain liquid crystal designs) where a low-frequency alternating current (AC) bias is used. The *power consumption* of the tuning technique and the *maximum bias level* are especially important for portable devices (such as laptops and cell phones), where the battery size and the recharging interval will be directly impacted.

Tuning speed is strongly dependent on the type of tuning method used. Integrated electronic devices usually offer the fastest tuning, while mechanical techniques are considerably slower. Tuning speed is an important consideration in applications where it is necessary to be able to quickly hop between frequency ranges. In other applications, where the frequency range needs to be adjusted only occasionally, tuning speed may not be a critical parameter.

The *maximum RF power* that the tuning configuration can handle will be an important consideration for certain high-power applications. One example would be an FAA used at a cellular base station. Antennas for these applications typically must handle several hundreds of watts, and the tuning mechanisms would need to support such power levels.

One aspect of the tuning mechanism that is difficult to quantify is its complexity. Challenges in the fabrication of the required devices and the relative difficulty in integrating the tuning mechanism with the antenna are factors that contribute to complexity. One indication of complexity is the number of components required. In general, the higher the number of components, the more complicated the biasing network becomes. A *component count* could thus be an appropriate quantity to indicate the complexity of the tuning method.

To get a sense of the potential complexity of an FAA, Figure 2.7 compares a passive printed dipole and a PIN diode-loaded printed dipole with frequency agility. The passive printed dipole (to be discussed further in Chapter 3) shown in Figure 2.7(a) is fed by a microstrip line and will operate over a narrow band whose center frequency is determined by the length L of the dipole. One method to make the printed dipole frequency agile is first to segment the length of the dipole and then integrate a set of PIN diodes across the gaps, as shown in Figure 2.7(b). Applying the appropriate set of DC bias voltages (V_1 to V_4) across the diodes will result in them being either open circuit or short circuit. With the diodes on either arm of the dipole biased in pairs, this leads to a dipole whose length can be switched between L_1 and L_4, resulting in four frequency bands. In order to prevent the DC bias signals from entering the RF source, a blocking capacitor needs to be integrated into the microstrip feed. Similarly, RF chokes (inductors) are required in all the bias lines to prevent RF signals from entering the DC source. The bias lines that lead up to the dipole arms have to be designed with care in order to minimize their impact on the electromagnetic performance of the antenna. The component count for this design is 15, which, as can be seen, corresponds to a fairly complex antenna.

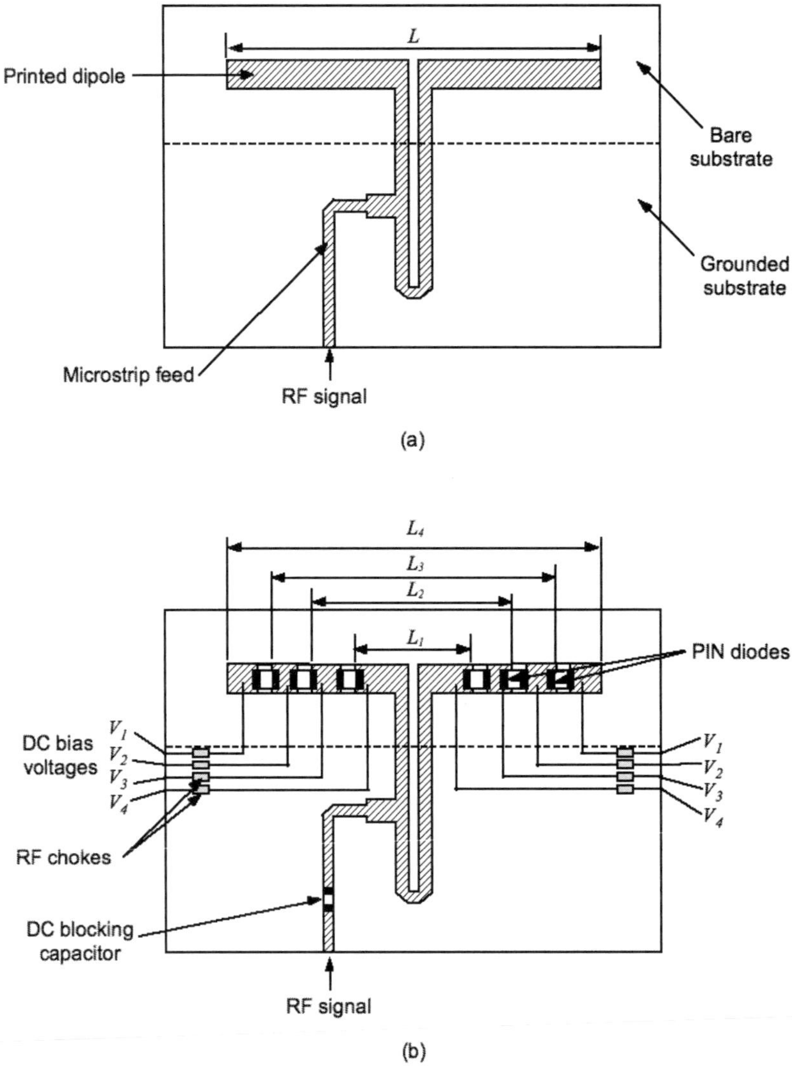

Figure 2.7 Comparison of (a) a passive printed dipole and (b) a tunable PIN-diode-loaded printed dipole.

Nonelectrical parameters such as component reliability and lifetime, thermal properties, temperature sensitivity, and the overall cost, size, and weight of the FAA are also important aspects that need to be included when evaluating the viability of the antenna for use in a particular application. In general, tuning mechanisms that do not make use of any moving parts will be more reliable and have a longer lifetime than those relying on physical movement or on MEMS devices. Thermal properties relate to the heat generated by the tuning circuitry, which, if too high, will require passive or active cooling. The temperature sensitivity of the tuning mechanism might be an issue in FAAs that have to function over a wide range of temperatures, which is the case for many outdoor applications. Minimizing the cost of the FAA is always of interest, and in most consumer applications, it is the driving factor. Size and weight are also critical parameters for handheld, laptop, or other mobile communication applications. Often the real estate available to fit an antenna for these applications is extremely limited, and it is a difficult challenge to come up with a viable design.

Rarely are all these parameters addressed in FAA literature. This is understandable since most publications are presenting prototype designs that have not necessarily been optimized but are intended to demonstrate the potential capability of the technique. In Chapters 5–8 where several examples are surveyed, as many of the metrics as possible will be provided in order to give the reader a sense of the relative performance of the various techniques.

2.3 FAA CLASSIFICATION

In this book, the FAAs will be classified by their tuning mode (discrete or continuous) and by the type of method used to achieve frequency agility. Although there are numerous techniques for frequency tuning, these can be grouped into one of three broad categories: mechanical actuation, tunable substrates, and active device integration. Mechanical actuation includes methods where there is a physical movement or a deformation of the antenna element and includes mechanisms such as piezoelectric actuators, voltage-controlled membranes, and MEMS devices. Tunable substrates are ones whose permittivity, permeability, or conductivity can be altered by the application of a bias field and include ferrites, ferroelectrics, liquid crystals, and semiconductors. Active device integration includes the use of various electronic components such as diodes, transistors, and optical switches for achieving frequency agility. Chapter 4 will examine the various tuning techniques in these three categories.

Chapter 3

Antenna Elements

In theory, almost any antenna could be made capable of frequency agility, but in practice certain antenna elements are more amenable than others. This chapter examines various antennas that are considered the most suited for portable or mobile wireless communications. Most of the antennas share the common feature that they are low-gain radiators (typically about 5 dB or less) and have maximum dimensions of under one wavelength. The basic principles of operation are discussed for each antenna, and equations are provided for determining appropriate dimensions for a given frequency of operation. The antennas presented in this chapter include dipoles, microstrip patch antennas, slot antennas, planar inverted-F antennas, dielectric resonator antennas, and helical antennas. Frequency-agile versions of these antennas will be examined in Chapters 5–8.

3.1 WIRE DIPOLES

A wire dipole is one of the oldest types of antennas consisting, in its simplest form, of a pair of thin colinear wires fed with an out-of-phase (or differential) signal. It has been used as an individual radiator or as an array element in numerous applications from short-wave communications to radio and television reception. There are also many variants of the simple dipole (including folded dipoles, biconical antennas, bowtie antennas, and printed dipoles) designed for reasons such as enhancing input impedance, widening operating bandwidth, or facilitating integration with printed circuit technology. Although wire dipoles are not easily modified into FAAs, several closely related antennas (like the printed dipole, the microstrip dipole, and the slot dipole) are well-suited for frequency agility. Since the radiation patterns of these various dipoles share many similarities, it is instructive to first examine the behavior of a simple wire dipole before proceeding to these variants.

The geometry of a wire dipole is depicted in Figure 3.1. In this instance, the two arms of the dipole are of equal length, and a signal is applied across the small central gap. If the radius a of the wire is much smaller than the free-space wavelength λ_o, and if the dipole has a total length L, then for the dipole aligned parallel to the z-axis, the far-field radiation pattern is given by

$$U(\theta) = \frac{\eta}{8}\left|I_o\right|^2 \left[\frac{\cos\left[\dfrac{\pi L}{\lambda_o}\cos\theta\right] - \cos\left[\dfrac{\pi L}{\lambda_o}\right]}{\pi \sin\theta} \right]^2 \tag{3.1}$$

where I_o is the maximum current on the dipole and η = 377Ω is the intrinsic impedance of free space. Equation (3.1) represents an omnidirectional pattern with a constant value in planes parallel to the z = 0 plane (azimuth planes). Elevation (ϕ = constant) plane radiation patterns of a dipole for various lengths of L are shown in Figure 3.2, where the peak value for each pattern is normalized to 0 dB. The half-power beamwidth in the elevation plane and the directivity of the dipole are plotted in Figure 3.3 as a function of dipole length. As the length increases from $L = 0.1\lambda_o$ to $L = 1.5\lambda_o$ the elevation beamwidth narrows from 89° to 19°. By applying curve-fitting, the following expression can be used to estimate the half-power beamwidth Θ_d of the dipole

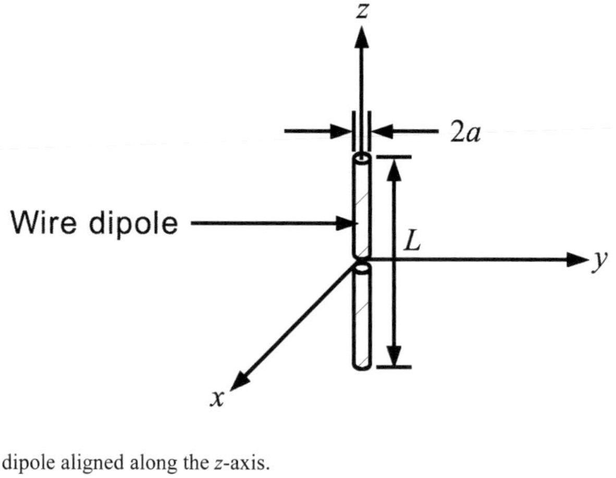

Figure 3.1 Wire dipole aligned along the z-axis.

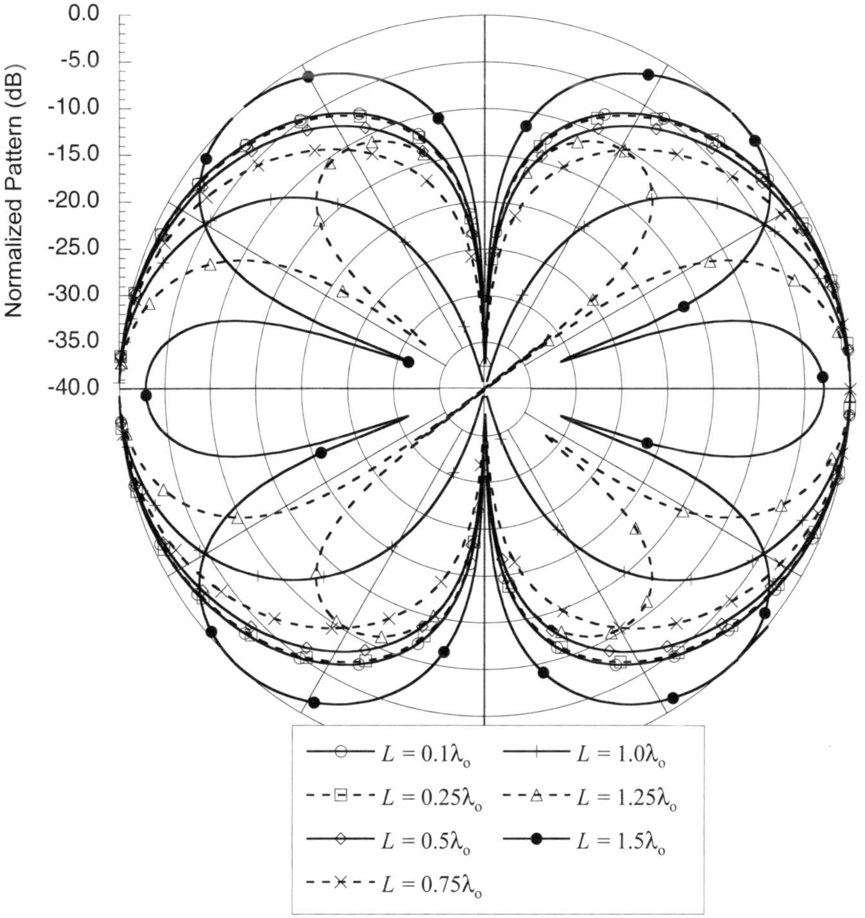

Figure 3.2 Normalized radiation patterns of a thin wire dipole aligned along the *z*-axis.

$$\Theta_d = 89.014 + 10.097\left[\frac{L}{\lambda_o}\right] - 78.618\left[\frac{L}{\lambda_o}\right]^2 + 27.441\left[\frac{L}{\lambda_o}\right]^3 \qquad (3.2)$$

The directivity increases from $1.76\ \mathrm{dB_i} \leq D \leq 5.2\ \mathrm{dB_i}$ as the dipole length increases from $0.1\lambda_o \leq L \leq 1.25\lambda_o$ and then drops as L increases further, due to the appearance and growth of sidelobes. An approximate expression for the directivity of the dipole (valid for $L \leq \lambda_o$) is given by

$$D = 1.7106 + 0.45761\left[\frac{L}{\lambda_o}\right] - 0.0244\left[\frac{L}{\lambda_o}\right]^2 + 1.6565\left[\frac{L}{\lambda_o}\right]^3 \qquad (3.3)$$

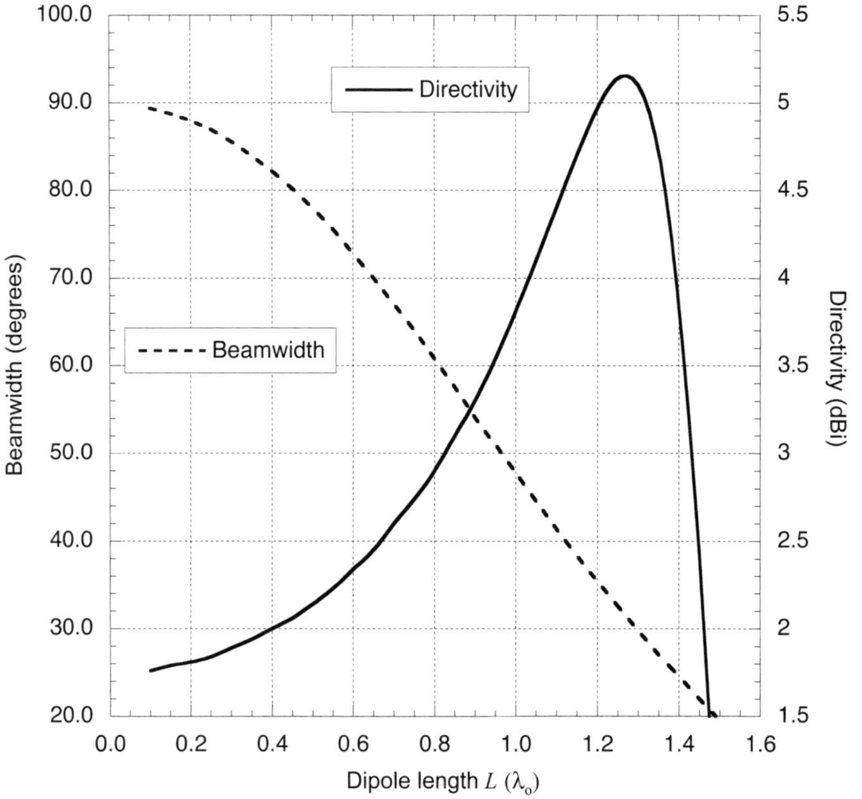

Figure 3.3 Beamwidth and directivity of a thin wire dipole as a function of dipole length.

Wire dipoles are typically designed to have a length of slightly less than $L = 0.5\lambda_o$, since at this length, the input impedance is purely real, with a value of approximately 73Ω, which is relatively easy to match to the characteristic impedance of a coaxial transmission line. For thin wire dipoles whose lengths are no longer than about $L = \lambda_o/2$, the input impedance ($Z_{in} = R_{in} + jX_{in}$) can be approximated by [1]

$$R_{in} = -0.4787 + 7.3246\left[\frac{\pi L}{\lambda_o}\right] + 0.3963\left[\frac{\pi L}{\lambda_o}\right]^2 + 15.613\left[\frac{\pi L}{\lambda_o}\right]^3 \tag{3.4}$$

$$X_{in} = g_L - 0.4456 + 17.0082\left[\frac{\pi L}{\lambda_o}\right] - 8.6793\left[\frac{\pi L}{\lambda_o}\right]^2 + 9.6031\left[\frac{\pi L}{\lambda_o}\right]^3 \tag{3.5}$$

where

$$g_L = 120\left[\ln\left(\frac{L}{2a}\right) - 1\right]\cot\left[\frac{\pi L}{\lambda_o}\right] \tag{3.6}$$

For a wire dipole having a rectangular cross-section of width w and height h, the radius a in (3.6) can be replaced by an equivalent radius a_e [1], where

$$a_e = 0.34h + 0.25w \tag{3.7}$$

To excite the dipole, a wire-pair transmission line could be used, but this is not practical due to the exposed currents on the transmission line; therefore a shielded coaxial cable is usually preferred. In addition to matching the impedance of the dipole to that of the cable to minimize reflections, a balun (balanced-to-unbalanced transformer) is needed to prevent currents from being excited along the outer conductor of the coaxial cable. An example of a coaxial balun feeding a wire dipole antenna is shown in Figure 3.4. The quarter wavelength, short-circuited section forms a stub line with the coaxial cable and presents a high impedance (open circuit) at the feed point, preventing currents from flowing on the outer conductor of the coaxial feed line. A printed version of the balun can be included in certain printed dipole designs, as described in Section 3.3.

3.2 MONOPOLE ANTENNAS

Closely related to the wire dipole is the wire monopole antenna. It consists of a thin wire of radius a protruding vertically by a height H from an electrically large ground plane, as shown in Figure 3.5. When the ground plane is sufficiently large (at least several wavelengths), it can be approximated by an infinitely large conductor, and the radiation pattern of the monopole approaches that of a

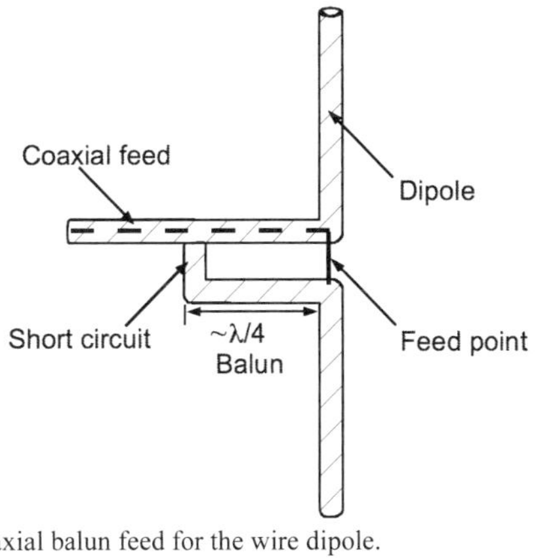

Figure 3.4 Example of a coaxial balun feed for the wire dipole.

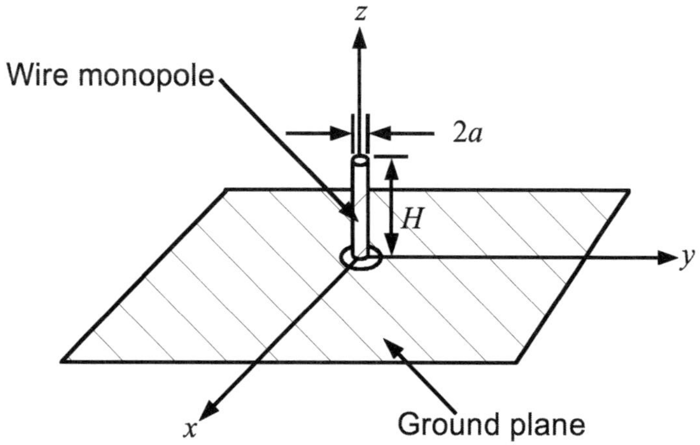

Figure 3.5 Wire monopole antenna.

corresponding dipole of twice its length ($L = 2H$) for the hemisphere above the ground plane ($0° \leq |\theta| \leq 90°$), as shown in Figure 3.6 for the case of a monopole with $H = 0.25\lambda_o$. This corresponds to the upper half of the normalized dipole pattern for $L = 0.5\lambda_o$ in Figure 3.2. Below the ground plane it is assumed that there is no radiation. Since all the radiated power is concentrated in the upper hemisphere, the directivity of the monopole will be twice that of the corresponding dipole, while its radiation resistance will be half that of the dipole. For finite-size ground planes, diffraction from the edges of the ground plane will cause ripples in the patterns as well as spillover below the ground plane and a reduction in directivity. As an example, the radiation pattern of the same quarter-wave monopole on a square ground plane of dimensions $1.25\lambda_o \times 1.25\lambda_o$ is also shown in Figure 3.6.

The monopole antenna has the attractive feature that it can be excited directly from a coaxial cable without the need of a balun. The center conductor of the coaxial cable is connected to the monopole while the outer conductor is connected to the ground plane. To avoid excessive mismatch loss, the characteristic impedance of the coaxial line needs to be close to the input impedance of the monopole.

Wire monopoles have a relatively narrow impedance bandwidth. This bandwidth can be dramatically improved by replacing the wire with a flat metal plate having any one of numerous shapes. Examples of these flat monopoles are shown in Figure 3.7, many of which have been designed for ultrawideband (UWB) applications, with percentage bandwidths of up to 10.7:1 [2-7]. Flat monopoles can be fabricated either from a thin metal sheet or by printing a conductor on a thin dielectric substrate.

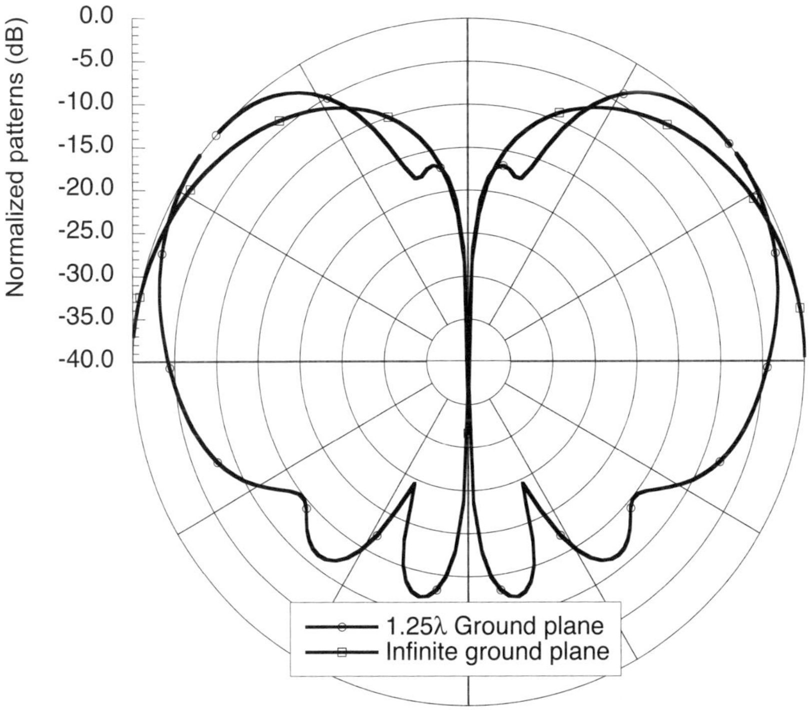

Figure 3.6 Normalized patterns of a monopole on a finite and infinite ground plane.

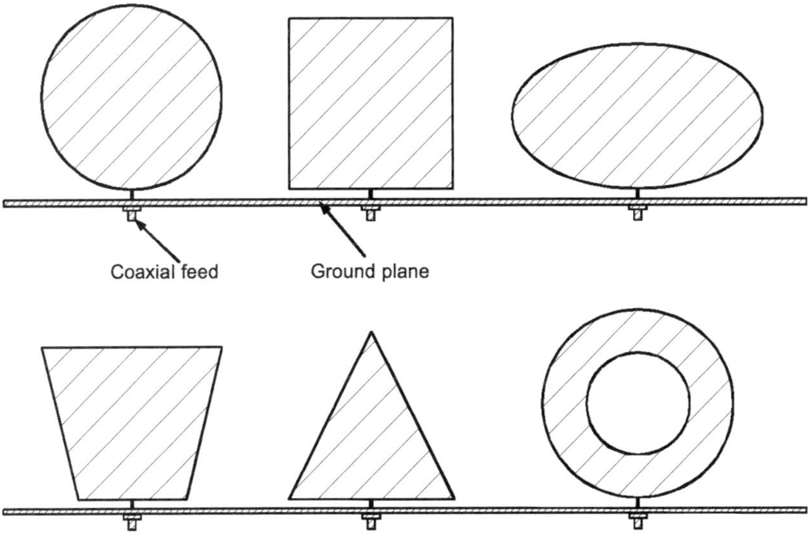

Figure 3.7 Examples of wideband flat monopole antennas.

Closely related to the flat monopole is the coplanar printed monopole antenna, where the monopole emerges parallel from the ground plane instead of perpendicular. Examples of coplanar monopole shapes are shown in Figure 3.8 [8–16]. Many publications do not make a distinction between flat and coplanar monopoles, and the term printed monopole is often used to describe either configuration. Coplanar monopoles have the advantage that they can be more readily integrated with microstrip or CPW technology. Both flat and coplanar monopoles are amenable to frequency agility and examples will be presented in later chapters.

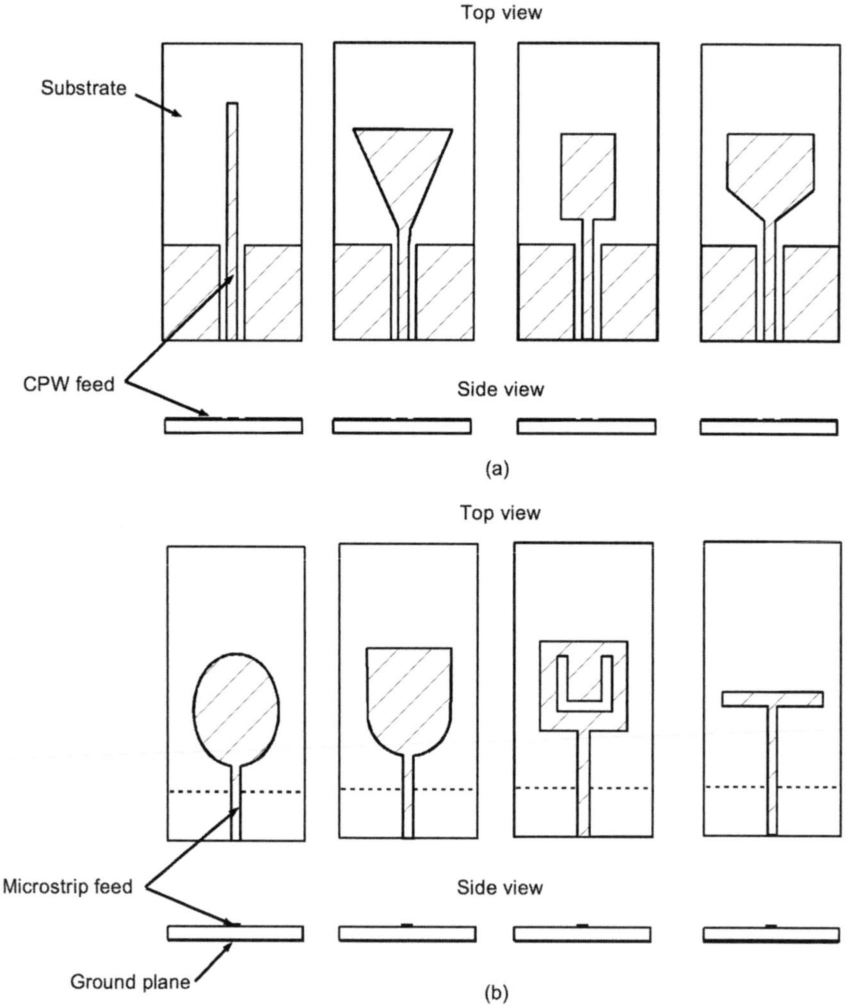

Figure 3.8 Examples of coplanar monopole antennas: (a) CPW-fed monopoles and (b) microstrip-fed monopoles.

3.3 PRINTED DIPOLES

Printed dipoles are similar to coplanar printed monopoles and can be considered variants of the simple wire dipole, where the wire is replaced by a metal trace printed on a thin dielectric substrate. The two arms of the dipole can be printed on the same side or on opposite sides of the substrate, as shown in Figure 3.9(a,b). For a thin dielectric substrate having a low dielectric constant, the radiation pattern

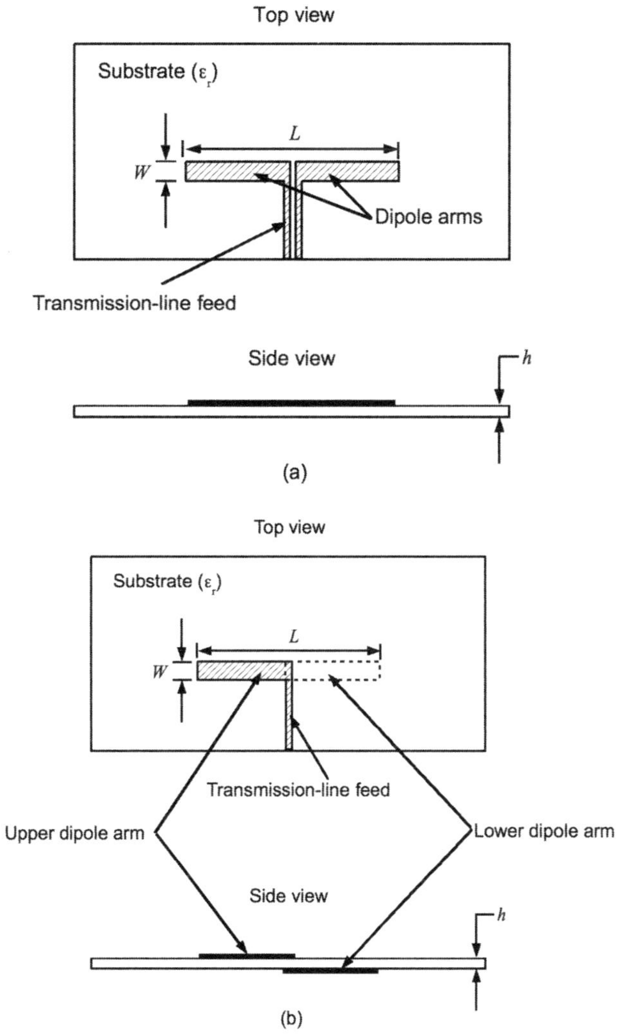

Figure 3.9 Printed dipole antennas: (a) single-sided printed dipole and (b) double-sided printed dipole.

of the printed dipole will be similar to that of a wire dipole, where the printed dipole width w is related to the wire dipole radius a by $w = 4a$ [17]. Although there is little difference in the radiation pattern of the single-sided or double-sided printed dipole, there is a significant implication on the feeding mechanism. For the double-sided case, the printed transmission line feeding the dipole will naturally provide the correct phasing to properly excite the dipole. In many cases the transmission line is connected to a microstrip line (examined in Section 3.4) via a matching transition section. Figure 3.10 shows several variations of the double-

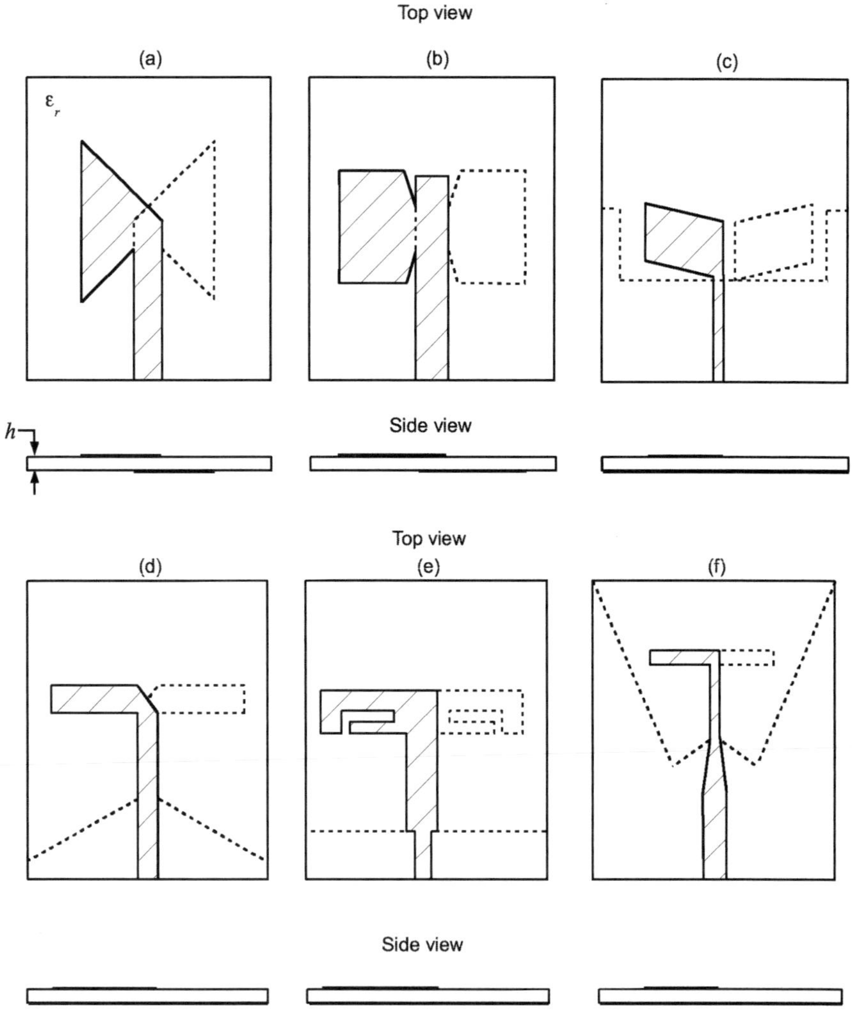

Figure 3.10 Printed dipole antenna examples. ((a) After [18]; (b) after [19]; (c) after [20]; (d) after [21]; (e) after [22]; and (f) after [23].)

sided printed dipole, which have been designed for wideband [Figure 3.10(a)–(d)], dual-band [Figure 3.10(e)], or increased directivity [Figure 3.10(f)] [18–23]. For these more complex designs, there are no closed-form expressions for calculating the input impedance, and full-wave electromagnetic simulations are required to accurately predict this parameter.

For the single-sided printed dipoles, the feed design is somewhat more complicated, since the two arms of the dipole need to be fed out of phase. There are two common approaches for achieving the proper feed design. The first is illustrated in Figure 3.11. The signal from a microstrip transmission line is split equally in two and a 180° phase delay is introduced in the path feeding one of the dipole arms. Transformers are also necessary to match between the impedance of the dipole arms to that of the microstrip lines.

The second design approach for feeding a single-sided printed dipole is illustrated in Figure 3.12; its equivalent circuit is shown in Figure 3.13 [25]. The configuration consists of an (unbalanced) microstrip feed printed on one side of the substrate and a pair of (balanced) microstrip coupled lines on the opposite side, forming the balun. The dipole arms are then integrated with the coupled lines. This configuration is analogous to the coaxial balun of a wire dipole shown in Figure 3.4. The microstrip feed line of characteristic impedance Z_a sees an

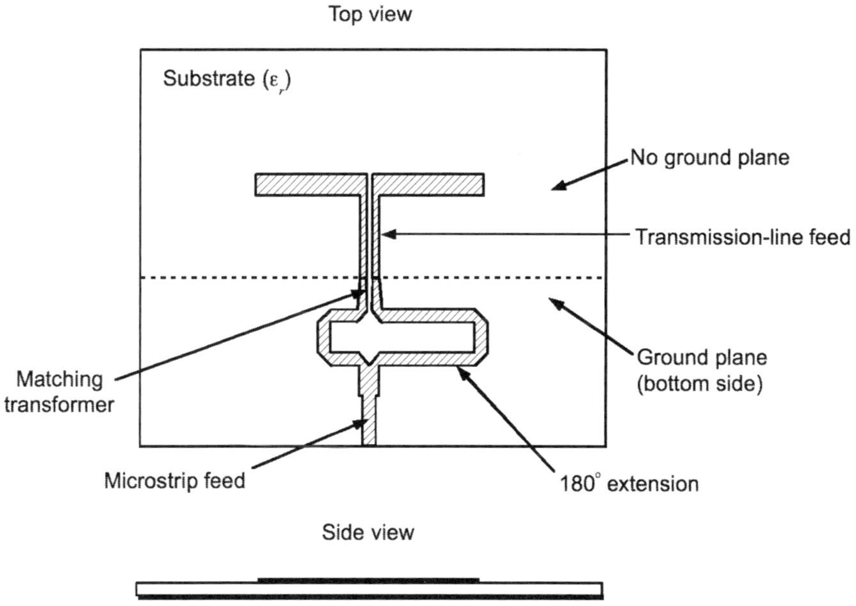

Figure 3.11 Single-sided printed dipole antenna with split microstrip feed. (After [24].)

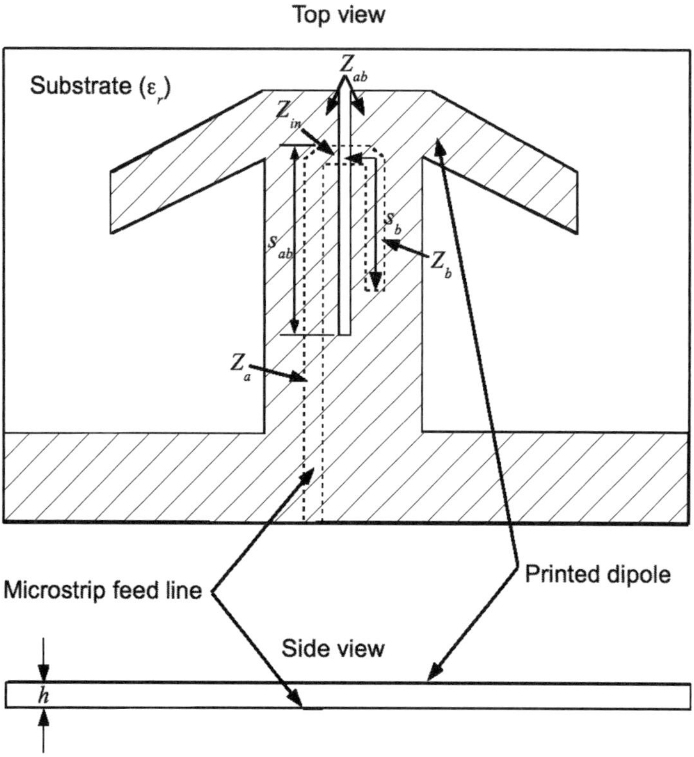

Figure 3.12 Printed dipole antenna with balun feed. (After [25].)

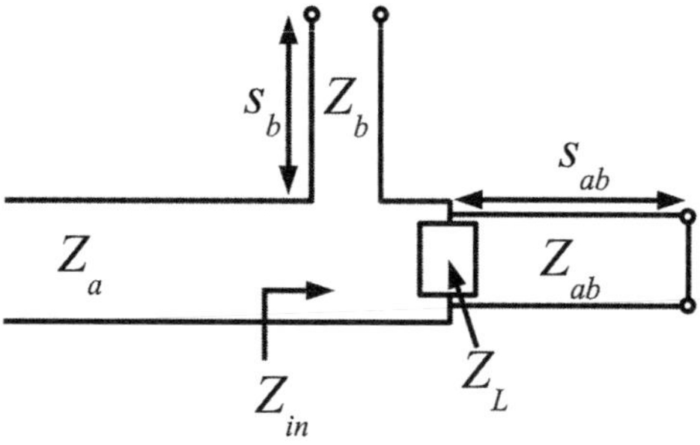

Figure 3.13 Equivalent circuit of the microstrip feed shown in Figure 3.12. (After [25].)

input impedance Z_{in} that is formed by a series open-circuit stub of length s_b having a characteristic impedance Z_b, and a short-circuit parallel stub of length s_{ab} with a characteristic impedance Z_{ab} and the dipole impedance Z_L. The input impedance Z_{in} can be calculated using [25]

$$Z_{in} = -jZ_b \cot(\beta_b s_b) + \frac{jZ_L Z_{ab} \tan(\beta_{ab} s_{ab})}{Z_L + jZ_L Z_{ab} \tan(\beta_{ab} s_{ab})} \tag{3.8}$$

where β_b and β_{ab} are the guided wave numbers for the stubs of characteristic impedance Z_b and Z_{ab}, respectively. The electrical lengths (θ_b and θ_{ab}) of the two stubs, where $\theta_b = s_b\ \beta_b$ and $\theta_{ab} = s_{ab}\ \beta_{ab}$, can each be adjusted from the nominal value of 90° to improve the impedance bandwidth performance. Section 3.4 contains equations for estimating the characteristic impedances of microstrip lines used to calculate values for Z_a and Z_b while expressions for the characteristic impedance (Z_{ab}) of coupled lines can be found in [26, Appendix B]. Examples of printed dipoles with frequency agility will be presented in later chapters.

3.4 MICROSTRIP ANTENNAS

Microstrip circuit technology was introduced in the 1960s, and microstrip antennas followed in the mid 1970s. The technology consists of a thin low-loss dielectric sheet (substrate), clad on both sides with a thin metallic layer (usually copper). The metal on one side of the dielectric serves as the ground plane and usually remains untouched. The other side is covered with a photo-resist film, exposed to ultraviolet light through a photo-mask containing a negative of the desired pattern. A chemical etchant is then applied to remove the unexposed metal, leaving the desired pattern which forms the microstrip circuit or antenna. (The metal pattern can also be designed by mechanical etching using a computer-controlled milling machine.) Figure 3.14 shows examples of various microstrip circuits and antennas.

Microstrip antennas offer several attractive features including a low profile, ease of fabrication, and relatively low manufacturing costs. Microstrip technology can also accommodate antenna designs over a wide range of frequencies from as low as about 100 MHz up to nearly 100 GHz. More complicated antenna designs using multiple layers of dielectric sheets, each having an independent thickness and dielectric constant, have also been developed to provide improved performance and greater design flexibility.

Microstrip patch antennas are resonant antennas that can be designed using almost any arbitrary shape. Historically, simple shapes such as rectangles, ellipses, and triangles have been most common due to the relative ease of analysis, and many closed-form expressions have been derived to predict the resonant frequency, impedance bandwidth, and radiation patterns of these antennas. With the advent of fast computers, large memory capacity, and commercial full-wave electromagnetic simulation software, complicated structures consisting of multiple layers and complex shapes can now also be readily analyzed. However, for many

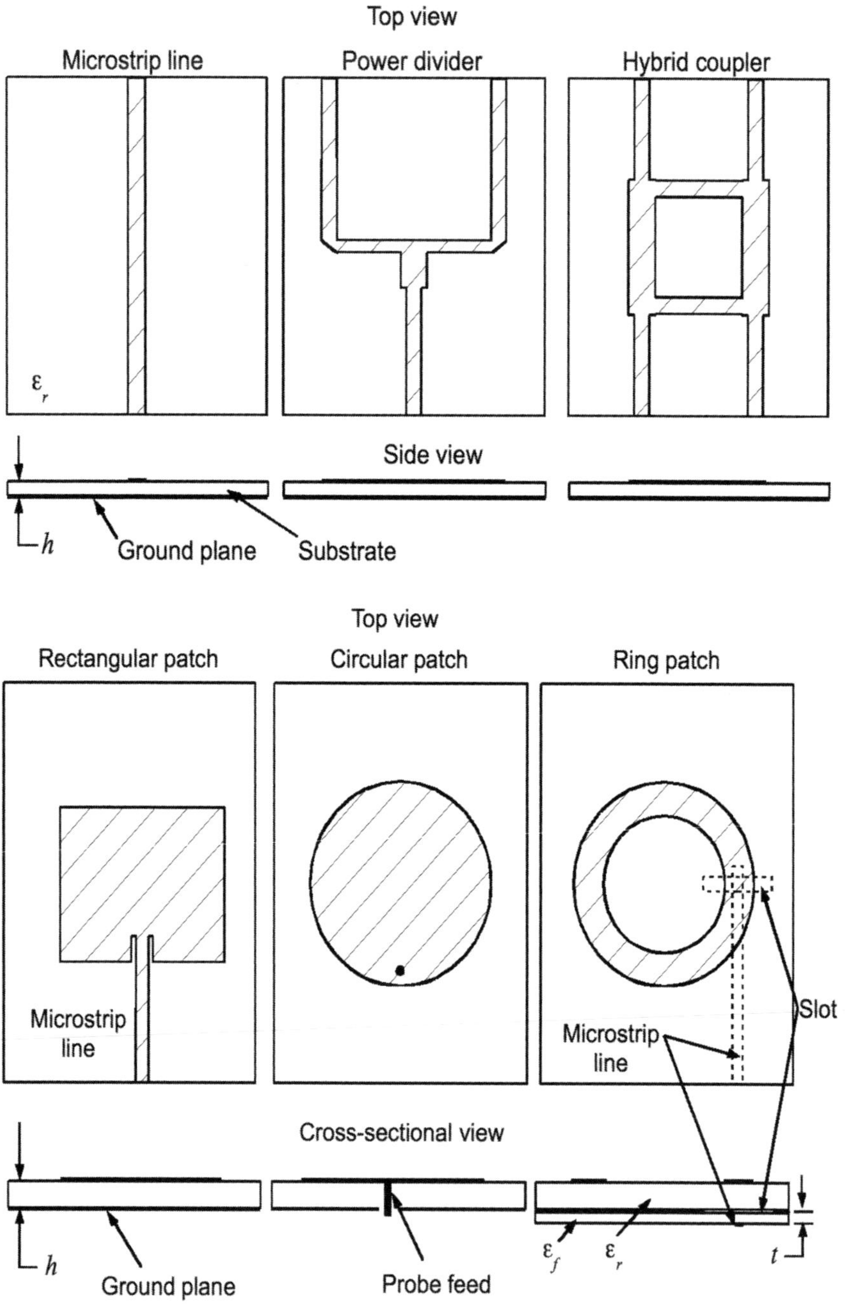

Figure 3.14 Example of microstrip circuits and antennas.

applications, simple-shaped microstrip patch designs based on the closed-form expressions are often adequate. A large body of knowledge exists on microstrip antennas, summarized in numerous textbooks, and several references are provided for the reader who wishes to delve deeper into microstrip antenna technology [26–31].

In this section, the basic design principles are given for a few of the most common microstrip patch antennas: the rectangular patch, the microstrip dipole, the circular patch, and the ring patch. The performance of each of these shapes will be found to be fairly similar, with some shapes having certain advantages in terms of size, bandwidth, or efficiency. The shape of the microstrip patch chosen by the designer will ultimately depend on the intended application.

Before examining the microstrip patch antennas, it is instructive to first consider the analysis of a typical microstrip line of width w, printed on a grounded dielectric substrate of thickness h and dielectric constant ε_r, as shown in Figure 3.15(a). A guided wave traveling along the microstrip line propagates in a quasi-TEM mode, with a cross-sectional field distribution sketched in Figure 3.15(b). The electrical parameters of interest for this microstrip line are its characteristic impedance Z_o and the wavelength λ_g of the guided wave. Since the fields of the wave are traveling in an inhomogeneous medium (partially in the

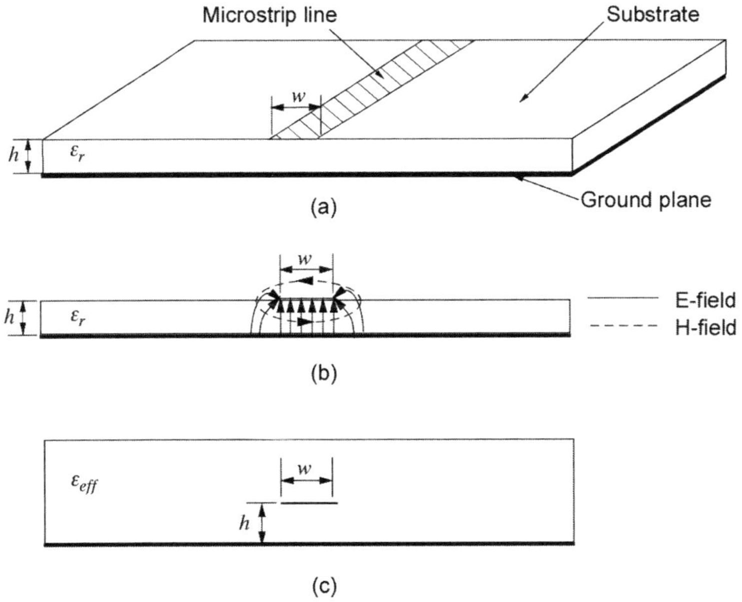

Figure 3.15 Microstrip line: (a) microstrip line geometry, (b) field configuration, and (c) equivalent model.

substrate and partially in air), a direct electromagnetic analysis of the microstrip line is difficult. Instead, an equivalent model is analyzed, shown in Figure 3.15(c), where the inhomogeneous medium is replaced with a homogeneous one having an effective dielectric constant ε_{eff} whose value lies between 1 and ε_r and is given by [32]

$$\varepsilon_{eff} = \begin{cases} \dfrac{\varepsilon_r + 1}{2} + \dfrac{\varepsilon_r - 1}{2}\left[1 + 12\dfrac{h}{w}\right]^{-1/2}, & \text{for } \dfrac{w}{h} > 1 \\ \dfrac{\varepsilon_r + 1}{2} + \dfrac{\varepsilon_r - 1}{2}\left[\left(1 + 12\dfrac{h}{w}\right)^{-1/2} + 0.04\left(1 - \dfrac{w}{h}\right)^2\right], & \text{for } \dfrac{w}{h} \le 1 \end{cases} \tag{3.9}$$

where it is assumed that the microstrip line has zero thickness. The guided wavelength is then determined using

$$\lambda_g = \frac{\lambda_o}{\sqrt{\varepsilon_{eff}}} \tag{3.10}$$

Equation (3.9) for ε_{eff} was based on a quasi-static model and assumes that ε_{eff} is independent of frequency. The accuracy of (3.9) is approximately 5% or better for frequencies up to about 30 GHz. A frequency-dependent function for ε_{eff} has been developed in [33] that provides a more accurate approximation and has been found to offer better than 1% accuracy for frequencies of up to 60 GHz.

The characteristic impedance of the microstrip is also a function of the effective dielectric constant, along with the line width and substrate thickness and can be calculated using [32]

$$Z_o = \begin{cases} \dfrac{120\pi}{\sqrt{\varepsilon_{eff}\left[\dfrac{w}{h} + 1.393 + 0.667\ln\left(\dfrac{w}{h} + 1.444\right)\right]}}, & \text{for } \dfrac{w}{h} > 1 \\ \dfrac{60}{\sqrt{\varepsilon_{eff}}}\ln\left[\dfrac{8h}{w} + \dfrac{w}{4h}\right], & \text{for } \dfrac{w}{h} \le 1 \end{cases} \tag{3.11}$$

(Again the microstrip line is assumed to have zero conductor thickness.) Convenient expressions also exist for determining the width of the microstrip line for a given substrate to achieve the desired characteristic impedance [26, Appendix B]

$$\frac{w}{h} = \begin{cases} \dfrac{2}{\pi}\left[B - 1 - \ln(2B - 1) + \dfrac{\varepsilon_r - 1}{2\varepsilon_r}\left(\ln(B - 1) + 0.39 - \dfrac{0.61}{\varepsilon_r}\right)\right], & \text{for } Z_o \le \dfrac{89.91}{\sqrt{\varepsilon_{eff}}} \\ \dfrac{8e^A}{e^{2A} - 2}, & \text{for } Z_o > \dfrac{89.91}{\sqrt{\varepsilon_{eff}}} \end{cases} \tag{3.12}$$

where

$$A = \frac{Z_o}{60}\left[\frac{\varepsilon_r + 1}{2}\right]^{1/2} + \frac{\varepsilon_r - 1}{\varepsilon_r + 1}\left[0.23 + \frac{0.11}{\varepsilon_r}\right] \quad (3.13a)$$

$$B = \frac{60\pi^2}{Z_o\sqrt{\varepsilon_r}} \quad (3.13b)$$

The equations for effective dielectric constant, characteristic impedance and guided wavelength of microstrip lines have been used in the design of microstrip patch antennas, which will be examined in Sections 3.4.1–3.4.4.

3.4.1 Rectangular Microstrip Patch Antenna

The rectangular microstrip patch antenna, shown in Figure 3.16, consists of a rectangular metal patch of length L and width W, printed on a grounded substrate having relative permittivity of ε_r, relative permeability of μ_r, and thickness h. Most substrates used are pure dielectric (with $\varepsilon_r > 1, \mu_r = 1$), but ferrite substrates (with $\varepsilon_r > 1$, $\mu_r > 1$) have also been used and will be discussed in more detail in subsequent chapters. In this section, the substrate will be considered to be a pure dielectric. The antenna depicted in Figure 3.16 is excited at a point located along the centerline of its width and inset from the edge by a distance s.

One of the advantages of microstrip patch antennas is the variety of feed mechanisms that can be used for excitation. The most common feeds are shown in

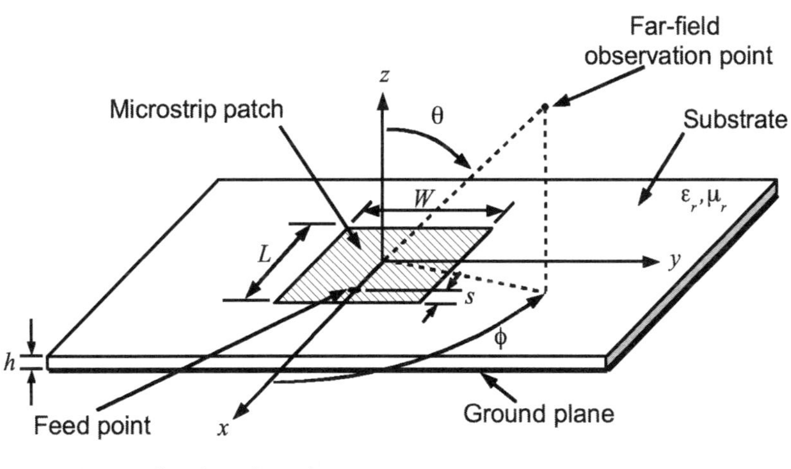

Figure 3.16 Rectangular microstrip patch antenna.

Figure 3.17. The probe-fed technique [Figure 3.17(a)] is one of the simplest, and can be implemented by drilling a hole through the substrate and attaching a commercial microwave connector (such as an N-type or SMA). The center pin of the connector is soldered to the patch and the connector ground is attached to the

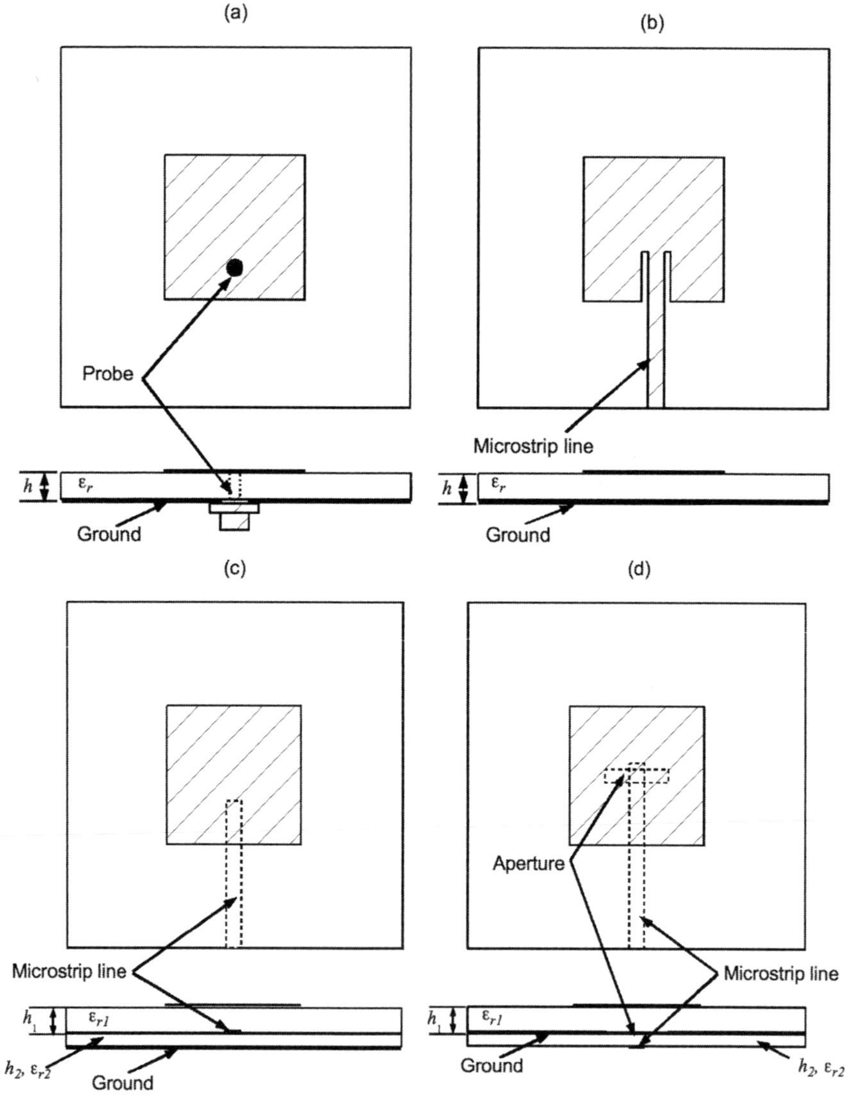

Figure 3.17 Examples of feeds for rectangular microstrip patch antennas: (a) probe-fed, (b) direct-fed, (c) proximity-fed, and (d) aperture-fed.

the ground plane of the microstrip patch antenna. The probe feed is suitable for lower frequency designs but may not be practical at higher frequencies, where the patch dimensions are on the order of a few millimeters. Another simple feed mechanism, which is more suitable at higher frequencies, is through a direct connection with a microstrip line [Figure 3.17(b)]. Both the probe-fed and direct microstrip-fed excitations only require a single substrate. The drawback with the direct-feeding approach is the potential for spurious radiation from the feed line, which can adversely affect the performance of the microstrip antenna. By moving the feed line to a second substrate [Figure 3.17(c)], unwanted radiation can be reduced since the height h_2 and dielectric constant ε_{r2} of the feed substrate can be optimized to minimize spurious radiation with minimal impact on the radiation performance of the patch. To fully shield the feed line from the microstrip patch, a ground plane can be introduced between the two substrates, and the coupling between microstrip line and antenna is achieved through a slot aperture in the ground plane [Figure 3.17(d)]. The selection of which feed excitation to implement will depend on several factors including the frequency of operation, the electric performance (such as impedance match and radiation cross-polarization), and cost requirements.

Numerous modes can be excited in the rectangular microstrip patch, each with its own characteristic radiation pattern. This antenna is usually excited with the lowest order mode, known as the transverse magnetic TM_{10} mode. The subscripts refer to the number of half-cycle variations of the E-field along the patch length and width, respectively. The E-fields of the TM_{10} mode are sketched in Figure 3.18.

The resonant frequency of the TM_{10} mode of the rectangular microstrip patch can be estimated using various methods with varying degrees of complexity and accuracy. One of the simplest approaches that nonetheless provides reasonable accuracy is based on treating the rectangular microstrip patch antenna as a section of microstrip transmission line having an effective length L_{eff} given by [34]

$$L_{eff} = L + 2\Delta L \tag{3.14}$$

where

$$\Delta L = 0.421h \frac{\varepsilon_{eff} + 0.3}{\varepsilon_{eff} - 0.258} \frac{\dfrac{W}{h} + 0.264}{\dfrac{W}{h} + 0.813} \tag{3.15}$$

The effective length is somewhat longer than the physical length L of the antenna since it accounts for the effect of the fringing E-fields, as shown in Figure 3.18. Using the effective length, and determining the effective dielectric constant by applying (3.9) where the width w of the microstrip line is substituted with the

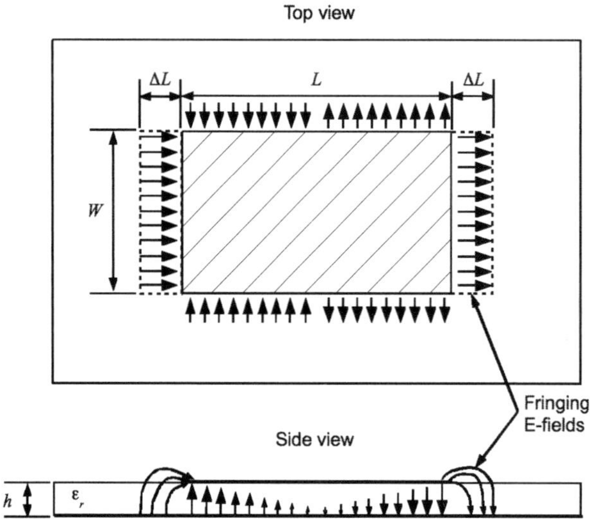

Figure 3.18 Sketch of the E-fields of the TM_{10} mode for the rectangular microstrip patch.

microstrip patch width *W*, the resonant frequency f_o is determined by [31, Ch. 5]

$$f_o = \frac{c}{2L_{eff}\sqrt{\varepsilon_r}} \tag{3.16}$$

where c is the speed of light in free space. From (3.16) it can be observed that the physical length L of the patch is the dominant parameter controlling the resonant frequency. The patch width W and substrate thickness h have less of an effect, since they modify the effective dielectric constant and hence the effective length. The width, dielectric constant, and substrate thickness all play a more important role when it comes to antenna impedance, bandwidth, and radiation efficiency, to be discussed shortly. In designing a rectangular microstrip patch antenna at a desired resonant frequency, for a given substrate thickness and dielectric constant, a useful starting value for the patch width to ensure a reasonably efficient radiator is

$$W = \frac{c}{2f_o}\sqrt{\frac{2}{\varepsilon_r + 1}} \tag{3.17}$$

The effective dielectric constant can then be determined using (3.9), and L_{eff} can be found by rearranging (3.16). The physical length L of the patch can then be resolved by applying (3.14) and (3.15).

Based on the transmission line model, the radiation from the rectangular microstrip patch antenna is modeled as that of an array of two thin rectangular slots of length W and width ΔL, separated by a center-to-center distance of $L + \Delta L$ and placed on an infinite ground plane. Ignoring the effect of the dielectric, and applying array theory, the far-field radiation pattern of the antenna can be derived to be [26, Ch. 4]

$$U = |E_\theta|^2 + |E_\phi|^2 \tag{3.18}$$

where

$$E_\theta = -A \cos\phi \, G_1 \, G_2 \, G_3 \tag{3.19a}$$

$$E_\phi = A \cos\theta \sin\phi \, G_1 \, G_2 \, G_3 \tag{3.19b}$$

$$A = jk_o V_o W \tag{3.19c}$$

$$G_1 = \frac{\sin\left[\frac{1}{2} k_o h \sin\theta \cos\phi\right]}{\frac{1}{2} k_o h \sin\theta \cos\phi} \frac{\sin\left[\frac{1}{2} k_o W \sin\theta \sin\phi\right]}{\frac{1}{2} k_o W \sin\theta \sin\phi} \tag{3.19d}$$

$$G_2 = 2\cos\left[\frac{1}{2} k_o L \sin\theta \cos\phi\right] \tag{3.19e}$$

$$G_3 = \frac{e^{-jk_o r}}{4\pi r} \tag{3.19f}$$

and V_o is the amplitude of the voltage excitation. These equations are fairly accurate for thin substrates ($h \leq 0.08\lambda_o$). More accurate (but more complicated) expressions can be found to account for the substrate effects [31, Ch. 5]. The normalized radiation pattern of a typical rectangular patch antenna based on (3.18) is shown in two principal planes in Figure 3.19. Referring to the rectangular patch orientation shown in Figure 3.16, the E-plane corresponds to $\phi = 0°$ while the H-plane corresponds to $\phi = 90°$.

The input impedance of the rectangular microstrip patch antenna at resonance can also be estimated using the transmission line model, where the antenna is treated as a transmission line, loaded at both ends by a conductance G_s. At resonance, the input impedance will be real and is given (for thin substrates, $h < 0.1\lambda_o$) by [35, Ch. 14]

$$R_{in} = \frac{1}{2G_s} \cos^2\left[\frac{\pi s}{L}\right] \tag{3.20}$$

where

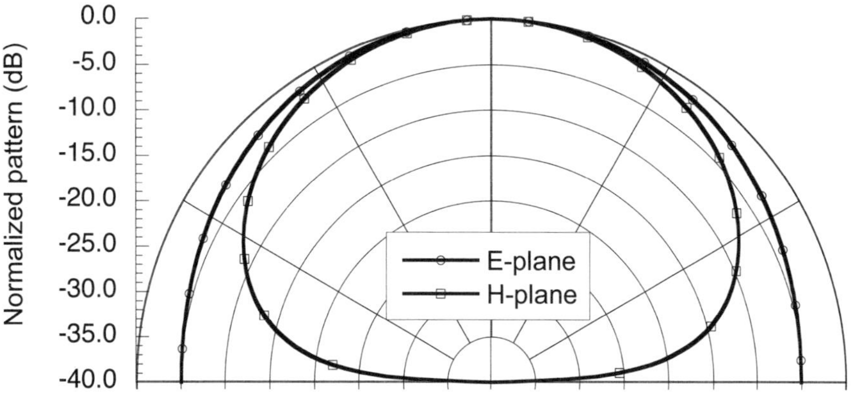

Figure 3.19 Normalized radiation pattern of the TM_{10} mode for the rectangular microstrip patch (with $L = 0.283\lambda_o$, $W = 0.395\lambda_o$, $h = 0.05\lambda_o$, and $\varepsilon_r = 2.2$).

$$G_s = \frac{W}{120\lambda_o}\left[1 - \frac{1}{24}(k_o h)^2\right] \quad (3.21)$$

and $k_o = 2\pi f_o / c$. The input impedance is a function of the offset distance s from the edge of the patch. It has a maximum value at the edge ($s = 0$) and drops to zero at the center of the patch ($s = L/2$).

The quality factor, radiation efficiency, and a more accurate approximation of the input impedance of the rectangular microstrip patch can be obtained by carrying out an analysis based on a cavity model of the antenna. An in-depth derivation of the following equations has been carried out in [31, Ch. 5], and only a brief summary and the resultant equations are presented here.

The total quality (Q) factor of the antenna can be expressed as the sum of four terms associated with conductor loss (Q_c), dielectric loss (Q_d), radiation into a space wave (Q_r), and radiation into a surface wave (Q_s):

$$Q = \left[\frac{1}{Q_c} + \frac{1}{Q_d} + \frac{1}{Q_r} + \frac{1}{Q_s}\right]^{-1} \quad (3.22)$$

The conductor loss quality factor of the microstrip antenna is related to its surface resistance R_s, by

$$Q_c = \frac{120\pi^2 f}{c}\frac{h}{R_s} \quad (3.23)$$

The surface resistance is a function of the conductivity σ of the metal patch through the relation

$$R_s = \sqrt{\frac{\pi\mu_o f}{\sigma}} \tag{3.24}$$

where μ_o is the permeability of free space ($\mu_o = 4\pi\ 10^{-7}$ H/m).

The dielectric loss quality factor is inversely related to the loss tangent (tanδ) of the dielectric substrate; that is,

$$Q_d = \frac{1}{\tan\delta} \tag{3.25}$$

The quality factor associated with the amount of power radiated into space is given by

$$Q_r = \frac{3}{16}\frac{\varepsilon_r^3}{p\left(\varepsilon_r^2 - \varepsilon_r + 0.4\right)}\frac{L_e}{W_e}\frac{\lambda_o}{h} \tag{3.26}$$

where

$$W_e = W + 2h\frac{\ln 4}{\pi} \tag{3.27a}$$

$$p = 1 + a_1 + a_2 + a_3 + a_4 \tag{3.27b}$$

$$a_1 = -0.016605\left[2\pi\frac{W_e}{\lambda_o}\right]^2 \tag{3.27c}$$

$$a_2 = 0.00023\left[2\pi\frac{W_e}{\lambda_o}\right]^4 \tag{3.27d}$$

$$a_3 = -0.01828\left[2\pi\frac{L_e}{\lambda_o}\right]^2 \tag{3.27e}$$

$$a_4 = 0.000217\left[2\pi\frac{W_e}{\lambda_o}\right]^2\left[2\pi\frac{L_e}{\lambda_o}\right]^2 \tag{3.27f}$$

and $\lambda_o = c/f_o$.

Finally, the quality factor associated with the surface wave is determined using

$$Q_s = \frac{P_r}{P_s}Q_r \tag{3.28}$$

where P_r represents the power radiated into space, and P_s represents the power radiated into surface waves of a horizontal electric dipole on a lossless dielectric substrate. These two powers can be calculated using the following expressions

$$P_r = 320h^2\left[\frac{\pi}{\lambda_o}\right]^4\left[\frac{\varepsilon_r^2 - \varepsilon_r + 0.4}{\varepsilon_r^2}\right] \tag{3.29}$$

$$P_s = 480\pi h^3 \left[\frac{\pi}{\lambda_o}\right]^5 \left[1 - \frac{1}{\varepsilon_r}\right]^3 \tag{3.30}$$

The radiation efficiency e_{rad} of the microstrip patch can then be expressed as the ratio of the total Q-factor over the space-wave Q-factor

$$e_{rad} = \frac{Q}{Q_r} \tag{3.31}$$

The fractional bandwidth (BW) of the microstrip patch can also be expressed as a function of the total Q-factor by the relationship

$$\text{BW} = \frac{\text{VSWR} - 1}{Q\sqrt{\text{VSWR}}} \cdot 100\% \tag{3.32}$$

The input impedance ($Z_{in} = R_{in} + jX_{in}$) of the microstrip patch at the resonant frequency when fed by a probe with a radius a at a distance s from the edge of the patch is given by

$$R_{in} = 480Q \frac{L_e}{W_e} \frac{h}{\lambda_o} \cos^2 \left[\frac{\pi(s + \Delta L)}{L_e}\right] \tag{3.33}$$

$$X_{in} = 120\pi \frac{h}{\lambda_o} \left[-\gamma + \ln\left(\frac{\lambda_o}{\pi a \sqrt{\varepsilon_r}}\right)\right] \tag{3.34}$$

where $\gamma = 0.577216$ is Euler's constant.

As a design example, for a microstrip patch printed on a substrate of thickness $h = 1.5$ mm and dielectric constant $\varepsilon_r = 2.2$, the width of the patch is selected using (3.17) to be $W = 11.86$ mm for a desired resonance at 10 GHz. Using (3.14) to (3.16), the required physical length is $L = 8.5$ mm. The total quality factor Q is then determined using (3.22) by calculating the individual quality factors (Q_d, Q_c, Q_r, and Q_s). The resultant value is $Q = 10.1$ which, for a VSWR = 2, corresponds to BW = 7%. The radiation efficiency is found to be $e_{rad} = 82.6\%$ using (3.31). Finally, for a probe of radius $a = 0.165$ mm located at the edge ($s = 0$) of the patch, the input impedance calculated using (3.33) and (3.34) is $Z_{in} = 175.5 + j58.4\Omega$.

3.4.2 Microstrip Dipoles

The microstrip dipole, shown in Figure 3.20, can be considered as a rectangular microstrip patch whose width W is significantly smaller than its length L and is typically smaller than $0.05\lambda_o$. The printed dipole has been the subject of in-depth study in the early 1980s [36–45]. The longitudinal currents of the fundamental

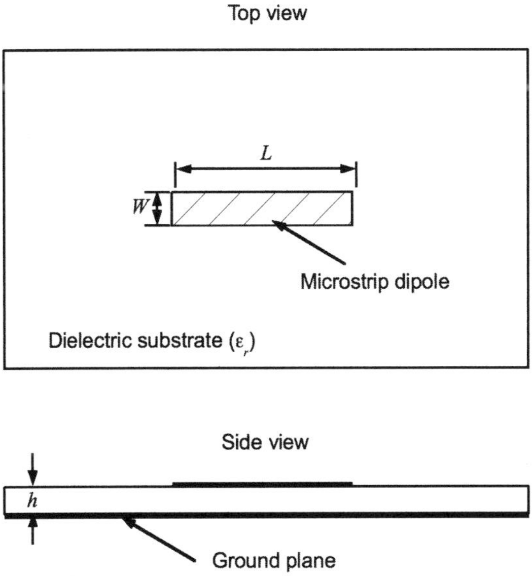

Figure 3.20 Microstrip dipole.

mode of the microstrip dipole are similar to those of the microstrip patch, and so the radiation patterns and gain of these two antennas are similar. The resonant lengths of the two antennas will be somewhat different since the effective dielectric constant ε_{eff} is a function of the antenna width W. The equations for resonant frequency and radiation patterns of the microstrip patch presented in Section 3.3.1 can be applied to predicting the behavior of the microstrip dipole. The input impedance and bandwidth of the microstrip dipole, however, will differ widely from that of the microstrip patch for the following reason. Due to the small width, the radiation resistance of the dipole is high (potentially thousands of ohms), and it is usually excited near the center, using proximity coupling, as shown in Figure 3.21, or through aperture coupling, as shown in Figure 3.22. By center-feeding the dipole, the high edge impedance (which appears as a parallel circuit) located approximately $\lambda_g/4$ from the center is transformed into a low impedance appearing in series.

The input impedance ($Z_{in} = R_{in} + jX_{in}$) of the printed dipole can be estimated by modifying the equations for a wire dipole over a ground plane derived in [42]

$$R_{in} = R_i \left\{ 1 - \frac{3}{2} \left[\frac{\sin(2kh)}{2kh} + \frac{\cos(2kh)}{(2kh)^2} - \frac{\sin(2kh)}{(2kh)^3} \right] \right\} \quad (3.35)$$

$$X_{in} = -2Z_o \cot\left[\frac{k\,L_{eff}}{2} \right] \quad (3.36)$$

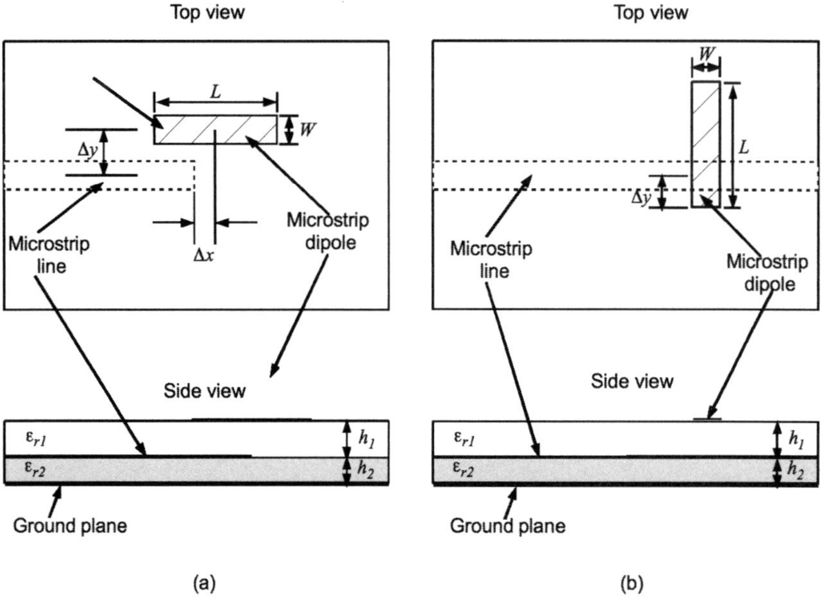

Figure 3.21 Microstrip dipole antennas excited by proximity coupling to microstrip lines: (a) after [36], and (b) after [43].

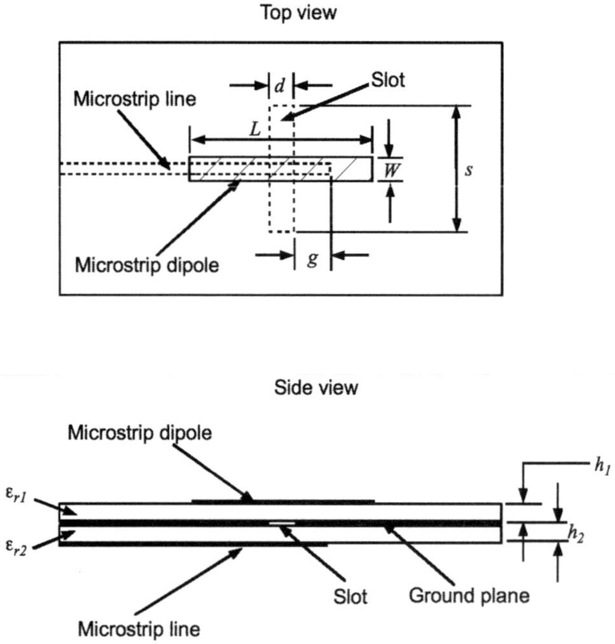

Figure 3.22 Microstrip dipole excited by aperture coupling to a microstrip line. (After [45].)

where

$$R_i = 60\left[\frac{k\,L_{eff}}{2}\right]^2 \frac{1-\dfrac{\sin\left(k\,L_{eff}\right)}{k\,L_{eff}}}{1-\cos\left(k\,L_{eff}\right)} \tag{3.37a}$$

$$k = \frac{2\pi\sqrt{\varepsilon_{eff}}}{\lambda_o} \tag{3.37b}$$

The quality factor, bandwidth, and radiation efficiency can be calculated using the same equations as for the microstrip patch antenna. Since it occupies less area than the microstrip patch, the printed dipole has found usage in array applications, as radiating elements in leaky wave antennas and in cases where a size constraint may preclude the use of a microstrip patch antenna.

A comparison can be made with the rectangular microstrip patch design example in Section 3.3.1. Using the same substrate (h = 1.5 mm and ε_r = 2.2) and choosing the width of the of the dipole to be W = 2 mm, the required length for resonance at 10 GHz is L = 8.85 mm, determined using (3.14) to (3.16). The dipole length is thus slightly longer than that of the corresponding microstrip patch (L = 8.5 mm) due to the lower effective dielectric constant arising from the smaller value of W. Using the equations in Section 3.4.1 for the various Q-factors of the microstrip patch, the total quality factor of the dipole is determined to be Q = 35, which translates to a fractional BW = 2% for a VSWR = 2. This is significantly smaller than the bandwidth for the corresponding microstrip patch antenna (BW = 7%) and is again due to the narrower width of the dipole. The radiation efficiency is determined using (3.31) to be e_{rad} = 79.7%, slightly lower than for the corresponding microstrip patch. Finally, the input impedance of the center-fed dipole at resonance will be purely real, and using (3.35) is determined to be 7.5Ω. The radiation pattern of this dipole is plotted in Figure 3.23, based on (3.18), being nearly identical to that of the corresponding rectangular microstrip antenna.

3.4.3 Circular Microstrip Patch Antennas

The circle or disk is another common shape for microstrip patch antennas [46–49]. As shown in Figure 3.24, it is characterized by a metallic disk of radius a printed on a dielectric substrate of thickness h and dielectric constant ε_r and fed at a point located at a radius s from the center. Like the rectangular microstrip patch, the circular patch can be excited in various modes using different feeding mechanisms such as a probe, a microstrip line, or an aperture. The radiation pattern of the fundamental mode (TM_{11}) is similar to the TM_{10} mode of the rectangular

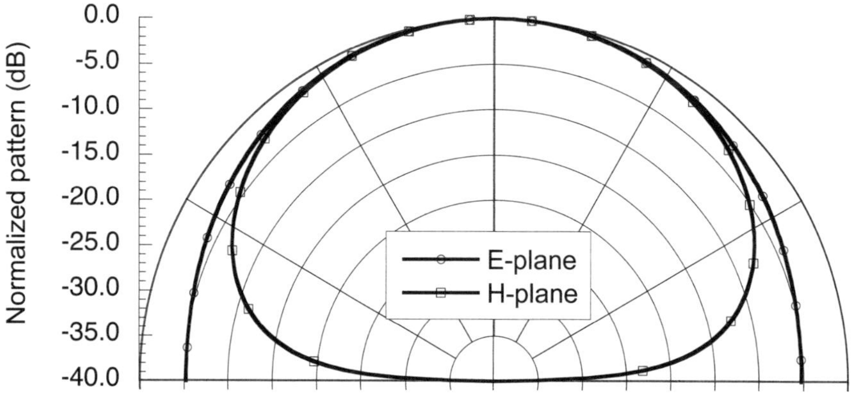

Figure 3.23 Microstrip dipole pattern (with $h = 1.5$ mm, $\varepsilon_r = 2.2$, $W = 2$ mm, $L = 8.85$ mm, and $f = 10$ GHz).

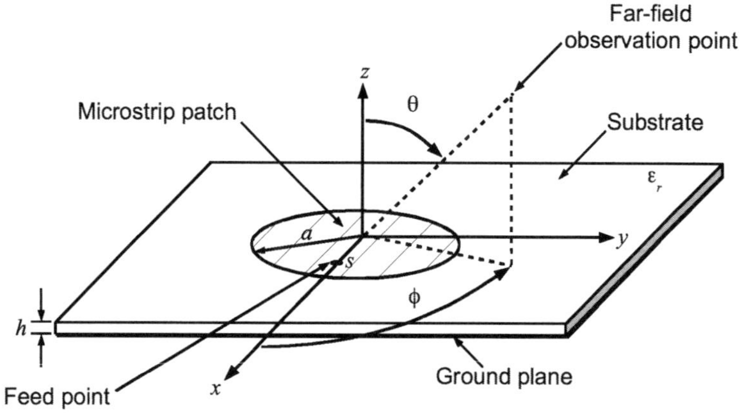

Figure 3.24 Circular microstrip patch antenna.

microstrip patch. The area of the circular microstrip patch is usually slightly smaller than that of the corresponding rectangular patch, making its use advantageous in array environments. The circular patch can also be readily excited in several higher order modes that produce conically shaped radiation patterns.

Using a cavity model analysis, the resonant frequency f_{mn} for the TM_{mn} mode of the circular microstrip antenna is determined by [48]

$$f_{mn} = \frac{\alpha_{mn} c}{2\pi a_e \sqrt{\varepsilon_r}} \tag{3.38}$$

where α_{mn} is the mth root of the derivative of the Bessel function of order n. For the TM_{11} mode, $\alpha_{mn} = 1.8411$. The effective radius a_e is somewhat larger than the physical radius to account for the fringing fields and is given by [46]

$$a_e = a\sqrt{\left\{1 + \frac{2h}{\pi a \varepsilon_r}\left[\ln\left(\frac{\pi a}{2h}\right) + 1.7726\right]\right\}} \tag{3.39}$$

The far-field electric field components for the TM_{mn} mode of the circular microstrip patch are determined using [29, Ch. 4]

$$E_\theta = \pi V_o a \frac{e^{-jk_o r}}{r} e^{-jn\pi/2} \cos n\phi \left[J_{n-1}(k_o a \sin\theta) - J_{n+1}(k_o a \sin\theta)\right] \tag{3.40}$$

$$E_\phi = -V_o \pi a \frac{e^{-jkr}}{r} e^{-jn\pi/2} \sin n\phi \cos\theta \left[J_{n-1}(k_o a \sin\theta) + J_{n+1}(k_o a \sin\theta)\right] \tag{3.41}$$

where $k_o a = \frac{\alpha_{mn}}{\sqrt{\varepsilon_r}}$ and $J_n(x)$ represent Bessel functions of the first kind of order n, and V_o is the amplitude of the voltage excitation. Since $k_o a = \frac{1.841}{\sqrt{\varepsilon_r}}$ for the TM_{11} mode, the equations for the far-field E-fields simplify to

$$E_\theta = -j\pi V_o a \frac{e^{-jk_o r}}{r} \cos\phi \left\{J_o\left[\frac{1.841}{\sqrt{\varepsilon_r}}\right] - J_2\left[\frac{1.841}{\sqrt{\varepsilon_r}}\right]\right\} \tag{3.42}$$

$$E_\phi = -jV_o \pi a \frac{e^{-jkr}}{r} \sin\phi \cos\theta \left\{J_o\left[\frac{1.841}{\sqrt{\varepsilon_r}}\right] + J_2\left[\frac{1.841}{\sqrt{\varepsilon_r}}\right]\right\} \tag{3.43}$$

The normalized far-field patterns of the TM_{11} mode of a circular microstrip patch antenna are shown in Figure 3.25. The patterns are very similar to those of the rectangular microstrip patch discussed in Section 3.4.1.

The input resistance R_{in} of the circular microstrip patch at a feed point located at a radius s from the origin operating in the TM_{11} mode can be estimated using [26, Ch. 5]

$$R_{in} = R_r \frac{J_1^2(k_{11}s)}{J_1^2(k_{11}a)} = 0.336 R_r J_1^2(1.841 s/a) \tag{3.44}$$

where R_r is the radiation resistance of the antenna, and $k_{11}a = 1.841$. The radiation resistance can be derived from the radiated power P_r of the antenna using

$$R_r = \frac{V_o^2}{2P_r} \tag{3.45}$$

where V_o is the amplitude of the voltage excitation. The radiated power is determined by

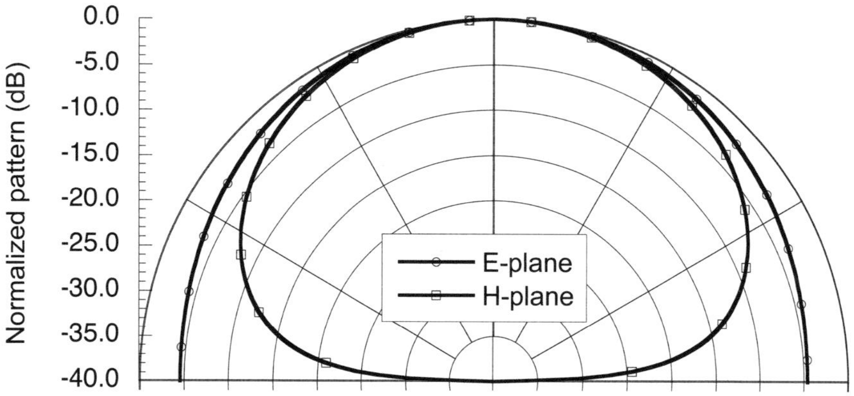

Figure 3.25 Normalized far-field patterns of the TM_{11} mode of circular microstrip antenna (with $a = 5.2$ mm, $h = 1.5$ mm, $\varepsilon_r = 2.2$, and $f_{11} = 10$ GHz).

$$P_r = \frac{V_o^{\,2}}{70.81\varepsilon_r} \int_0^{\pi/2} \left[J_0^2(X) - 0.59\varepsilon_r J_1^2(X) + J_2^2(X) \right] \sin\theta \, d\theta \tag{3.46}$$

where $X = \frac{1.841}{\sqrt{\varepsilon_r}} \sin\theta$. The integral in (3.46) needs to be evaluated numerically in order to determine the radiated power.

The radiation quality factor Q_r of the TM_{11} mode of the circular microstrip patch is obtained using

$$Q_r = \frac{2\pi f_{11} E_s}{P_r} \tag{3.47}$$

where E_s is the stored energy, given by

$$E_s = \frac{2.389 V_o^2}{16\pi h \, \mu_o f_{11}} \tag{3.48}$$

The quality factors associated with conductor loss (Q_c) and dielectric loss (Q_d) and surface wave loss (Q_s) are obtained using (3.23), (3.25), and (3.28) respectively. The radiation efficiency e_{rad} and BW can then be obtained using (3.31) and (3.32), respectively.

As a comparison to the rectangular microstrip patch of Section 3.4.1, a circular microstrip antenna printed on a substrate with thickness h = 1.5 mm and a dielectric constant of $\varepsilon_r = 2.2$ would require a radius of $a = 5.2$ mm to resonate at f_{11} = 10 GHz (for the TM_{11} mode), based on (3.38) and (3.39). The radiation quality factor (Q-factor) is found to be Q_r = 10.68 using (3.47) while the other values for the Q-factor are the same as for the corresponding rectangular patch, resulting in a total Q-factor of $Q = 9.0$. This translates to a radiation efficiency of $e_{rad} = 84.3\%$ and BW = 7.9% for a VSWR = 2. These results are close to those of the corresponding rectangular microstrip patch, but the area of the circular patch is about 84% that of the rectangular patch. Using (3.45), the radiation resistance of the circular microstrip patch is determined to be $R_r = 338\Omega$, which is significantly higher than that of the rectangular patch (due to the smaller area of the circular antenna).

3.4.4 Annular Microstrip Patch Antennas

The annular or ring microstrip patch antenna has also found widespread use. As shown in Figure 3.26, the antenna consists of a ring with inner radius a and outer radius b, fed at a point located at a radius s from the center, and printed on a substrate of thickness h and dielectric constant ε_r. The analysis of the annular microstrip patch antenna follows closely that of the disk antenna and a cavity model has been applied by various authors [50–54]. The outer diameter of the resonant annular microstrip antenna is smaller than the corresponding circular microstrip antenna, since for the former it is the mean circumference of the ring, which is equal to the guided wavelength. For the TM_{11} mode, however, the input impedance is higher and the impedance BW is lower than for the disk. One advantage of the annular microstrip antenna is the extra degree of freedom provided by the inner radius. By controlling the ratio of the outer to inner radii, the resonant frequencies of the higher order modes can be shifted, which may be of significant practical importance for multiband applications.

The resonant frequency of the TM_{n1} mode can be derived by expressing the wave number k_{1n} as a function of the average radius

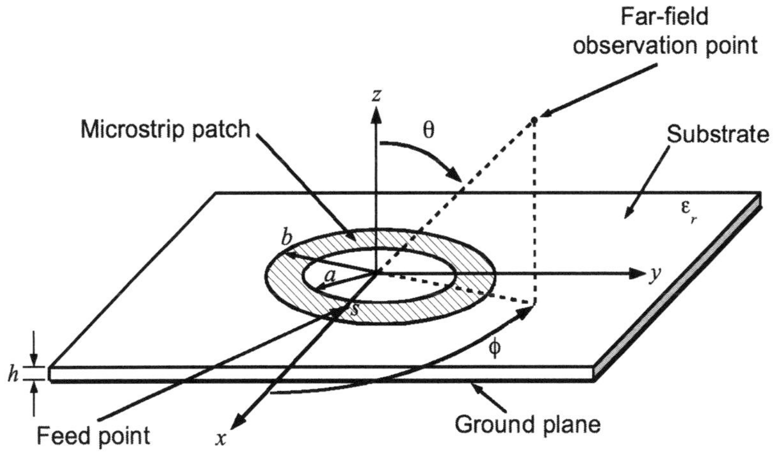

Figure 3.26 Ring microstrip patch antenna.

$$k_{n1} = \frac{2n}{a+b} \tag{3.49}$$

Expressing this in terms of the resonant frequency f_{1n},

$$f_{n1} = \frac{n\,c}{\pi(a+b)\sqrt{\varepsilon_{eff}}} \tag{3.50}$$

where c is the speed of light and ε_{eff} is the effective dielectric constant, which can be approximated using (3.9) by setting $w = b - a$. Reasonably accurate values for k_{nl} are obtained for $n \leq 5$ and $(b - a) / (b + a) < 0.35$ [27]. A more accurate value for the resonant frequency can be determined by applying the cavity model and solving for the following characteristic equation resulting from the application of the appropriate boundary conditions [54]

$$J_n'(kb)Y_n'(ka) - J_n'(ka)Y_n'(kb) = 0 \tag{3.51}$$

where $J_n(x)$ and $Y_n(x)$ are the Bessel functions of the first and second kind, respectively, of order n. If the roots of (3.51) are denoted by α_{nm}, then the resonant frequency is

$$f_{nm} = \frac{\alpha_{nm} c}{2\pi a\sqrt{\varepsilon_{eff}}} \tag{3.52}$$

Further improvements in accuracy can be obtained by replacing the physical radii a and b with their effective values

$$a_e = a - \frac{W_e - (b-a)}{2} \tag{3.53}$$

$$b_e = b + \frac{W_e - (b-a)}{2} \tag{3.54}$$

where

$$W_e = \frac{120\pi h}{Z_o\sqrt{\varepsilon_{eff}}} \tag{3.55}$$

The characteristic impedance Z_o is obtained using (3.11) by again setting $w = b - a$. Equation (3.53) is only valid if $(W_e - b + a)/2 < a$, otherwise the value for a_e becomes negative. Approximations to (3.53) and (3.54) are

$$a_e = a - \frac{3h}{4} \tag{3.56}$$

$$b_e = b + \frac{3h}{4} \tag{3.57}$$

The far-field E-fields of the annular microstrip antenna are given by

$$E_\theta = \frac{j^n 2E_o k_o h}{\pi k_{nm}} \frac{e^{-jk_o r}}{r}\left[J_n^{'}(k_o a\sin\theta) - \frac{J_n^{'}(k_{nm}a)}{J_n^{'}((k_{nm}b))} J_n^{'}(k_o b\sin\theta)\right]\cos n\phi \quad (3.58)$$

$$E_\phi = \frac{-nj^n 2E_o k_o h}{\pi k_{nm}} \frac{e^{-jk_o r}}{r}\left[\frac{J_n^{'}(k_o a\sin\theta)}{k_o a\sin\theta} - \frac{J_n^{'}(k_{nm}a)}{J_n^{'}((k_{nm}b))} \frac{J_n^{'}(k_o b\sin\theta)}{k_o b\sin\theta}\right]\sin n\phi\cos\theta$$

(3.59)

where $k_{nm}a = \alpha_{nm}$.

The far-field patterns of the TM_{11} mode of a ring microstrip patch antenna are shown in Figure 3.27 and are similar to both the rectangular and circular shaped microstrip patches already presented.

The radiation quality factor for the TM_{11} mode of the ring patch is determined using [28, Ch. 3]

$$Q_r = \frac{60\varepsilon_r}{f_{11}\,\mu_o h\, I_1}\left\{\frac{J_n^{'2}(k_{11}a_e)}{J_n^{'2}(k_{11}b_e)}\left[1 - \frac{1}{k_{11}^2 b_e^2}\right] - \left[1 - \frac{1}{k_{11}^2 a_e^2}\right]\right\} \quad (3.60)$$

where

$$\begin{aligned} I_1 = &\int_0^{\pi/2} \frac{\cos^2\theta}{k_o^2\sin\theta}\left\{\frac{J_1(k_o a_e\sin\theta)}{a_e} - \frac{J_1(k_o b_e\sin\theta)}{b_e}\frac{J_1^{'}(k_{11}a_e)}{J_1^{'}(k_{11}b_e)}\right\}^2 d\theta \\ &+ \int_0^{\pi/2} \sin\theta\left\{J_1^{'}(k_o a_e\sin\theta) - J_1^{'}(k_o b_e\sin\theta)\frac{J_1^{'}(k_{11}a_e)}{J_1^{'}(k_{11}b_e)}\right\}^2 d\theta \end{aligned} \quad (3.61)$$

Finally, the input impedance of the microstrip ring antenna fed by a coaxial probe at a radius ρ_o from the center can be determined by solving

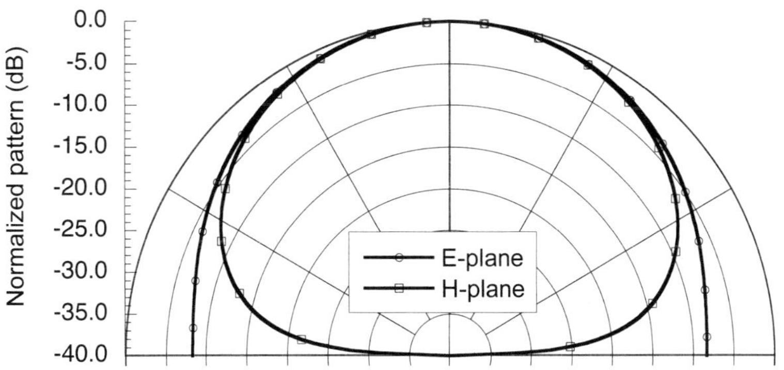

Figure 3.27 Normalized far-field patterns of the TM_{11} mode of ring microstrip antenna (with $a = 3.9$ mm, $b = 4.5$ mm, $h = 1.5$ mm, $\varepsilon_r = 2.2$, and $f_{11} = 10$ GHz).

$$Z_{in} = j2\pi f\mu_o h \cos^2 n\phi_o \sum_{n=0}^{\infty}\sum_{m=1}^{\infty} A_{mn} \tag{3.62}$$

where

$$A_{mn} = \frac{\pi k_{nm}^2 \left(\dfrac{\sin 2nw_f}{2nw_f}\right)^2 \left[J_n(k_{nm}\rho_o)Y_n'(k_{nm}a_e) - J_n'(k_{nm}a_e)Y_n(k_{nm}\rho_o)\right]^2}{2n\delta_n\left(k_{eff}^2 - k_{nm}^2\right)\left\{\dfrac{J_n'^2(k_{nm}a_e)}{J_n'^2(k_{nm}b_e)}\left[1-\dfrac{n^2}{k_{nm}^2 b_e^2}\right]-\left[1-\dfrac{n^2}{k_{nm}^2 a_e^2}\right]\right\}} \tag{3.63a}$$

$$k_{eff} = k_o\sqrt{\varepsilon_{eff}\left[1 - j\frac{1}{Q}\right]} \tag{3.63b}$$

$$\delta_n = \begin{cases} 1 & \text{for } n \neq 0 \\ 2 & \text{for } n = 0 \end{cases} \tag{3.63c}$$

The finite diameter of the probe is accounted for by modeling it as a uniform current ribbon of angular width $2w_f$ in (3.63a).

For the ring patch printed on a substrate with thickness h = 1.5 mm and dielectric constant of ε_r = 2.2, one design with the TM_{11} mode resonant at 10 GHz would require inner and outer radii of a = 3.9 mm and b = 4.5 mm, resulting in less area than the previously examined rectangular and circular patches, but more than the printed dipole. The radiation quality factor is found to be Q_r = 17.8 using (3.64) while the other values for Q-factor are the same as for the corresponding rectangular patch, resulting in a total quality factor of Q = 13.6. This translates to a radiation efficiency of e_{rad} = 76.7% and BW = 5.1% for VSWR = 2.

A comparison of the performance of the four microstrip patch antennas designed at 10 GHz is summarized in Table 3.1.

Table 3.1

Summary of the 10-GHz Microstrip Antenna Design Examples
(h = 1.5 mm, ε_r = 2)

Antenna	Dimensions (λ_o)	Area (λ_o^2)	Q	Bandwidth (VSWR = 2)	Radiation Efficiency
Rectangular patch	L = 0.283 W = 0.395	0.112	10.1	7%	82.6%
Microstrip dipole	L = 0.295 W = 0.067	0.02	35	2%	79.7%
Circular patch	a = 0.173	0.094	9.0	7.9%	84.3%
Annular patch	a = 0.13 b = 0.15	0.071	13.6	5.1%	76.7%

3.5 SLOT ANTENNAS

A thin rectangular slot antenna formed by cutting an aperture in an infinitely large but thin perfect metal conductor was analyzed in 1946 and was shown to exhibit complementary behavior to a corresponding wire dipole radiating in free space [55]. The slot produces the same radiation pattern as the dipole except that the E-fields and H-fields are exchanged (that is, the polarization of the slot pattern is orthogonal to that of the dipole). The terminal impedance Z_s of the slot and Z_d of the dipole are also related by the expression

$$Z_s = \frac{\eta^2}{4Z_d} \tag{3.64}$$

where η is the intrinsic impedance of free space.

Slot antennas also share many common attributes with printed antennas. They have a low profile, are relatively easy to fabricate, can be formed into several shapes (as shown in Figure 3.28), and can be excited using various feeding techniques, such as those shown in Figure 3.29. Slot antennas can be formed by cutting apertures in metal sheets or waveguide walls, or by using photo-etching techniques similar to those used for fabricating microstrip patch antennas. Slot antennas can be designed for bi-directional radiation [which is the case for the examples in Figure 3.29(a)–(c)] or hemispherical radiation as is the case for a slot backed by a metal cavity or by the slot in a waveguide shown in Figure 3.29(d). This section examines the behavior of the rectangular slot and the circular ring slot, two of the most common slot shapes.

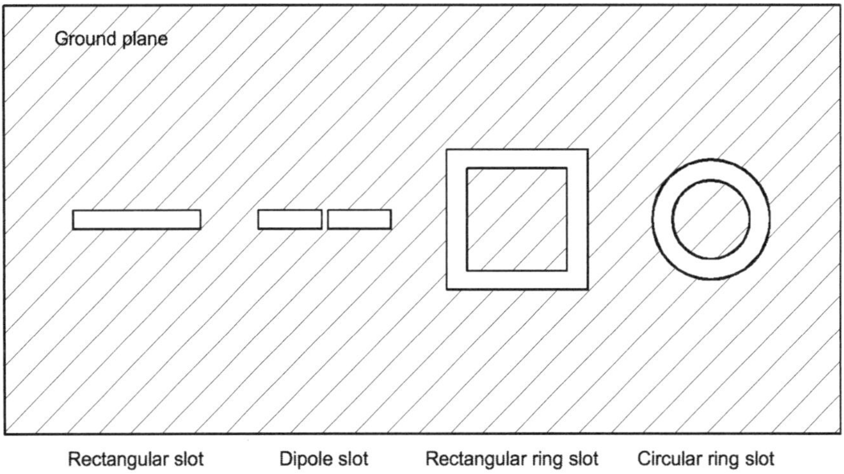

Figure 3.28 Examples of slot antennas.

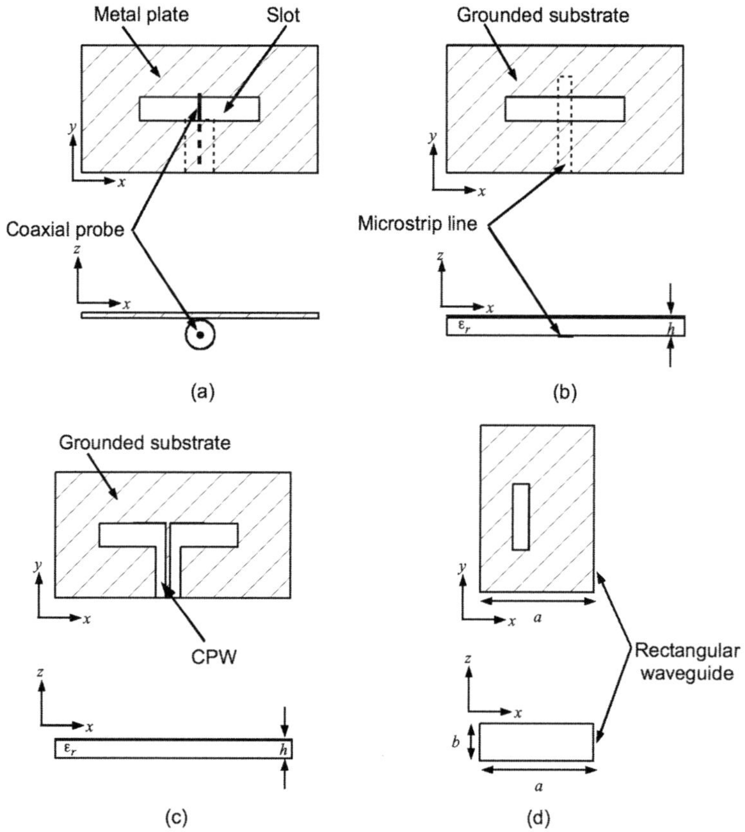

Figure 3.29 Typical methods of exciting a slot antenna: (a) coaxial feed, (b) microstrip feed, (c) CPW feed, and (d) slotted waveguide.

3.5.1 Rectangular Slot Antennas

The rectangular slot antenna is typically excited such that the amplitude of the electric field in the aperture of the slot exhibits a cosinusoidal behavior. Figure 3.30 sketches the E-field for the lowest order mode (TE_{10}) for a rectangular slot antenna of length L and width W. Assuming that the aperture E-field exists only within the slot itself and is zero elsewhere; that is,

$$E_a = \begin{cases} \hat{\mathbf{y}} E_o \cos\left(\frac{\pi}{L} x\right) & \text{for } -\frac{L}{2} \le x \le \frac{L}{2}, -\frac{W}{2} \le y \le \frac{W}{2} \\ 0 & \text{elsewhere} \end{cases} \tag{3.65}$$

then the far-field E-field can be represented by the following expressions [35, Ch. 12]

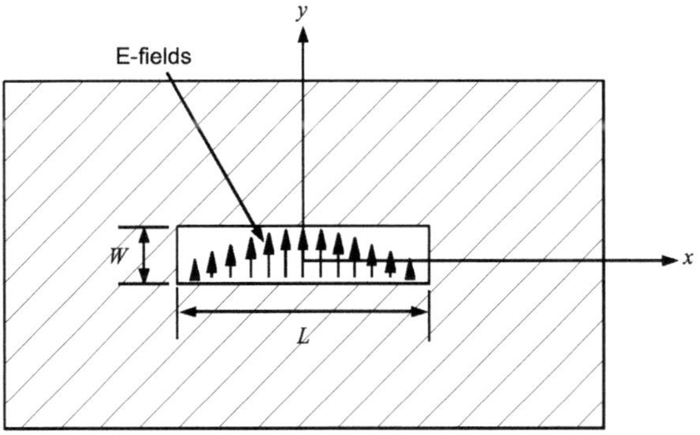

Figure 3.30 Rectangular slot antenna excited in the TE_{10} mode.

$$E_\theta = A\sin\phi \frac{\cos X}{X^2 - \left(\frac{\pi}{2}\right)^2} \frac{\sin Y}{Y} \tag{3.66}$$

$$E_\phi = A\cos\theta\cos\phi \frac{\cos X}{X^2 - \left(\frac{\pi}{2}\right)^2} \frac{\sin Y}{Y} \tag{3.67}$$

where

$$X = \frac{kL}{2}\sin\theta\cos\phi \tag{3.68a}$$

$$Y = \frac{kW}{2}\sin\theta\sin\phi \tag{3.68b}$$

$$A = -j\frac{kWLE_o}{4}\frac{e^{-jkr}}{r} \tag{3.68c}$$

The normalized patterns of a rectangular slot with dimensions $L = \lambda_o/2$, $W = \lambda_o/5$ based on (3.66) and (3.67) are shown in Figure 3.31. The slot radiates as if it were a magnetic dipole; that is, its radiation pattern is the same as that of an electric dipole except that the E- and H-planes are exchanged. The back radiation pattern of the slot antenna can be suppressed by adding a metal back plate or enclosing the backside in a metal cavity.

3.5.2 Annular Slot Antennas

The circular ring or annulus is another commonly used shape for a slot antenna, producing similar patterns to those of the circular ring microstrip patch for the case where a metal plate or cavity backing is used. The geometry of the annular slot

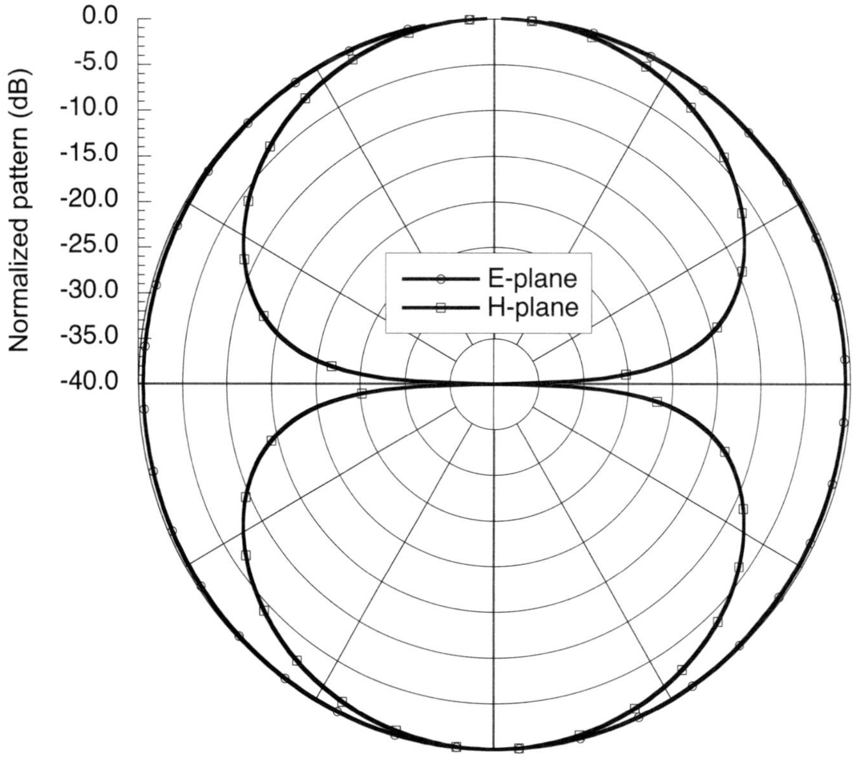

Figure 3.31 Normalized radiation pattern of a rectangular slot antenna excited in the TE_{10} mode (with $L = \lambda_o/2$, and $W = \lambda_o/5$).

antenna is shown in Figure 3.32. For a thin slot ($b - a << \lambda_o$), the radial E-field in the slot can be approximated by $E_\rho = E_o\cos(n\phi')$. The resultant far-field E-field can then be determined using [26, Ch. 7]

$$E_\theta = -j^n \frac{E_o a k_o (b-a)}{2} \frac{e^{-jk_o r}}{r} \cos(n\phi) J'_n(ak_o \sin\theta) \tag{3.69}$$

$$E_\phi = j^n n E_o (b-a) \frac{e^{-jk_o r}}{r} \sin(n\phi) \cot\theta J_n(ak_o \sin\theta) \tag{3.70}$$

If a constant E-field can be maintained within the slot (i.e., $n = 0$ and thus $E_\rho = E_o$), then the slot will radiate the dipole-like pattern shown in Figure 3.33. For the $n = 1$ mode, the slot will radiate a broadside pattern, as shown in

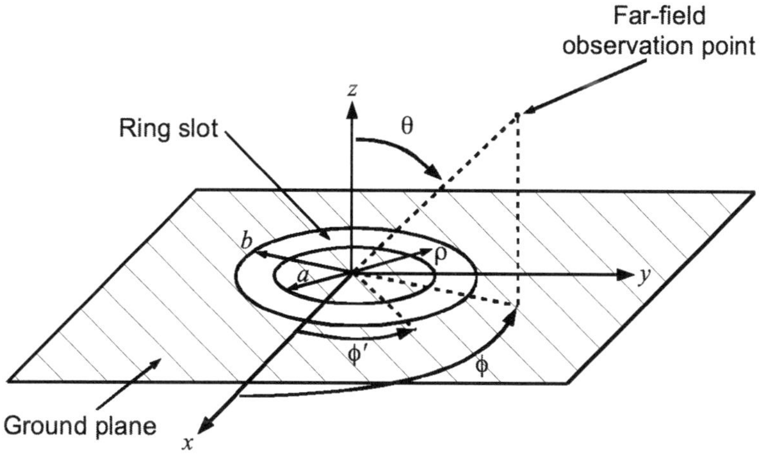

Figure 3.32 Annular slot antenna.

Figure 3.34. Again, the patterns in Figures 3.33 and 3.34 could be made hemispherical by enclosing the underside of the slot antennas with a metal cavity. Examples of frequency-agile slot antennas will be examined in subsequent chapters.

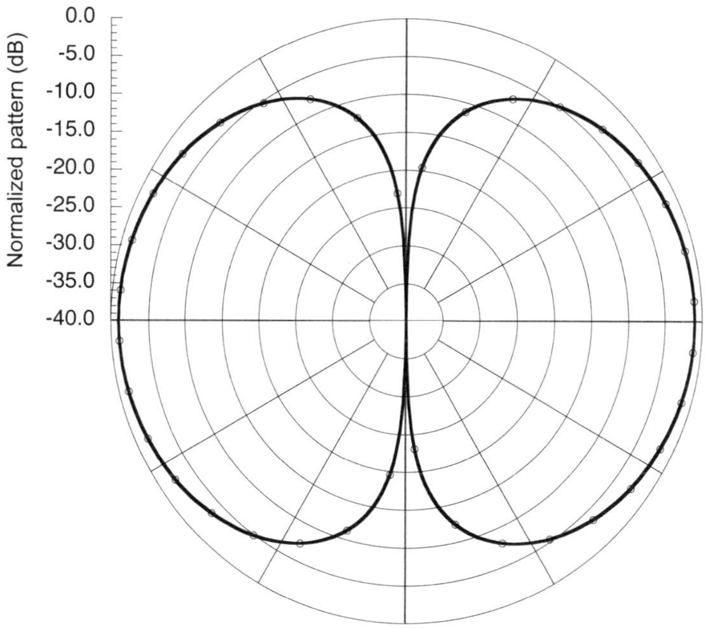

Figure 3.33 Normalized radiation pattern of a ring slot antenna ($n = 0$, $ak_o = 1$).

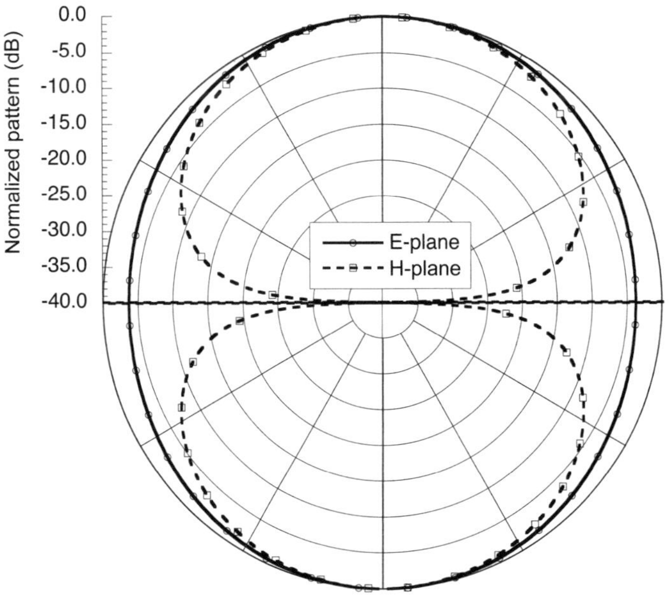

Figure 3.34 Normalized radiation pattern of a ring slot antenna (n = 1, ak_o = 1).

3.6 IFAs

The IFA belongs to a class of low-profile bent wire antennas originally proposed for use as telemetry antennas on missiles [56]. The IFA, as shown in Figure 3.35, consists of a wire of radius a and length L placed horizontally at a height h above a ground plane. The wire is shorted to the ground plane at one end and is fed at a distance L_1 from the short. Printed versions of the IFA have also been designed, such as that shown in Figure 3.36, where the IFA is coplanar with the ground plane. Other examples of printed IFAs are found in [57–65]. The length L of the IFA is typically chosen to be near a quarter wavelength, while the height h is a small fraction of a wavelength ($h \sim 0.1\lambda_o$).

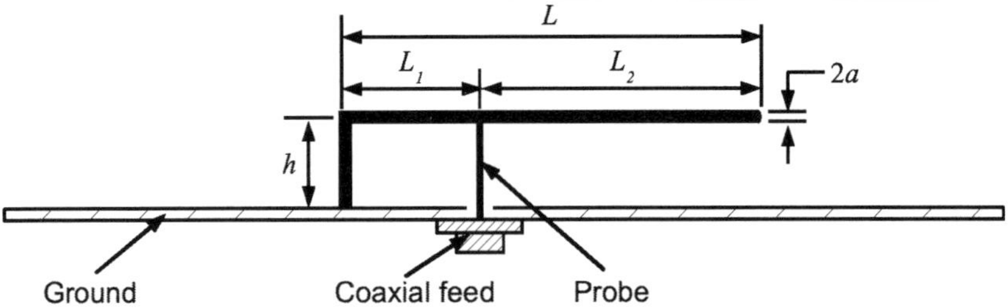

Figure 3.35 IFA geometry.

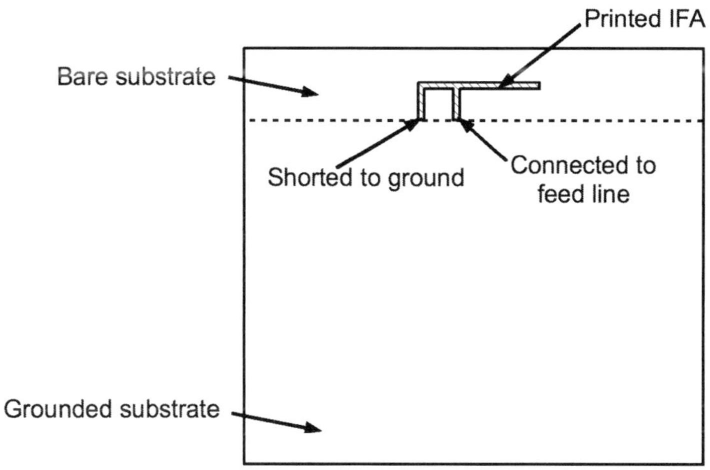

Figure 3.36 Printed IFA.

The radiation resistance and input impedance of the wire IFA have been estimated by modeling the antenna as a parallel wire transmission line, whose generalized circuit model is shown in Figure 3.37, where an infinite ground plane is assumed [56]. This circuit model allows for the more general case where the short circuit and open circuit ends of the IFA are replaced with two arbitrary load impedances, Z_1 and Z_2.

If the transmission line is excited by a voltage source V_o that generates a current I_o, then the currents I_1 to the left of the generator and I_2 to the right of the generator can be determined from [56]

$$I_1(x) = -I_o\left[\frac{\cosh(\gamma L_2 + \theta_2)\sinh(\gamma x + \theta_1)}{\sinh(\gamma L + \theta_1 + \theta_2)}\right] \quad \text{for } 0 \le x \le L_1 \tag{3.71}$$

$$I_2(w) = I_o\left[\frac{\cosh(\gamma L_1 + \theta_1)\sinh(\gamma w + \theta_2)}{\sinh(\gamma L + \theta_1 + \theta_2)}\right] \quad \text{for } 0 \le w \le L_2 \tag{3.72}$$

where

$$\theta_1 = \rho_1 + j\phi_1 = \coth^{-1}\left[\frac{Z_1}{Z_o}\right] \tag{3.73a}$$

$$\theta_2 = \rho_2 + j\phi_2 = \coth^{-1}\left[\frac{Z_2}{Z_o}\right] \tag{3.73b}$$

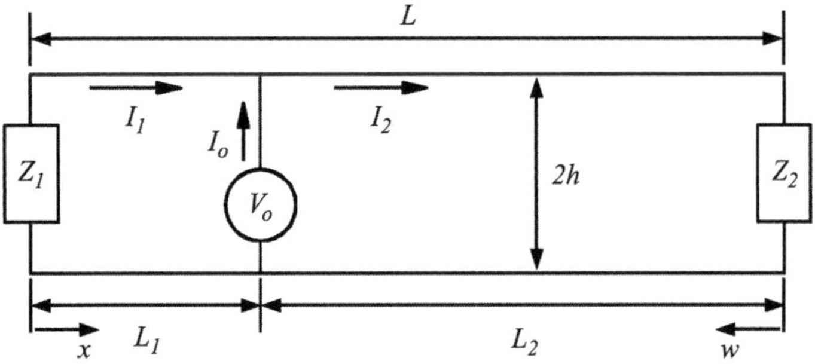

Figure 3.37 Generalized equivalent circuit of an IFA. (After [56].)

$$Z_o = \frac{120}{\pi}\cosh^{-1}\left[\frac{h}{a}\right] \tag{3.73c}$$

In (3.71) and (3.72), $x = 0$ and $w = 0$ represent the location of loads Z_1 and Z_2, respectively (with $w = L\text{-}x$). Also, $\gamma = \alpha + j\beta$ is the complex propagation constant where α is the attenuation constant and β the guided wave number ($\beta = 2\pi/\lambda_o$). For low-loss lines, $\gamma \approx j\beta$, $\theta_1 \approx j\phi_1$, and $\theta_2 \approx j\phi_2$.

The input impedance Z_{in} of the generalized circuit model as seen across the voltage generator is given by

$$Z_{in} = R_{in} + jX_{in} = R_{rad} + j\left[\frac{X_1X_2}{X_1 + X_2}\right] \tag{3.74}$$

where

$$X_1 = -Z_o\cot(\beta L_1 + \phi_1) \tag{3.75a}$$

$$X_2 = -Z_o\cot(\beta L_2 + \phi_2) \tag{3.75b}$$

The radiation resistance R_{rad} is expressed as

$$R_{rad} = 120(\beta h)^2\left[0.5\left(1 + K_1^2 + K_2^2\right) - A - B\right] \tag{3.76}$$

where

$$A = K_2\sin\phi_1\frac{\sin\beta L_1}{\beta L_1} + K_1\sin\phi_2\frac{\sin\beta L_2}{\beta L_2} \tag{3.77a}$$

$$B = K_1K_2\cos(\phi_1 + \phi_2)\frac{\sin\beta L}{\beta L} \tag{3.77b}$$

$$K_1 = \frac{\cos(\beta L_1 + \phi_1)}{\sin(\beta L + \phi_1 + \phi_2)} \quad (3.77c)$$

$$K_2 = \frac{\cos(\beta L_2 + \phi_2)}{\sin(\beta L + \phi_1 + \phi_2)} \quad (3.77d)$$

The general formula for R_{rad} in (3.76) can be greatly simplified for specific antenna cases.

(1) For an end-driven open-circuit line, corresponding to a simple bent monopole [Figure 3.38(a)] where $\phi_1 = \phi_2 = 0$, and $L_1 = 0$, then

$$R_{rad} = \frac{120(\beta h)^2}{\sin^2 \beta L}\left[1 - \frac{\sin 2\beta L}{2\beta L}\right] \quad (3.78)$$

(2) For an end-driven short-circuited line, corresponding to a half-square loop [Figure 3.38(b)] where $\phi_1 = 0$, $\phi_2 = \pi/2$, and $L_1 = 0$, then

$$R_{rad} = \frac{120(\beta h)^2}{\cos^2 \beta L}\left[1 - \frac{\sin 2\beta L}{2\beta L}\right] \quad (3.79)$$

(3) For the IFA with open-circuit line [Figure 3.38(c)]: ($\phi_1 = \pi/2$, $\phi_2 = 0$)

(a) If ($\cos\beta L \neq 0$), then

$$R_{rad} = \frac{120(\beta h)^2}{\cos^2 \beta L}\left[A_1 - A_2\right] \quad (3.80)$$

where

$$A_1 = 0.5\left(\cos^2 \beta L + \sin^2 \beta L_1 + \cos^2 \beta L_2\right) \quad (3.81a)$$

$$A_2 = \cos \beta L \cos \beta L_2 \frac{\sin \beta L_1}{\beta L_1} \quad (3.81b)$$

(b) If $\beta L_2 = \pi/2$, then

$$R_{rad} = 120(\beta h)^2 \quad (3.82)$$

(4) For the IFA with short-circuited line [Figure 3.38(d)]: ($\phi_1 = 0$, $\phi_2 = \pi/2$)

(a) If ($\sin\beta L \neq 0$), then

$$R_{rad} = \frac{120(\beta h)^2}{\sin^2 \beta L}\left[A_3 - A_4\right] \quad (3.83)$$

$$A_3 = 0.5\left(\sin^2 \beta L + \sin^2 \beta L_1 + \sin^2 \beta L_2\right) \quad (3.84a)$$

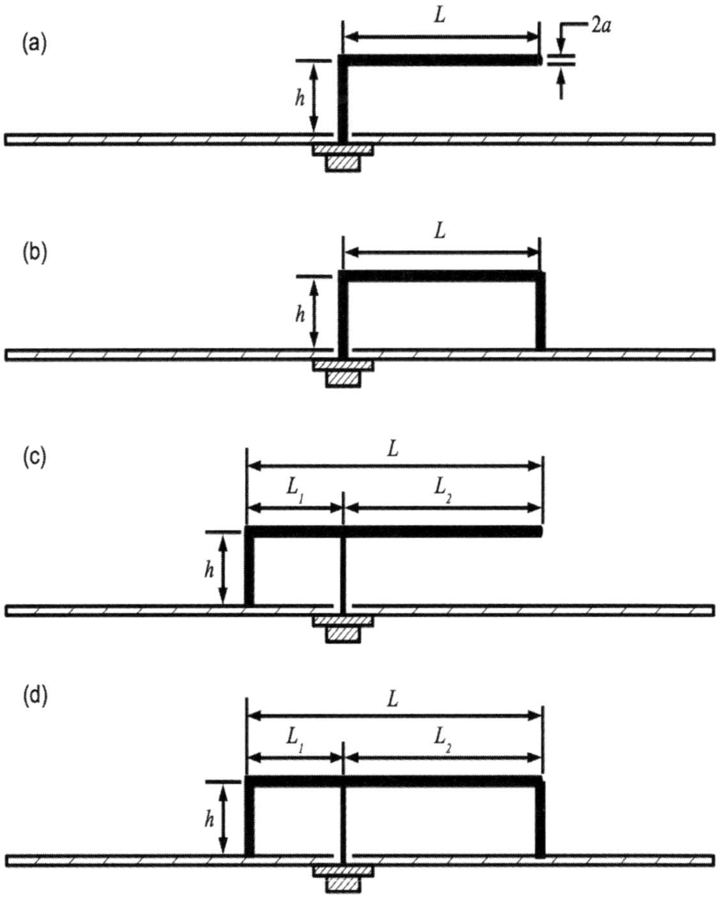

Figure 3.38 Different configurations of bent wire antennas.

$$A_4 = \frac{L_1^2 + L_1 L_2 + L_2^2}{\beta L L_1 L_2} \sin \beta L \sin \beta L_1 \sin \beta L_2 \qquad (3.84b)$$

(b) If $\beta L_2 = \pi/2$, then (3.82) applies.

The radiation patterns of the IFA can be estimated based on the currents in (3.71) and (3.72) by using a combination of image theory and the auxiliary vector potential method, which is valid for the case of an infinite ground plane [35, Ch. 4]. In the presence of a small ground plane, diffraction effects need to be included, and the radiation pattern is best determined using an electromagnetic numerical method. Figure 3.39 shows the normalized far-field E-field components in two orthogonal planes ($\phi = 0^\circ$ and $\phi = 90^\circ$), determined using the method of

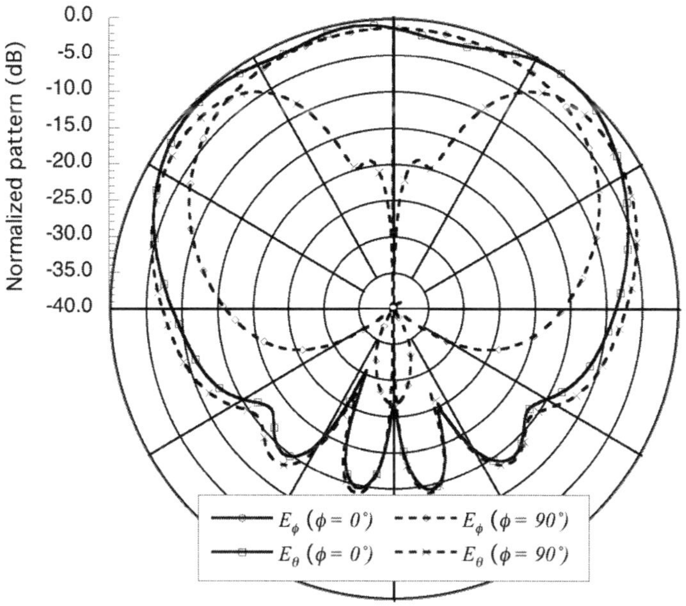

Figure 3.39 Normalized E-fields of a bent monopole (h = 0.1λ_o, L = 0.25λ_o) on finite ground (1.25λ_o x 1.25λ_o).

moments [66] of the bent wire antenna of Figure 3.8(a) with $h = 0.1\lambda_o$ and $L = 0.25\lambda_o$ placed near the center of a finite square ground plane ($1.25\lambda_o$ x $1.25\lambda_o$). The combined far-fields would thus form a nearly hemispherical pattern, with both orthogonal polarizations being strongly excited. The significant amount of spillover radiation below the finite ground plane is evident in Figure 3.39.

In recent years, printed IFAs, similar to the one in Figure 3.36, have been used as internal antennas in cellular phones and other portable wireless devices. In these applications, the small size of the ground plane and surrounding electronics has a significant impact on the IFA performance. Full-wave electromagnetic simulations are generally required to accurately predict the input impedance and radiation patterns for the printed IFAs used in such applications. Typically the resultant radiation patterns have a fairly low gain (~1 dBi) with almost equal power radiated in the E_θ and E_ϕ far-field components. As an example, an IFA mounted on a small ground plane with the dimensions of a typical cell phone, as shown in Figure 3.40, is analyzed. The radiation pattern is again determined using the method of moments, and the results are shown in Figure 3.41. For this configuration, the radiation is nearly isotropic, which is desirable for handheld wireless communications, where the orientation and position of the device can be continuously changing. In real applications, blockage from a user's hand or head can cause significant distortion in the radiation pattern, and multiple antennas might be used to improve performance.

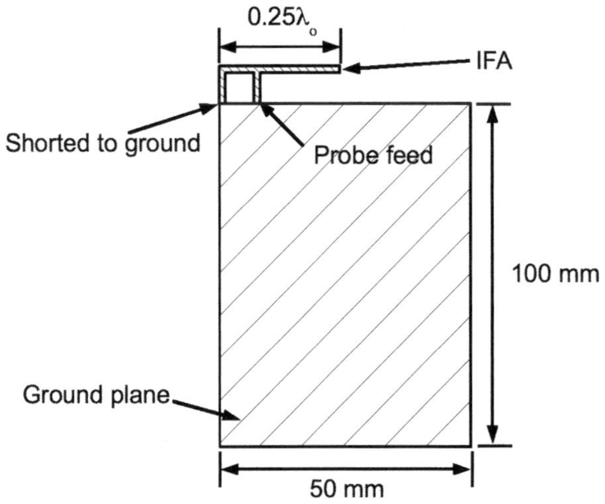

Figure 3.40 IFA mounted on cell-phone-sized ground plane.

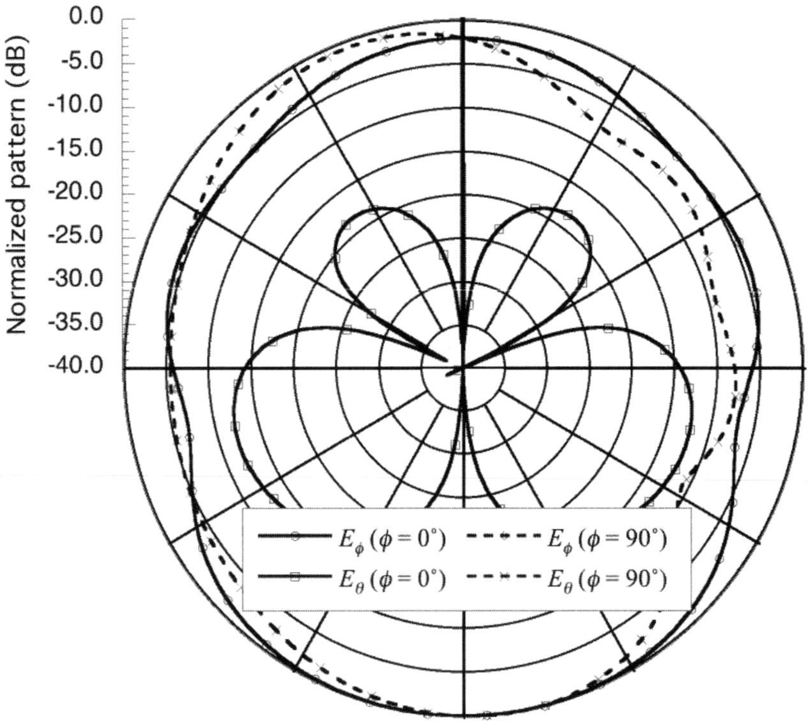

Figure 3.41 Normalized E-fields of the IFA of Figure 3.40.

3.7 PIFAs

The PIFA consists of a small rectangular plate of length L and width W, placed on a dielectric block of height h and relative permittivity ε_r, all located on a ground plane (usually of small electrical size). The plate is connected to ground at one end by a vertical wall of width s and is excited by a probe located at a distance d from the shorting wall. In many designs, the dielectric block is absent, and the metal plate is suspended in air above the ground plane. A sketch of the PIFA is shown in Figure 3.42. The PIFA can be considered as an extension of the IFA where the horizontal wire of the IFA is replaced by the rectangular plate. Examples of PIFAs are found in [67–71]. The resonant frequency f_0 of the PIFA is approximated by

$$f_o = \frac{c}{4\sqrt{\varepsilon_r}\left(L + W - s\right)} \tag{3.85}$$

where c is the speed of light. Equation (3.85) shows that the width s of the shorting wall has a significant impact on the resonant frequency of the PIFA. If the shorting plate width is equal to the plate width ($s = W$), the resonant frequency of the PIFA is similar to a short-circuited quarter-wave microstrip patch antenna. However, by reducing s, this resonant frequency can be lowered, resulting in a more compact antenna design.

PIFAs are generally designed to have a low profile, resulting in a value of h, which is a small fraction of a wavelength. The input impedance of the PIFA can be controlled by adjusting the separation distance d between the feed probe and the short circuit. The real part of the input impedance decreases toward zero as the feed is moved closer to the shorting wall (that is, as d approaches zero). The inductance of the probe can be reduced by introducing a capacitive gap between the feed probe and the top plate of the PIFA [70]. The bandwidth can be enhanced for wider band or multiband operation by modifying the PIFA shape, often by

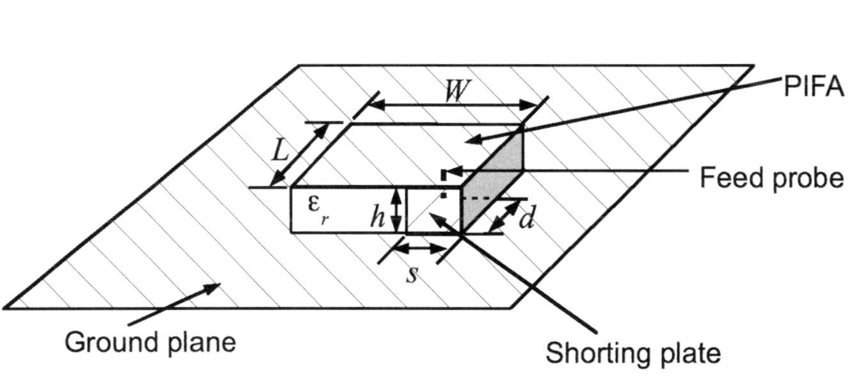

Figure 3.42 PIFA geometry.

introducing slots, or nested elements [72–75]. PIFAs have also found widespread use in handsets and portable devices, and, like IFAs, their impedance and radiation characteristics will be considerably affected by the small ground plane and circuitry of the device. They typically produce low-gain patterns (1–2 dBi) with nearly equal power radiated in both orthogonal polarizations. Frequency-agile versions of the PIFA will be examined in Chapters 7 and 8.

3.8 DRAs

Dielectric resonators, typically cylindrical in shape and composed of low-loss dielectric material with a high relative permittivity, have been used in microwave circuits for many years due to their compact size, high Q-factor, and ease of integration with printed circuits [76]. It was not until the early 1980s that an in-depth investigation was carried out on the radiation characteristics of dielectric resonators [77, 78]. This initial and subsequent work highlighted the potential benefits of DRAs especially in terms of their design flexibility, making them a viable alternative to microstrip technology. Detailed analysis and design information on DRAs have been compiled in two books [79, 80] and a summary of the recent state of the art of DRA technology can be found in [81]. Many DRA shapes have been proposed for improving impedance bandwidth, generating circular polarization, or achieving higher gain; however, this section will focus only on the two most common and simple DRA shapes: cylindrical and rectangular DRAs. Even with these two simple shapes, a high degree of design flexibility can be achieved, making them useful for a wide range of applications.

3.8.1 Cylindrical DRAs

The cylindrical DRA, as shown in Figure 3.43, is formed by a dielectric material of relative permittivity ε_r, having a height h and radius a. Although not strictly required, the DRA is usually placed on a large ground plane to facilitate excitation. As was the case for microstrip antennas, DRAs can be excited using numerous feeding techniques, with the two most common ones being a probe or a slot aperture. The resonant frequency and bandwidth of the cylindrical DRA are a function of the height-to-radius (h/a) ratio and the relative permittivity ε_r of the dielectric material. The shape of the radiation pattern of the cylindrical DRA depends on the mode being excited. Two modes that are commonly excited are the $TM_{01\delta}$ and the $HEM_{11\delta}$.

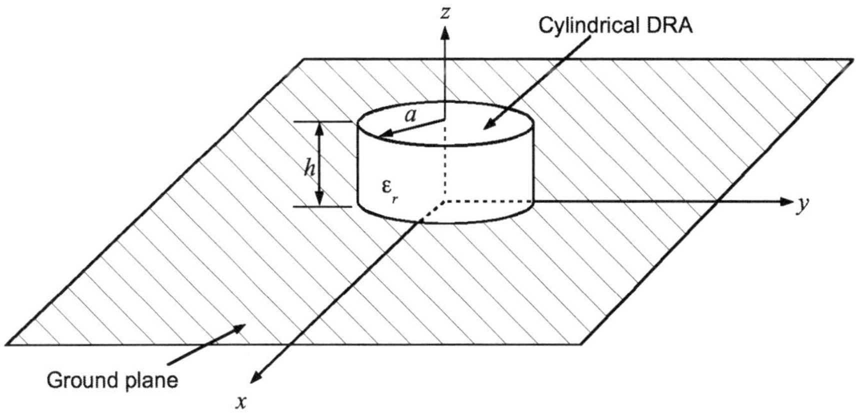

Figure 3.43 Cylindrical DRA.

The H-fields of the $TM_{01\delta}$ mode within the cylindrical DRA are sketched in Figure 3.44. This mode is typically excited using a probe located at the center of the DRA. The $TM_{01\delta}$ mode produces a conical pattern similar to that of a short monopole, as sketched in Figure 3.44. The resonant frequency f_o of the cylindrical DRA operating in the $TM_{01\delta}$ mode is given by [82]

$$f_o = \frac{c\sqrt{3.83^2 + (\pi/2)^2 (a/h)^2}}{2\pi a\sqrt{\varepsilon_r + 2}} \tag{3.86}$$

where c is the speed of light and where (3.86) is valid for $0.4 \le a/h \le 6$. The radiation Q-factor for this mode is a more complicated function of aspect ratio and can be approximated by the following curve-fitted expression [83]

$$\begin{aligned} Q_r &= 0.008721\varepsilon_r^{0.888413} e^{0.0397475\varepsilon_r} \left\{1 - \left[0.3 - 0.2\frac{a}{h}\right]\left[\frac{38-\varepsilon_r}{28}\right]\right\} \\ &\times \left\{9.498186\frac{a}{h} + 2058.33\left[\frac{a}{h}\right]^{4.322261} e^{-3.50099\left(\frac{a}{h}\right)}\right\} \end{aligned} \tag{3.87}$$

The radiation efficiency of DRAs can approach 100% when they are fabricated from low-loss microwave dielectrics, since conductor losses are negligible and

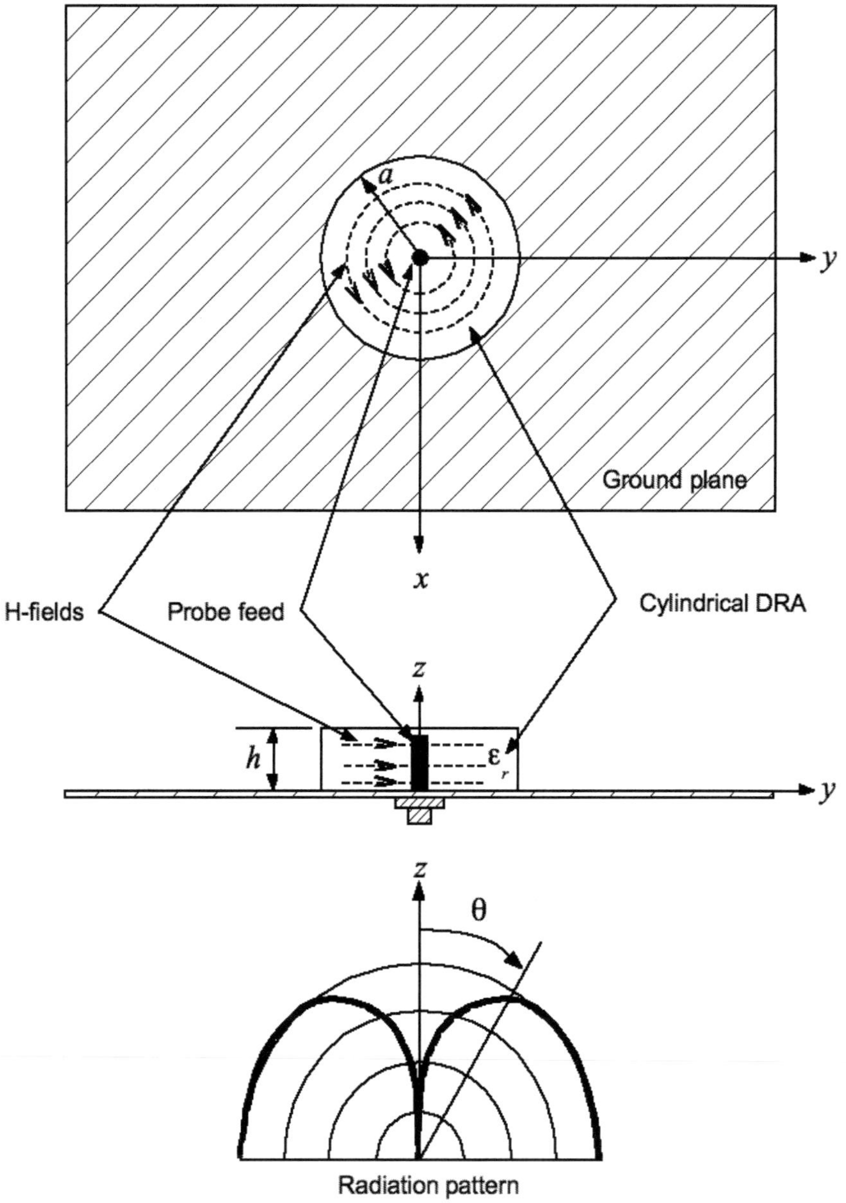

Figure 3.44 $TM_{01\delta}$ mode of the cylindrical DRA.

surface waves are nonexistent. The bandwidth of the DRA can thus be well approximated by substituting the radiation Q-factor of (3.87) into (3.31).

There are no simple expressions for determining the input impedance of the DRA. Full-wave electromagnetic solvers are usually used to analyze the DRA and to determine the input impedance, which can be controlled to some extent by adjusting the probe height and diameter.

The H-fields of the $HEM_{11\delta}$ mode are sketched in Figure 3.45. This mode can be excited by a slot, as shown in Figure 3.45, or by a probe located close to the outer wall of the cylinder. The $HEM_{11\delta}$ mode produces a radiation pattern similar to that of a short horizontal magnetic dipole, as sketched in Figure 3.45. The resonant frequency and radiation Q-factor for this mode can be estimated by the following expressions [82]

$$f_o = \frac{6.324c}{2\pi a\sqrt{\varepsilon_r + 2}}\left\{0.27 + 0.36\frac{a}{2h} + 0.02\left(\frac{a}{2h}\right)^2\right\} \tag{3.88}$$

$$Q_r = 0.01007\varepsilon_r^{1.3}\frac{a}{h}\left\{1 + 100e^{-2.05\left(\frac{a}{2h} - \frac{1}{80}\left(\frac{a}{h}\right)^2\right)}\right\} \tag{3.89}$$

where (3.88) is valid over the range of $0.4 \leq a/h \leq 6$ and (3.89) is valid for $0.4 \leq a/h \leq 5$. The cylindrical DRA operating in the $HEM_{11\delta}$ mode can also be designed to radiate circular polarization by using a cross-shaped slot or two probes fed 90° out of phase.

Table 3.2 lists dimensions of various cylindrical DRAs all designed to resonate at 10 GHz for the $TM_{01\delta}$ and $HEM_{11\delta}$ modes, along with the accompanying Q-factors and corresponding bandwidths. One of the advantages of the cylindrical DRA over the circular microstrip patch is the ability to vary the DRA aspect ratio (a/h) for a given dielectric constant while still maintaining the same resonant frequency. This allows the designer flexibility in accommodating physical constraints that may be imposed by certain applications such as maintaining a low profile or a compact footprint. A much wider range of values for the dielectric constant can also be chosen for DRAs than for microstrip patches, since there is no power lost in surfaces waves and thus little impact on radiation efficiency. There are, however, implications on bandwidth as can be seen by the examples in Table 3.2.

3.8.2 Rectangular DRAs

The rectangular DRA is characterized by a height h, a width w, a depth d, and a dielectric constant ε_r, as shown in Figure 3.46. Like the cylindrical DRA, it is usually mounted on a large ground plane and can be excited using numerous

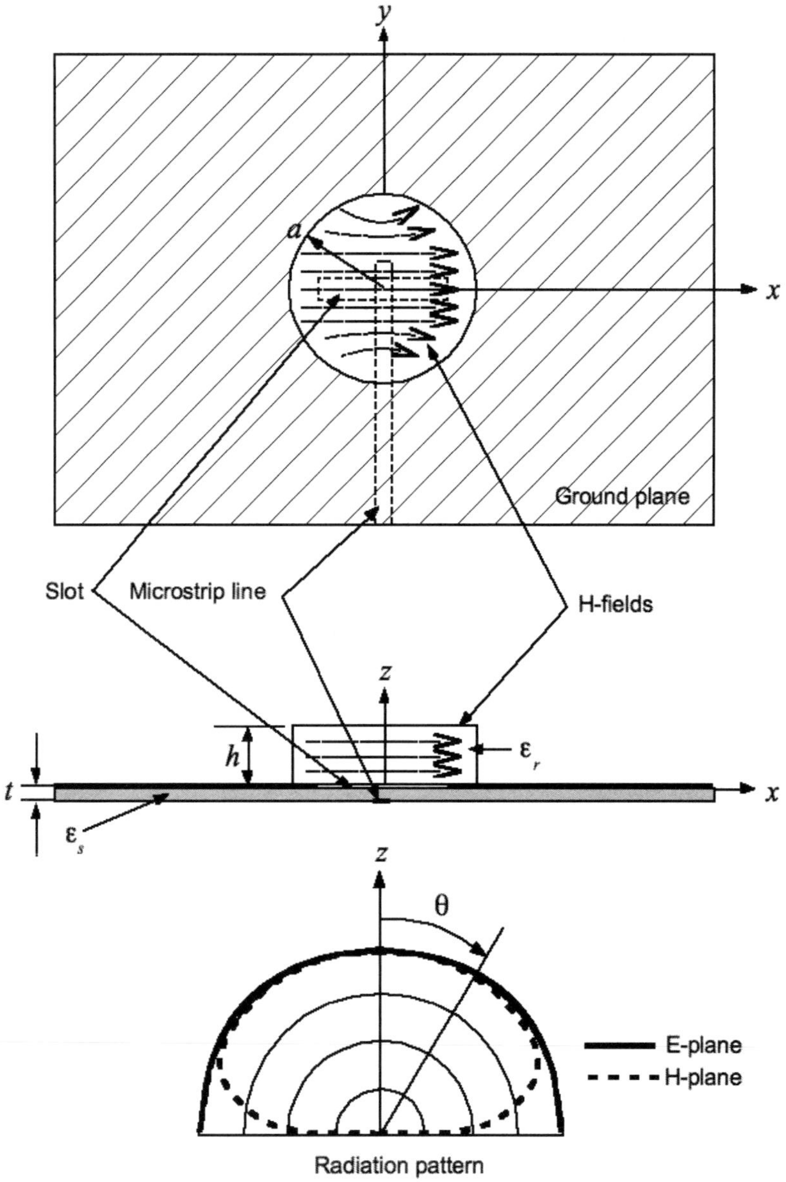

Figure 3.45 $HEM_{11\delta}$ mode of the cylindrical DRA.

Table 3.2

Sample Designs of Cylindrical DRAs Resonant at 10 GHz

a (mm)	h (mm)	ε_r	Q_r	Bandwidth
		$TM_{01\delta}$ Mode		
10.5	2.5	10	6.23	11.3%
6.7	3.5	10	6.53	10.8%
6.0	4.5	10	7.73	9.1%
5.0	2.5	20	16.6	4.3%
4.3	3.5	20	21.2	3.35
4.35	1.5	40	39.5	1.8%
3.15	2.5	40	90.7	0.8%
		$HEM_{01\delta}$ Mode		
a (mm)	**h (mm)**	**ε_r**	**Q_r**	**Bandwidth**
7.3	2.5	10	4.2	16.7%
4.35	3.5	10	7.5	9.4%
3.6	4.5	10	7.4	9.6%
3.3	2.5	20	18.3	3.95
2.6	3.5	20	17.8	4.0%
3.0	1.5	40	37.2	1.9%
1.9	2.5	40	44.0	1.6%

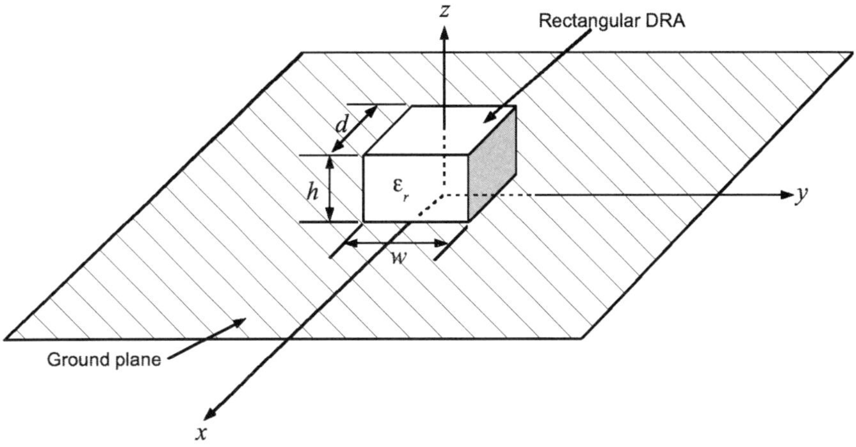

Figure 3.46 Rectangular DRA.

feeding mechanisms including slot apertures and probes. The rectangular shape offers an extra degree of freedom over the cylindrical shape, since the ratios *w/h* and *w/d* can be chosen independently for a given resonant frequency and dielectric constant. There is therefore a greater amount of flexibility in designing rectangular DRAs to achieve the desired profile and bandwidth characteristics.

With the rectangular DRA oriented as shown in Figure 3.46, the lowest order mode is the $TE^{x}_{\delta 11}$, where the H-fields are sketched in Figure 3.47, and whose radiation pattern is similar to that of the cylindrical DRA operating in the $HEM_{11\delta}$ mode. For the $TE^{x}_{\delta 11}$ mode, the resonant frequency, f_o, is estimated by solving the following transcendental equation [84]:

$$k_x \tan(k_x d/2) = \sqrt{(\varepsilon_r - 1)k_o^2 - k_x^2} \tag{3.90}$$

where

$$k_o = \frac{2\pi}{\lambda_o} = \frac{2\pi f_o}{c}, \; k_y = \frac{\pi}{w}, \; k_z = \frac{\pi}{2h}, \text{ and } k_x^2 + k_y^2 + k_z^2 = \varepsilon_r k_o^2$$

The radiation Q-factor of the rectangular DRA is determined using [84]

$$Q = \frac{2\omega W_e}{P_{rad}} \tag{3.91}$$

where W_e and P_{rad} are the stored energy and radiated power, respectively, and $\omega = 2\pi f_o$. These quantities are given by

$$W_e = \frac{\varepsilon_o \varepsilon_r whdA^2}{16}\left[1 + \frac{\sin(k_x d)}{k_x d}\right]\left(k_y^2 + k_z^2\right) \tag{3.92}$$

and

$$P_{rad} = 10k_o^4 |\boldsymbol{p}_m|^2 \tag{3.93}$$

where $\boldsymbol{p}_m$ is the magnetic dipole moment of the DRA expressed as

$$\boldsymbol{p}_m = \frac{-j\omega 8\varepsilon_o(\varepsilon_r - 1)A}{k_x k_y k_z}\sin(k_x d/2)\hat{\mathbf{x}} \tag{3.94}$$

and A is an arbitrary constant related to the maximum amplitude of the fields.

Table 3.3 lists a few examples of rectangular DRAs designed to be resonant at 10 GHz. The rectangular DRA has even greater flexibility than the cylindrical DRA in that it has two degrees of freedom. That is, different values for (w/d) and (h/d) can be selected for a given dielectric constant while resonating at the same frequency. To limit the possibilities, the designs in Table 3.3 were selected with $w = d$, but this is not a requirement and numerous other combinations are equally viable. A rectangular DRA can thus be designed to better meet any physical constraints of a given application.

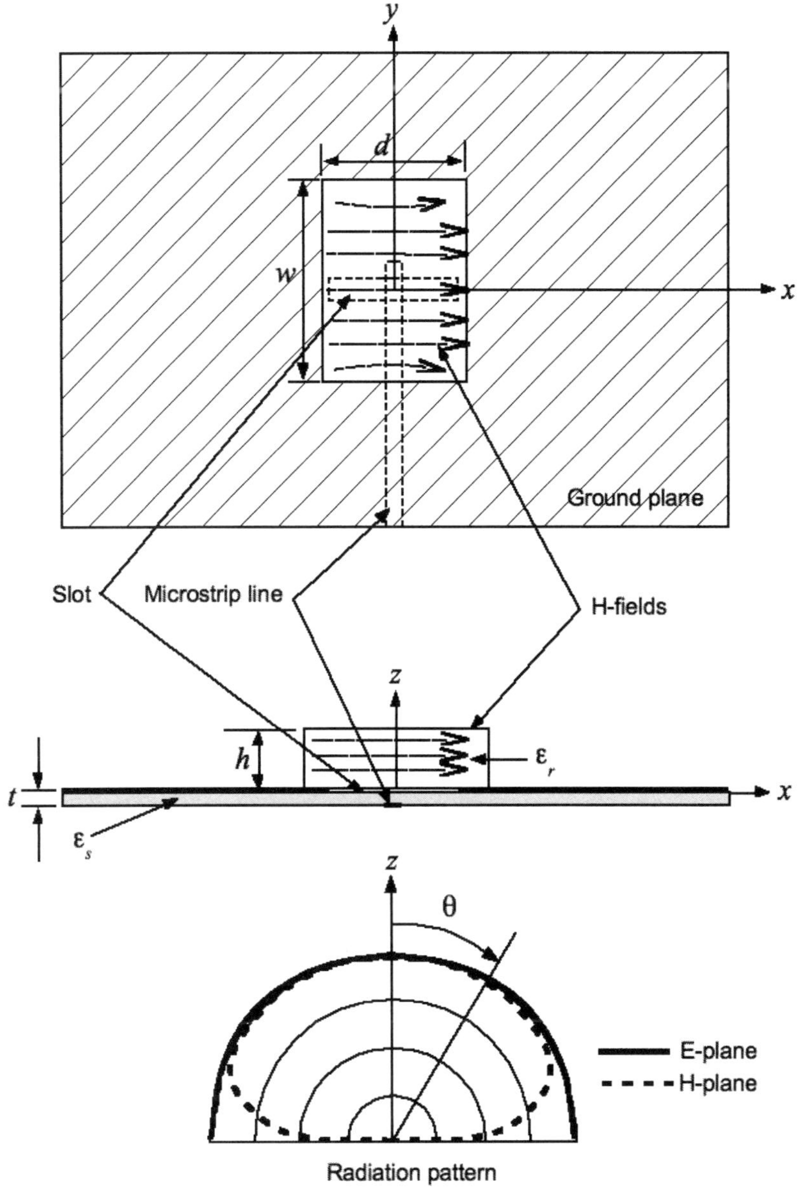

Figure 3.47 $TE^{x}_{\delta 11}$ mode of the rectangular DRA.

Table 3.3

Sample Designs of Rectangular DRAs Resonant at 10 GHz

w (mm)	d (mm)	h (mm)	ε_r	Q_r	Bandwidth
18	18	2.5	10	2.4	29.0%
7.9	7.9	3.5	10	7.6	9.4%
6.7	6.7	4.5	10	7.6	9.2%
5.5	5.5	2.5	20	19.2	3.7%
4.6	4.6	3.5	20	18.5	3.8%
3.2	3.2	2.5	40	48.8	1.4%
3.0	3.0	3.5	40	39.2	1.8%

3.9 HELICAL ANTENNAS

The helical antenna is formed by winding a wire around in a corkscrew fashion, as shown in Figure 3.48, and is usually attached to a ground plane. The antenna was first reported by Krauss in the 1950s [85]. The number of turns (N) in the winding, the spacing (s) between turns, the diameter (d) of the helix, and the size and shape of the ground plane are the parameters that will affect the radiation pattern. The helix is a traveling-wave antenna and as such, is broadband in nature. Due to its corkscrew shape, the helix is inherently suited for radiating circular polarized patterns. Helical antennas are used in numerous applications that require broadband circular polarization, such as in small mobile terminals for satellite communications, as feeds for spacecraft reflectors and satellite dishes, and as elements in arrays for radio astronomy.

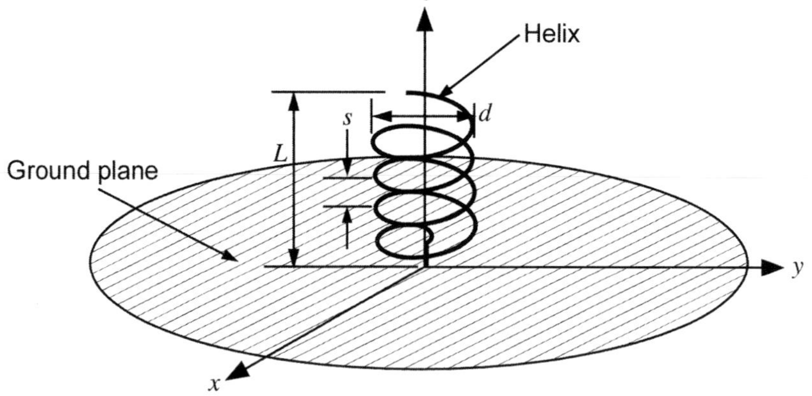

Figure 3.48 Geometry of a helical antenna.

The helical antenna can operate in many modes, the most common of which are the normal mode (or broadside) and the axial (or end-fire) mode. In the normal mode, the field radiated by the helical antenna is maximum in the plane normal to the helix axis and minimum along its axis, as shown in Figure 3.49(a). The pattern is similar to that of a short dipole or small loop antenna. To operate in the normal mode, the dimensions of the helical antenna are generally small compared to the wavelength of operation. In this mode, the helical antenna can be thought of as consisting of N small loops and N short dipoles connected together in series, as shown in Figure 3.50(b).

Since the length of the helical antenna operating in the normal mode is small compared to a the wavelength ($L << \lambda$), the far-field pattern will be independent of the number of loops and dipoles, so that a fairly accurate description can be obtained by simply summing the patterns of a single short dipole of length s and single small loop antenna of diameter D. The resultant normalized far-field electric field of the helix excited by a uniform current I_o can be approximated by [35, Ch.10]

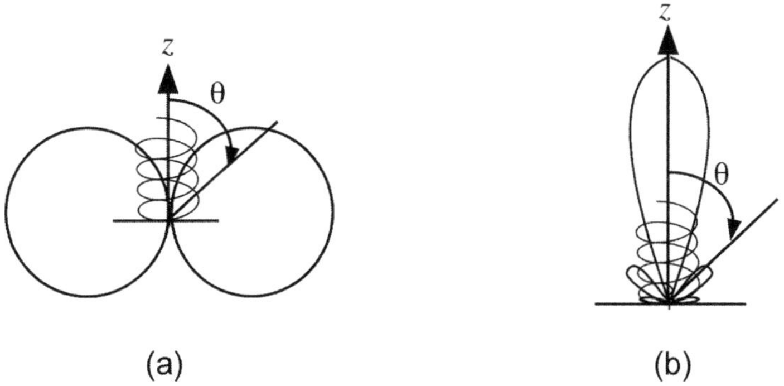

Figure 3.49 Common modes of the helical antenna: (a) normal mode and (b) axial mode.

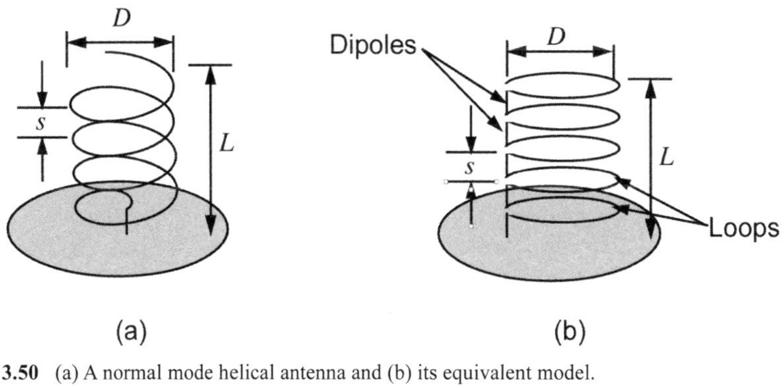

Figure 3.50 (a) A normal mode helical antenna and (b) its equivalent model.

$$E = \left[j\frac{s}{2\lambda}\hat{\theta} + \left(\frac{\pi d}{2\lambda}\right)^2 \hat{\phi} \right] I_o \eta \sin\theta \frac{e^{-jkr}}{r} \tag{3.95}$$

The helical antenna will radiate a circularly polarized wave by selecting the dimensions of s and d such that

$$s = \frac{(\pi d)^2}{2\lambda} \tag{3.96}$$

Because the dimensions of the helical antenna must be very small compared to the wavelength for normal-mode operation, the bandwidth of this mode is quite narrow, and the radiation efficiency is typically low.

Most helical antennas are designed for axial-mode operation, where the radiation pattern produced has a major lobe in the end-fire direction, as shown in Figure 3.46(b). To operate in this mode, the helix diameter and spacing between windings are a significant fraction of a wavelength. The circumference ($C = \pi d$) of the helix is typically designed to be approximately one wavelength, while the spacing is approximately one-quarter wavelength. The pitch angle ($\alpha = \tan^{-1}(s/C)$) is usually between 12° and 18°. Helical antennas are most often fed with a coaxial line and are mounted on a ground plane having a diameter of at least a half wavelength. The normalized far-field electric field of an N-turn axial mode helical antenna can be approximated using [35, Ch. 10]

$$E \approx \sin\frac{\pi}{2N}\cos\theta \frac{\sin(\psi N/2)}{\sin(\psi/2)} \frac{e^{-jkr}}{r} \tag{3.97}$$

where

$$\psi = \frac{2\pi}{\lambda}\left[s\cos\theta - \frac{\sqrt{C^2 + s^2}}{p} \right] \tag{3.98}$$

and p depends on the phase progression of the current along the windings of the helix. There are various approximations for p, including one for the ordinary end-fire condition

$$p = \left[\sin\alpha + \frac{\lambda\cos\alpha}{C} \right]^{-1} \tag{3.99}$$

Normalized patterns of an axial-mode helix for various numbers of turns are plotted in Figure 3.51. Examples of frequency-agile helical antennas will be presented in Chapters 5 and 7.

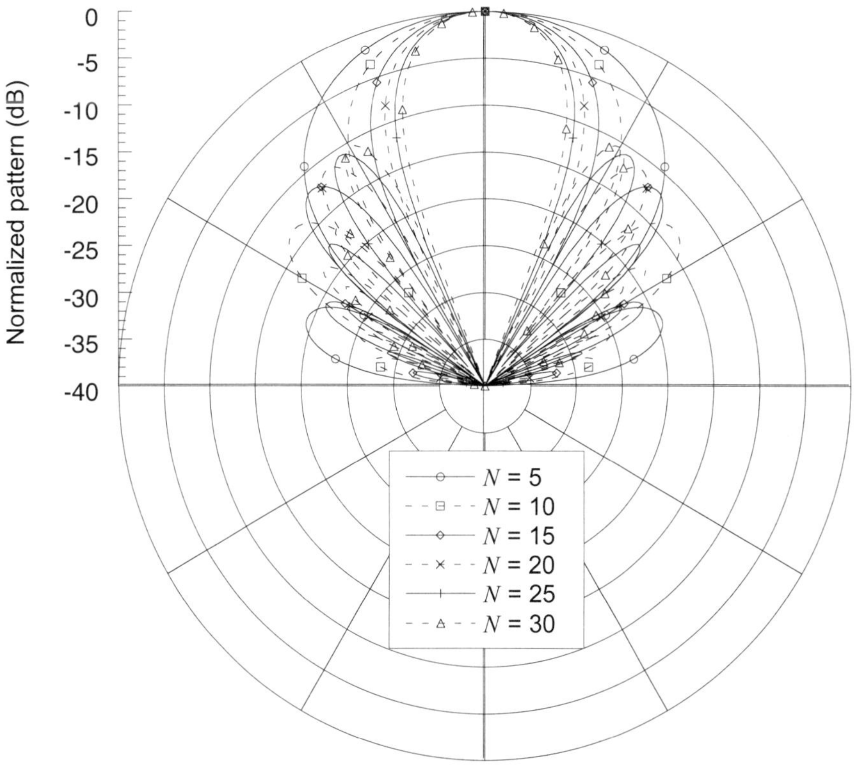

Figure 3.51 Normalized patterns of an axial-mode helical antenna.

References

[1] Tai, C. T., and S. A. Long, "Dipoles and Monopoles," in *Antenna Engineering Handbook*, Ch. 4, Third Edition, R. Johnson (ed.), New York, NY: McGraw Hill, 1992.

[2] Honda, S., et al., "A Disk Monopole Antenna with 1:8 Impedance Bandwidth and Omnidirectional Radiation Pattern," *Proc. International Symposium on Antennas and Propagation ISAP'92,* Saporo, Japan, September 22-25, 1992, pp. 1145-1148.

[3] Agrawall, N. P., G. Kumar, and K. P. Ray, "Wideband Planar Monopole Antennas," *IEEE Transactions on Antennas and Propagation,* Vol. 46, No. 2, February 1998, pp. 294-295.

[4] Evans, J. A., and M. J. Ammann, "Planar Trapezoidal and Pentagonal Monopoles with Impedance Bandwidths in Excess of 10:1," *Proc. IEEE International Symposium on Antennas and Propagation,* Orlando, Florida, July 11-16, 1999, pp. 1558-1561.

[5] Lee, E., P. S. Hall, and P. Gardner, "Compact Wideband Planar Monopole Antenna," *Electronics Letters,* Vol. 35, No. 25, December 1999, pp. 2157-2158.

[6] Chen, Z. N., et al., "Annular Planar Monopole Antennas," *IEE Proceedings on Microwaves, Antennas, and Propagation,* Vol. 149, No. 4, August 2002, pp. 200-203.

[7] Wong, K. L., C. H. Wu, and S. W. Su, "Ultrawide-Band Square Planar Metal-Plate Monopole Antenna with a Trident-Shaped Feeding Strip," *IEEE Transactions on Antennas and Propagation,* Vol. 53, No. 4, April 2005, pp. 1262-1269.

[8] Chung, K., T. Yun, and J. Choi, "Wideband CPW-Fed Monopole Antenna with Parasitic Elements and Slots," *Electronics Letters,* Vol. 40, No. 17, August 2004, pp. 1038-1040.

[9] Liang, X. L., S. S. Zhong, and F. W. Yao, "Compact UWB Tapered-CPW-Fed Planar Monopole Antenna," *Proc. Asia Pacific Microwave Conference,* Suzhou, China, December 4-7, 2005.

[10] Jan, J. Y., and T. M. Kuo, "CPW-Fed Wideband Planar Monopole Antenna for Operations in DCS, PCS, 3G, and Bluetooth Bands," *Electronics Letters,* Vol. 41, No. 18, September 2005, pp. 991-993.

[11] Alshehri, A. A., and A. R. Sebak, "A Novel UWB Planar Patch Antenna for Wireless Communications," *Proc. IEEE International Symposium on Antennas and Propagation,* San Diego, CA , July 5-11, 2008.

[12] Zhang, H., H. Xin, and R. W. Ziolkowski, "Design of Novel Printed Elliptical Monopole Antenna for UWB Applications," *Proc. IEEE International Symposium on Antennas and Propagation,* San Diego, CA, July 5-11, 2008.

[13] Pillalamarri, R., and G. S. B. Rao, "Analysis on Size Miniaturization in Printed Circular Disc Monopole Antennas for UWB Communications," *Proc. International Conference on Ultra Modern Telecommunications & Workshops,* St. Petersburg, Russia, October 12-14, 2009.

[14] Chu, Q. X., and L. H. Ye, "Design of Compact Dual-Wideband Antenna with Assembled Monopoles," *IEEE Transactions on Antennas and Propagation,* Vol. 58, No. 12, December 2010, pp. 4063-4066.

[15] Sun, M., Y. P. Zhang, and Y. Lu, "Miniaturization of Planar Monopole Antenna for Ultrawideband Radios," *IEEE Transactions on Antennas and Propagation,* Vol. 58, No. 7, July 2010, pp. 2420-2425.

[16] Brzezina, G., L. Roy, and L. MacEachern, "Planar Antennas in LTCC Technology with Transceiver Integration Capability for Ultra-Wideband Applications," *IEEE Microwave Theory and Techniques,* Vol. 54 No. 6, June 2006, pp. 2830-2839.

[17] Butler, C. M., "The Equivalent Radius of a Narrow Conducting Strip," *IEEE Transactions on Antennas and Propagation,* Vol. 30, No. 4, July 1982, pp. 755-758.

[18] Wilkinson, W. C., "A Class of Printed Circuit Antennas," *Proc. IEEE International Symposium on Antennas and Propagation,* Atlanta, GA, June 10-12, 1974, pp. 270-273.

[19] Ashkenazy, J., E. Levine, and D. Treves, "Conformal Arrays of Double-Sided Printed Dipoles," *Proc. IEEE International Symposium on Antennas and Propagation,* Philadelphia, PA, June 8-13, 1986, pp. 553-556.

[20] Chiu, K. Y., J. Y. Jan, and H. M. Chen, "Broadband Printed Dipole Antenna with a Pair of Sleeves for 2-6 GHz WiMax Applications," *Proc. IEEE International Symposium on Antennas and Propagation,* Charleston, SC, June 1-5, 2009.

[21] Chen, G. Y., and J. S. Sun, "Low Profile Planar Dipole Antenna," *Electronics Letters,* Vol. 40, No. 13, June 2004, pp. 790-792.

[22] Chen, H. M., et al., "Microstrip-Fed Printed Dipole Antenna for 2.4/5.2 GHz WLAN Operation," *Proc. IEEE International Symposium on Antennas and Propagation,* Monterey, CA, June 20-25, 2004, pp. 2584-2587.

[23] Petosa, A., S. Thirakoune, K. Levis, and A. Ittipiboon, "Microwave Holographic Antenna with Integrated Printed Dipole Feed," *Electronics Letters,* Vol. 40, No. 19, September 2004,

pp. 1162-1163.

[24] Qian, Y., et al., "Microstrip-Fed Quasi-Yagi Antenna with Broadband Characteristics," *Electronics Letters,* Vol. 34, No. 23, November 1998, pp. 2194-2196.

[25] Edward, B., and D. Rees, "A Broadband Printed Dipole with Integrated Balun," *Microwave Journal,* May 1987, pp. 339-344.

[26] Garg, R., P. Bhartia, I. Bahl, and A. Ittipiboon, *Microstrip Antenna Design Handbook,* Norwood, MA: Artech House, 2001.

[27] Bahl, I., and P. Bhartia, *Microstrip Antennas,* Dedham, MA: Artech House, Dedham, 1980.

[28] James, J. R., and P. S. Hall (eds.), *Handbook of Microstrip Antennas,* London: Peter Peregrinus, 1989.

[29] James, J. R., P. S. Hall, and C. Wood, *Microstrip Antennas: Theory and Design,* London: Peter Peregrinus, 1981.

[30] Bhartia, P., K. V. S. Rao, and R. S. Tomar, *Millimeter-Wave Microstrip and Printed Circuit Antennas,* Norwood, MA: Artech House, 1991.

[31] Lee, K. F., and W. Chen (eds.), *Advances in Microstrip and Printed Antennas,* New York, NY: John Wiley & Sons, 1997.

[32] Bahl, I. J., and R. Garg, "Simple and Accurate Formulas for a Microstrip with Finite Strip Thickness," *Proceedings of the IEEE,* Vol. 65, No. 11, November 1977, pp. 1611-1612.

[33] Kirshning, M., and R. Jansen, "Accurate Model for Effective Dielectric Constant of Microstrip With Validity Up to Millimeter-Wave Frequencies," *Electronics Letters,* Vol. 18, No. 6, March 1982, pp. 272-273.

[34] Hammerstad, O. E., "Equations for Microstrip Circuit Design," in *Proceedings 5th European Microwave Conference,* Hamburg, Germany, 1975, pp. 268-272.

[35] Balanis, C. A., *Antenna Theory: Analysis and Design,* Second Ed., New York, NY: John Wiley & Sons, 1997.

[36] Oltman, H. G., and D. A. Huebner, "Electromagnetically Coupled Microstrip Dipoles," *IEEE Transactions on Antennas and Propagation,* Vol. 29, No. 1, January 1981, pp. 151-157.

[37] Rana, I. E., and N. G. Alexopoulous, "Current Distribution and Input Impedance of Printed Dipoles," *IEEE Transactions on Antennas and Propagation,* Vol. 29, No. 1, January 1981, pp. 99-105.

[38] Katehi, P. B., and N. G. Alexopoulos, "On the Effect of Substrate Thickness and Permittivity on Printed Circuit Dipole Properties," *IEEE Transactions on Antennas and Propagation,* Vol. 31, No. 1, January 1983, pp. 34-39.

[39] Pozar, D. M., "Considerations for Millimeter Wave Printed Antennas," *IEEE Transactions on Antennas and Propagation,* Vol. 31, No. 5, September 1983, pp. 740-747.

[40] Elliot, R. S., and G. J. Stern, "The Design of Microstrip Dipole Arrays Including Mutual Coupling, Part I: Theory," *IEEE Transactions on Antennas and Propagation,* Vol. 29, No. 5, September 1981, pp. 757-760.

[41] Stern, G. J., and R. S. Elliot, "The Design of Microstrip Dipole Arrays Including Mutual Coupling, Part II: Experiment," *IEEE Transactions on Antennas and Propagation,* Vol. 29, No. 5, September 1981, pp. 761-765.

[42] Dubost, G., *Flat Radiating Dipoles and Applications to Arrays,* Chichester, England: Research Studies Press, 1981.

[43] Yang, H. Y., et al., "Design of Transversely Fed EMC Microstrip Dipole Arrays Including Mutual Coupling," *IEEE Transactions on Antennas and Propagation,* Vol. 38, No. 2, February 1990, pp. 145-151.

[44] Lepeltier, P. M., J. Citerne, and J. M. Floch, "On the EMC Dipole Feed-Line Parasitic Radiation," *IEEE Transactions on Antennas and Propagation,* Vol. 38, No. 6, June 1990, pp. 878-882.

[45] Ittipiboon, A., et al., "Slot-Coupled Excitation of Microstrip Dipole Antennas," *Electronics Letters,* Vol. 28, No. 2, January 1992, pp. 193-194.

[46] Shen, L. C., et al., "Resonant Frequency of a Circular Disc, Printed-Circuit Antenna," *IEEE Transactions on Antennas and Propagation,* Vol. 25, No. 4, July 1977, pp. 595-596.

[47] Long, S. A., L. C. Shen, and P. B. Morel, "Theory of the Circular-Disc Printed-Circuit Antenna," *Proceedings of the IEE,* Vol. 125, No. 10, October 1978, pp. 925-928.

[48] Derneryd, A. G., "Analysis of the Microstrip Disk Antenna Element," *IEEE Transactions on Antennas and Propagation,* Vol. 27, No. 5, September 1979, pp. 660-664.

[49] Shen, L. C., "Analysis of a Circular-Disc Printed-Circuit Antenna," *Proceedings of the IEE,* Vol. 126, No. 12, December 1979, pp. 1220-1222.

[50] Richards, W. F, J. D. Ou, and S. A. Long, "A Theoretical and Experimental Investigation of Annular, Annular Sector, and Circular Sector Microstrip Antennas," *IEEE Transactions on Antennas and Propagation,* Vol. 32, No. 8, August 1984, pp. 864-867.

[51] Shafai, L., "Characteristics of Printed Ring Microstrip Antennas," *Proc. Antenna Technology and Applied Electromagnetics Symposium, ANTEM'96*, Montreal, Canada, August 7-9, 1996, pp. 379-382.

[52] Mink, J. W., "Circular Ring Microstrip Antenna Elements," *Proc. IEEE International Symposium on Antennas and Propagation,* Québec, Canada, June 2-6, 1980, pp. 605-608.

[53] Bahl, I. J., S. S. Stuchly, and M. A. Stuchly, "A New Microstrip Radiator for Medical Applications," *IEEE Transactions on Microwave Theory and Techniques,* Vol. 28, No. 12, December 1980, pp. 1464-1468.

[54] Bahl, I. J., and S. S. Stuchly, "Closed-Form Expressions for Computer-Aided Design of Microstrip Ring Antennas," *International Journal of Microwave and Millimeter-Wave Computer-Aided Engineering,* Vol. 2, No. 3, 1992, pp. 144-154.

[55] Booker, H. G., "Slot Aerials and Their Relation to Complementary Wire Aerials," *Journal of the Institute of Electrical Engineers,* Part III A, 1946, pp. 620-626.

[56] King, R., C. W. Harrison, and D. H. Denton, "Transmission-Line Missile Antennas," *IRE Transactions on Antennas and Propagation,* Vol. 8, No. 1, January 1960, pp. 88-90.

[57] Ali, M., and G. Hayes, "Analysis of Integrated Inverted F Antennas for Bluetooth Applications," *Proc. IEEE AP-S Conference on Antennas and Propagation for Wireless Communications,* Waltham, MA, November 6-8, 2000, pp. 21-24.

[58] Kuo, Y. L., T. W. Chiou, and K. L. Wong, "A Novel Dual-Band Printed Inverted-F Antenna," *Microwave and Optical Technology Letters,* Vol. 31, No. 5, December 2001, pp. 353-355.

[59] Ammann, M. J., and L. E. Doyle, "A Loaded Inverted-F Antenna for Mobile Handset," *Microwave and Optical Technology Letters,* Vol. 28, No. 4, February 2001, pp. 226-228.

[60] Soras, C., et al., "Analysis and Design of an Inverted-F Antenna Printed on a PCMCIA Card for the 2.4 GHz ISM Band," *IEEE Antennas and Propagation Magazine,* Vol. 44, No. 1, February 2002, pp. 37-44.

[61] Ali, M., and G. J. Hayes, "Small Printed Integrated Inverted-F Antenna for Bluetooth Application," *Microwave and Optical Technology Letters,* Vol. 33, No. 5, June 2002, pp. 347-349.

[62] Karaboikis, M., et al., "Compact Dual-Printed Inverted-F Antenna Systems for Portable Wireless Devices," *IEEE Antennas and Wireless Propagation Letters,* Vol. 3, 2004, pp. 9-14.

[63] Wang, Y. S., M. C. Lee, and S. J. Chung, "Two PIFA-Related Miniaturized Dual-Band Antennas," *IEEE Transactions on Antennas and Propagation,* Vol. 55, No. 3, March 2007, pp. 805-811.

[64] Chan, P. W., H. Wong, and E. K. N. Yung, "Dual-Band Printed Inverted-F Antenna for DCS, 2.4 GHz WLAN Applications," *Proc. Loughborough Antenna & Propagation Conference LAPC 2008,* Loughborough, England, March 17-18, 2008, pp. 185-188.

[65] Oh, I. Y., et al., "Multi-Radiating-Element Printed Inverted-F Antenna with Independent Resonant Frequencies for Bandwidth Enhancement," *IEEE Antennas and Wireless Propagation Letters,* Vol. 9, 2010, pp. 926-929.

[66] Harrington, R. F., *Field Computation by Moment Methods,* New York, NY: Macmillan, 1968.

[67] Taga, T., and K. Tsunekawa, "Performance Analysis of a Built-In Planar Inverted F Antenna for 800 MHz Band Portable Radio Units," *IEEE Journal of Selected Areas in Communications,* Vol. 5, No. 5, June 1987, pp. 921-929.

[68] Tsunekawa, K., "Diversity Antennas for Portable Telephones," *IEEE Vehicular Technology Conference,* San Francisco, CA, May 1989, pp. 50-56.

[69] Tsunoda, K., and T. Taga, "Analysis of Planar Inverted F Antenna Using Spatial Network Method," *Proc. IEEE International Symposium on Antennas and Propagation,* Dallas, TX, May 7-11, 1990, pp. 664-667.

[70] Rowell, C. R., and R. D. Murch, "A Capacitively Loaded PIFA for Compact Mobile Telephone Handsets," *IEEE Transactions on Antennas and Propagation,* Vol. 45, No. 5, May 1997, pp. 837-842.

[71] Kashiwa, T., N. Yoshida, and I. Fukai, "Analysis of Radiation Characteristics of Planar Inverted-F Type Antenna on Conductive Body of Hand-Held Transceiver by Spatial Network Method," *Electronics Letters,* Vol. 25, No. 16, August 1989, pp. 1044-1045.

[72] Liu, Z. D., P. S. Hall, and D. Wake, "Dual-Frequency Planar Inverted-F Antenna," *IEEE Transactions on Antennas and Propagation,* Vol. 45, No. 10, October 1997, pp. 1451-1458.

[73] Salonen, P., M. Keskilammi, and M. Kivikovski, "Single-Feed Dual-Band Planar Inverted-F Antenna with U-Shaped Slot," *IEEE Transactions on Antennas and Propagation,* Vol. 48, No. 8, August 2000, pp. 1262-1264.

[74] Kim, B. C., J. H. Yun, and H. D. Choi, "Small Wideband PIFA for Mobile Phones at 1800 MHz," *Proc. IEEE Vehicular Technology Conference,* Milan, Italy, May 17-19, 2004, pp. 27-29.

[75] Nashaat, D. M., H. A. Elsadek, and H. Ghali, "Single Feed Compact Quad-Band PIFA Antenna for Wireless Communications Applications," *IEEE Transactions on Antennas and Propagation,* Vol. 53, No. 8, August 2005, pp. 2631-2365.

[76] Kajfez, D., and P. Guillon (eds.), *Dielectric Resonators,* Norwood, MA: Artech House, 1986.

[77] Long, S. A., M. W. McAllister, and L. C. Shen, "The Resonant Cylindrical Dielectric Cavity Antenna," *IEEE Transactions on Antennas & Propagation,* Vol. 31, No. 3, March 1983, pp. 406-412.

[78] McAllister, M. W., S. A. Long, and G. L. Conway, "Rectangular Dielectric Resonator Antenna,"

Electronics Letters, Vol. 19, No. 6, March 1983, pp. 218-219.

[79] Luk, K. M., and K. W. Leung (eds.), *Dielectric Resonator Antennas,* Hertfordshire, England: Research Studies Press, 2003.

[80] Petosa, A., *Dielectric Resonator Antenna Handbook,* Norwood, MA: Artech House, 2007.

[81] Petosa, A., and A. Ittipiboon, "Dielectric Resonator Antennas: A Historical Review and the Current State of the Art," *IEEE Antennas and Propagation Magazine,* Vol. 52, No. 5, October 2010, pp. 91-116.

[82] Mongia, R. K., and P. Bhartia, "Dielectric Resonator Antennas, A Review and General Design Relations for Resonant Frequency and Bandwidth," *International Journal of Microwave and Millimeter-Wave Computer-Aided Engineering,* Vol. 4, No. 3, 1994, pp. 230-247.

[83] Kishk, A. A., A. W. Glisson, and G. P. Junker, "Study of Broadband Dielectric Resonator Antennas," *Proc. 1999 Antenna Applications Symposium,* Allerton Park, Monticello, Il, September 1999, pp. 45-68.

[84] Mongia, R. K., and A. Ittipiboon, "Theoretical and Experimental Investigations on Rectangular Dielectric Resonator Antennas," *IEEE Transactions on Antennas & Propagation,* Vol. 45, No. 9, September 1997, pp. 1348-1356.

[85] Krauss, J. D., *Antennas,* Second Edition, New York, NY: McGraw-Hill, 1988.

Chapter 4

Frequency Tuning Techniques

This chapter reviews various methods that have been or could potentially be used for altering the operating frequency of an antenna. These techniques are grouped into three categories, encompassing methods that do one of the following:

- Physically change one or more antenna dimensions;
- Alter the material properties of the antenna;
- Integrate switching or tunable electronic devices.

Representative examples from each category will be presented, along with a brief description of their basic mode of operation and their main characteristics. Potential advantages and disadvantages for each of these methods will also be highlighted. Section 4.1 examines various types of actuators (including electromechanical, piezoelectric, hydraulic, and microactuators) that could be used to physically alter one or more antenna dimensions, causing a shift in operating frequency. Section 4.2 covers four classes of materials (ferrites, ferroelectrics, liquid crystals, and semiconductors) whose permeability, permittivity, or conductivity can be altered by an applied bias. These materials, once incorporated within an antenna design, could be biased to cause a shift in the operating frequency of the antenna. Section 4.3 deals with various electronic devices including PIN diodes, field effect transistors (FETs), MEMS RF switches, and varactors. These devices can be integrated into antenna designs and used as electronic switches or as variable reactive loads to alter the operating frequency. The specific ways in which these various techniques are combined with antenna elements to achieve frequency agility will be brought to light in subsequent chapters.

4.1 CHANGING PHYSICAL DIMENSIONS

Altering the size or shape of an antenna can affect its operating frequency. This approach is particularly suited for resonant type antennas, where a change in one or more dimensions will cause a shift in the resonant frequency. There are numerous ways in which such physical changes can be implemented, most of which make use of some form of actuator. Since an abundant amount of literature on the operating principles of various types of actuators is readily available (such as that in [1–13]), only the most basic concepts are briefly considered here. More attention will then be given to those types of actuators that are potentially suitable for FAAs, with various examples presented in Chapter 5.

Actuators come in a myriad of forms and sizes and can be designed for various applications ranging from the precision micron control of delicate optical components to the displacement of large masses. One way to categorize these actuators is by the type of method used for actuation. Table 4.1 lists the most common of these methods along with a brief description and some typical examples of their usage. The performance of these actuators can be quantified by numerous parameters. The major performance metrics are listed and described in Table 4.2. Sections 4.1.1–4.1.5 take a closer look at some of these actuators and at their potential use in developing FAAs.

4.1.1 Electromechanical Actuators

Electromechanical actuators typically consist of an electric motor attached to a screw in such a way as to convert the rotary motion of the motor into a linear displacement of the actuator. A simplified sketch depicting the basic operation of an electromechanical actuator is shown in Figure 4.1. A traveling nut is fastened around a lead screw, which is made to rotate by an electric motor. The traveling nut is attached to the shaft and is prevented from rotating. Rotation of the lead screw in one direction forces the nut up the screw, extending the shaft. Rotation of the screw in the opposite direction forces the nut down, retracting the shaft. The rotary motion of the motor is thus translated into a linear motion of the shaft. There are numerous variations of these actuators, each designed to focus on specific performance criteria such as mechanical efficiency, speed, load capacity, or actuator miniaturization. For use in FAAs, these actuators need to be compact and lightweight, have fairly fine precision of movement, and require low-voltage biasing. Miniature electromechanical actuators are now commercially available with maximum dimensions of under 15 mm and weights as light as 15 grams. Miniature electric motors are readily available with lengths on the order of 5 mm and diameters as small as 2 mm [14–16]. These actuators also require low operating power (less than half a watt) during actuator motion and no power to hold the actuator in a fixed position.

Electromechanical actuators are relatively inexpensive and can offer continuous frequency-tuning capabilities. Unlike some other types of actuators (like

Table 4.1

List of Common Actuator Types

Actuation Method	Description
Chemical	These actuators make use of chemical reactions to generate force and displacement. They can be used to design micropumps and microvalves for biochemical analysis and lab-on-a-chip applications.
Electric	Electric actuators make use of electrostatics to exert forces on deformable bodies. (This is a common technique for microactuators.) Piezoelectric materials or electroactive polymers can also be used in electric actuators where an applied voltage causes material deformation. They are used in applications requiring precision control and are typically limited to small displacements.
Electro-mechanical	Electromechanical actuators incorporate motors and a mechanical assembly to translate the rotary motion of the motor to a linear motion. They can be designed for numerous requirements ranging from precision movements to the displacement of heavy loads. They are used in many commercial and industrial applications requiring tilting, lifting, pivoting, and positioning.
Hydraulic	Hydraulic actuators are based on fluids, usually requiring a piston, spring, and hydraulic pump. They are commonly used for displacing heavy loads. They are also used for vibration damping, and commercial and industrial applications.
Magnetic	Magnetic actuators are operated by a magnetic field, usually generated using coils or permanent magnets. Also included in this category are magnetostrictive actuators, where material deformation is caused by an applied magnetic field. Applications include voice coils, vibration generators, synchronized circuit breakers, active vibration damping, microvalves, locking mechanisms, and shock absorbers. Magnetic actuators can be very light (weighing less than a gram) and can find applications in model flyers, microrobotics, optical communications, scanning-based display systems, and biomedical devices.
Optical	Optical actuators include devices that use photostrictive materials (which deform in response to light) or that use light for heating or for generating current or for changing resistivity/conductivity in photoresponsive materials. Potential applications include actuators for micromirrors, optical relay switches, micromachines and microrobotics, and optical servo systems.
Pneumatic	Pneumatic actuators use air to pressurize pistons. They are typically designed for moving light loads with applications including precise tensioning, positioning mechanisms supporting delicate materials, sensitive force control for holding or clamping fragile materials, radio-controlled models, medical or precise laboratory test equipment, vibration dampers, shock absorbers, or valve actuators.
Thermal	Thermal actuators make use of the difference in thermal expansion coefficients of different materials to cause a deformation. This category also includes actuators based on shape memory alloys, which deform to predefined shapes at certain temperatures. They are used in applications such as temperature-actuated valves, switches, latches, clamps, and control devices.

Table 4.2

Actuator Performance Measures

Metric	Description
Accuracy	A measure of how closely the actuator output approximates a desired calibrated scale.
Dimensions and weight	The size and weight of an actuator relative to its load-carrying capacity.
Drift	A measure of the changes of the actuator output with time when there is no control signal applied (the unbiased or zero-biased case).
Hysteresis	A measure of the difference in the actuator output when it is reached using inputs from two opposite directions. (Feedback control is usually required in order to determine the actual output.)
Linearity	The actuator output (expressed as a percentage of its maximum value) can be plotted against the input value (also expressed as a percentage of its maximum value). The resultant curve would be a straight diagonal line for an actuator that exhibits an ideal linear response. Linearity is a measure of the deviation of the actual response to this ideal curve.
Noise	A measure of the random fluctuations in actuator output (due to factors such as vibrations and thermal noise).
Load-bearing capability	A measure of the actuator output behavior when connected to a load, which is different from the unloaded behavior. (Requires knowledge of the mechanical characteristics of the load.)
Power efficiency	A measure of how efficiently the input power to the actuator is transformed into output power. (Output power is usually mechanical and is measured as the force generated by the actuator multiplied by the displacement.)
Precision	A measure of the exactness and the reproducibility of a desired actuator output.
Repeatability	A measure of the maximum difference in the actuator output value for a given input value when repeated over numerous cycles.
Resolution	The smallest increment in the value of the input that results in a detectable actuator output.
Sensitivity	The ratio of change in actuator output value to an incremental change in the input value.
Span	The full-scale operating range of the actuator output.
Speed	A measure of how quickly the actuator output can change by an incremental amount.
Step-response	A measure of how the actuator output responds to an abrupt (step) change in the input.
Threshold	A measure of the smallest initial increment in input (starting from zero input) that results in a detectable change in actuator output.

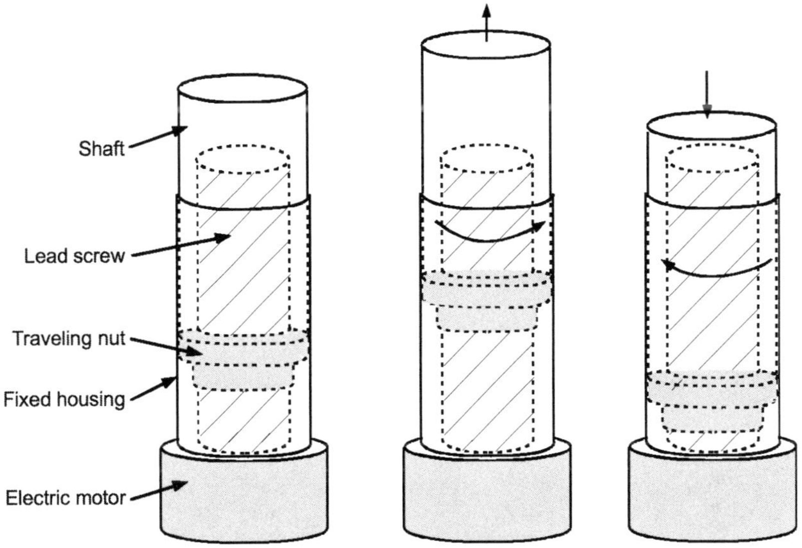

Figure 4.1 Example of an electromechanical linear actuator.

piezoelectric), electromechanical actuators do not suffer from hysteresis (see Table 4.2), which simplifies the control mechanism. These actuators can be made small enough to incorporate into certain FAA designs, especially at the lower end of the microwave frequency band, such as for cellular or wireless base station antennas.

One of the drawbacks of electromechanical actuators is that the moving parts are prone to wear, which would limit the lifetime of the FAA. The moving parts will also place a limit on the speed at which the frequency can be tuned, which may preclude some applications requiring fast frequency hopping. Also, the current commercially available dimensions of electromechanical actuators could restrict or prevent their use for small handheld devices or for antennas designed at higher frequencies where their dimensions are significantly smaller than those of the actuator. For these applications, microactuators (discussed in Section 4.1.4) could be considered.

4.1.2 Piezoelectric Actuators

The piezoelectric effect is the property of certain materials (for instance, certain crystals and ceramics) of generating an electric field in response to an applied mechanical stress. This electric field results in a voltage being developed across the material. The piezoelectric effect is reversible such that applying a voltage across the material will produce stress and/or strain. This behavior is illustrated in Figure 4.2 where a piezoelectric cylinder of length L and diameter D is polarized along its

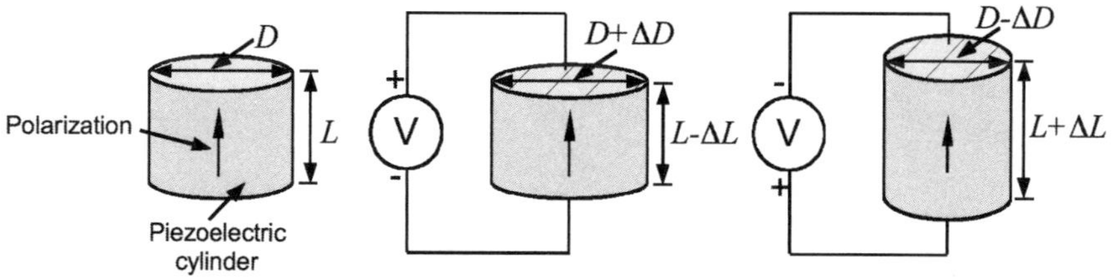

Figure 4.2 The piezoelectric effect.

length. Applying a bias voltage opposite to the direction of polarization results in a shrinkage along the cylinder length by ΔL and increase along the width by ΔD. Reversing the bias voltage has the opposite effect, where the cylinder length increases by ΔL and the width decreases by ΔD. The resultant change in shape in these materials is typically quite small (usually less than 0.1%) and requires high applied voltages. Piezoelectric materials typically have dielectric constants ranging from about 200 to 1,500.

Piezoelectric actuators can be used for very fine precision movements (microns or better). These actuators can be quite small, with dimensions of as low as 2 x 2 x 3 mm being commercially available. Piezoelectric actuators in the form of thin membranes (0.2-mm-thick) are also available [17]. These small actuators can typically deflect only a few microns and require several hundreds of volts for activation.

The advantages of piezoelectric actuators are their small size and precision control, making them suitable for continuous frequency-tuning applications. However, they require high activation voltages, and they exhibit hysteresis (i.e., the position is different in expansion or contraction), which makes it difficult to control their displacement in a repeatable manner. A position feedback control is thus required in order to repeatedly return to a specific position, increasing the complexity of these devices. Piezoelectric actuators are small enough that they could potentially be integrated into handheld devices; however, the high activation voltages that are required make them unlikely candidates for small portable devices where battery size and lifetime are of significant concern. High-deflection designs have also been developed for larger displacements (up to several centimeters) using stacked reduced and internally biased oxide wafer (Rainbow) actuators [9]. Rainbow actuators have been used in a frequency-agile stacked microstrip patch antenna design and will be discussed in Chapter 5.

4.1.3 Hydraulic and Pneumatic Actuators

Hydraulic (liquid-based) and pneumatic (air-based) actuators typically involve a hollow cylinder into which a piston is inserted. The two sides of the piston are alternately pressurized and depressurized using an external hydraulic or pneumatic pump to achieve precise linear displacement of the piston and, in turn, the device connected to the piston. A simplified schematic diagram of a hydraulic actuator is shown in Figure 4.3. Pumping the fluid or gas into the chamber pushes the piston forward, compressing the spring and displacing the shaft out of the casing. When the fluid is pumped out, the spring decompresses and the shaft moves back into the casing. Hydraulic actuators are typically designed to displace heavy loads and are relatively large. They are also susceptible to leakage and require feedback control if accurate positioning is needed. Pumps can also be designed at a small scale for microfluidics applications. As with other mechanical actuators, hydraulic or pneumatic actuators have relativley low switching speeds but do not require as high a bias level as piezoelectric or electrostatic actuators [18–22]. Examples of FAAs controlled by hydraulic or pneumatic actuators will be presented in Chapter 5.

4.1.4 Microactuators

Microactuators is a term that usually describes actuators whose maximum dimensions are a few millimeters or less. Scaled-down versions of the previously considered actuators would fall into this category. These actuators are usually fabricated using micromachining techniques. Common microactuator structures include cantilever beams, bridges, diaphragms, ratchets, torsion mirrors, pistons, and comb drives. Activation methods include electrostatic, thermal, chemical, or optical. For electrostatic actuators, the required fields are generated through an applied voltage bias (ranging from about 10–300V). Deflections are typically on the order of a few microns. In the field of optics, microactuators have been used to control miniature mirror arrays and have also been used for sensors, switches, and relays. At microwave frequencies, microactuators have been used for low-loss switches (discussed in Section 4.3), resonators, oscillators, and filters. As wireless applications move to higher frequencies, the dimensions of the required antenna will be small enough such that these microactuators could be an integral part of the antenna and used to alter dimensions to create FAAs, as examples in Chapter 5 will illustrate.

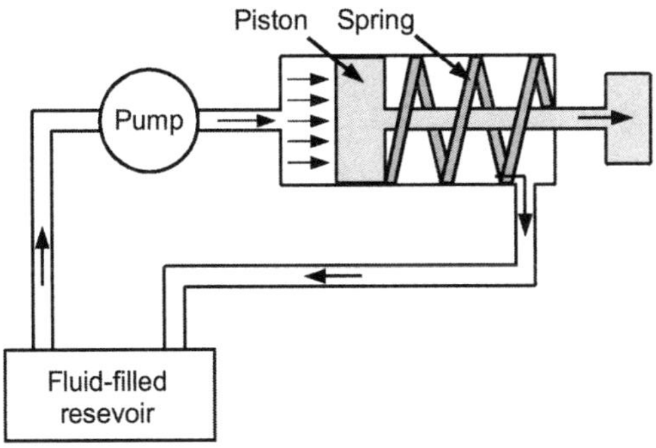

Figure 4.3 Schematic of a hydraulic actuator.

4.1.5 Inherently Deformable Antennas

Instead of using actuators that are integrated into the antenna to execute the desired change in dimensions, antennas could potentially be made up of inherently deformable material. Examples of such materials include electroactive polymers and shape memory alloys [12]. Electroactive polymers change size or shape in response to an applied electric field. They typically undergo much larger deformations (strains up to 380%) compared to ceramic or piezoelectric actuators but still require high actuation voltages (hundreds to thousands of volts).

Shape memory alloys are alloys that can "remember" their original shape. This effect is made possible through a phase change between two solid states (martensite and austenite) of the alloy. In the martensite phase, which exists at lower temperatures, the alloy is relatively soft and deformable. When heated to the austenite phase, it reverts to its original shape. In most shape memory alloys, a temperature change of only about 10°C is needed to begin the phase change. Shape memory alloys could be formed into wire antennas such as dipoles or helices with a change in shape resulting in a frequency shift. Cycle times of a few milliseconds are possible for microactuator designs [12]. The disadvantages of current shape memory alloys are their relatively high cost and their poor fatigue properties, which limit their lifetime. Further developments in materials science are therefore required before shape memory alloys can become more viable candidates for achieving FAAs.

4.2 CHANGING SUBSTRATE PARAMETERS

One of the main drawbacks of changing the dimension or the shape of an antenna as a mechanism for frequency tuning stems from the required physical displacement, implying a certain time delay as well as potential problems with reliability (due to repetitive stress, fatigue, and wear on moving components), which may adversely affect the antenna lifetime. The frequency-tuning techniques presented in this section avoid these potential problems, since there is no mechanical motion involved. Instead, this section examines various materials whose inherent electrical properties (permittivity, permeability, or conductivity) can be altered by an applied bias. These materials can be integrally incorporated within the antenna in such a manner that alteration of their electrical properties translates into a shift in operating frequency. Various materials that are suitable for use in FAAs are presented in this section.

4.2.1 Ferrites

Ferrites are ceramic-like materials made from sintering a mixture of metallic oxides. They are characterized by relative permeability values of several thousand (at low frequencies), dielectric constant values of 10 or greater, and high resistivity. Ferrites have found widespread applications as cores (used in inductors, transformers, and electromagnets), beads (for suppressing high-frequency noise in electronic components), and powders (used for magnetic recording tapes). At microwave frequencies, ferrites are used in such devices as isolators, circulators, and phase shifters. The behavior and properties of ferrites used at microwave frequencies have been well-documented in numerous publications (such as [23–34]), and this section provides only a brief overview of the main characteristics.

The magnetic properties of ferrites can be attributed primarily to the magnetic dipole moment arising from the spin of the electrons within the material. As sketched in Figure 4.4, when a uniform static DC magnetic field H_o is applied to the material, the electrons begin to precess around the axis of the applied field at a frequency known as the Larmor frequency, ω_o, given by

$$\omega_o = 2\pi\,\gamma H_o \tag{4.1}$$

where γ is the gyromagnetic ratio (γ = 2.8 MHz/Oe). The spinning electrons can be considered as a magnetic dipole and the precession is known as free precession. The ferrite is considered to be saturated when the applied static field is strong enough so that all the individual dipole moments in the material are aligned and behave as one large dipole moment. This condition is characterized by the saturation magnetization, M_s, which is a measure of the magnetic flux density required to align all the dipole moments in the ferrite. The saturation magnetization is a physical property of the ferrite material.

If a small-amplitude radio frequency (RF) magnetic field of frequency ω is then superimposed on the static field of a saturated ferrite, the magnetic moment undergoes a forced precession. The resultant permeability is a tensor quantity, known as the Polder permeability tensor, which has the general form

$$[\mu] = \begin{bmatrix} \mu_{xx} & \mu_{xy} & \mu_{xz} \\ \mu_{yx} & \mu_{yy} & \mu_{yz} \\ \mu_{zx} & \mu_{zy} & \mu_{zz} \end{bmatrix} \tag{4.2}$$

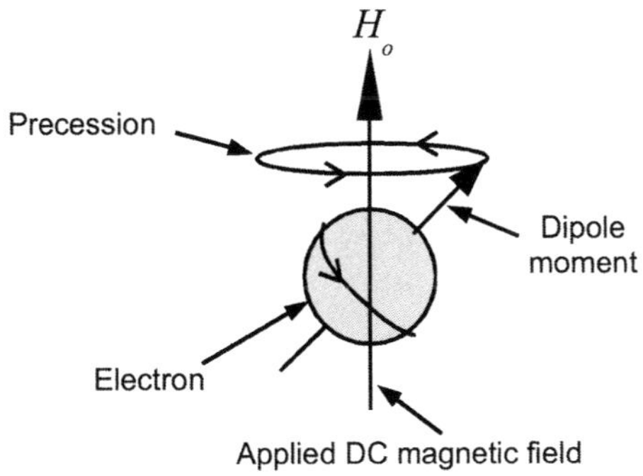

Figure 4.4 Electron precession around an applied static magnetic field.

If H_o is applied in the z-direction and the RF H-fields lie in the x- and y-directions (i.e., $\boldsymbol{h} = h_x\hat{x} + h_y\hat{y}$), the Polder permeability tensor reduces to

$$[\mu] = \mu_o\begin{bmatrix} \mu & j\kappa & 0 \\ -j\kappa & \mu & 0 \\ 0 & 0 & 1 \end{bmatrix} \tag{4.3}$$

(Expressions for the Polder tensor for x- and y-directed magnetic bias can be found in [32]). For the case of small magnetic losses,

$$\mu = \mu' - j\mu'' = 1 + \frac{\omega_m(\omega_o + j\omega_L)}{(\omega_o + j\omega_L)^2 - \omega^2} \tag{4.4}$$

$$\kappa = \kappa' - j\kappa'' = \frac{-\omega_m\omega_o}{(\omega_o + j\omega_L)^2 - \omega^2} \tag{4.5}$$

where $\omega = 2\pi f$, $\omega_m = 2\pi\,\gamma(4\pi M_s)$, $\omega_L = 1/T = 2\pi\ (\gamma\Delta H)/2$, T is the relaxation time, ΔH is the resonance line width, and f is the frequency of the microwave field.

Commercially available microwave ferrite materials include magnesium ferrites, nickel ferrites, lithium ferrites, and various yttrium garnets (YIGs). They are characterized by a saturation magnetization ranging from $300\text{G} \le 4\pi M_s \le 5{,}000\text{G}$, a line width ranging from $10\text{ Oe} \le \Delta H \le 900\text{ Oe}$, a dielectric constant ranging from $10 \le \varepsilon_r \le 20$, and a loss tangent of $tan\delta \le 0.0005$. These ferrite materials are available in various shapes including sheets and rods of circular or rectangular cross-section. Thin-film ferrites (whose thickness is on the order of microns) are also being considered in microwave applications. These materials (such as copper-, zinc-, manganese-, and barium-ferrites) are characterized by very high values of saturation magnetization ($4\pi M_s > 10{,}000\text{G}$) and line widths of $\Delta H \sim 1$ Oe. Such films, however, are not as readily commercially available.

It is convenient to use CGS units for the variables in (4.1) to (4.5). With H_o expressed in oersteds, $\gamma = 2.8$ MHz/Oe, and ($4\pi M_s$) expressed in gauss, the following frequencies will have units of megahertz

$$f_o = \omega_o/2\pi = \gamma\,H_o \tag{4.6}$$

$$f_m = \omega_m/2\pi = \gamma\,(4\,\pi\,M_s) \tag{4.7}$$

$$f_L = \omega_L/2\pi = \gamma\Delta H/2 \tag{4.8}$$

If the RF magnetic fields $\boldsymbol{h}$ and resultant magnetic flux density $\boldsymbol{b}$ are expressed in terms of clockwise (+) and counterclockwise (-) circular polarization, their relationship can be expressed as

$$\begin{bmatrix} b_+ \\ b_- \\ b_z \end{bmatrix} = \mu_o \begin{bmatrix} \mu_+ & 0 & 0 \\ 0 & \mu_- & 0 \\ 0 & 0 & 1 \end{bmatrix} \begin{bmatrix} h_+ \\ h_- \\ h_z \end{bmatrix} \tag{4.9}$$

where

$$\begin{aligned} b_+ &= \frac{1}{2}\left(b_x - jb_y\right) \quad & h_+ &= \frac{1}{2}\left(h_x - jh_y\right) \\ b_- &= \frac{1}{2}\left(b_x + jb_y\right) \quad & h_- &= \frac{1}{2}\left(h_x + jh_y\right) \\ \mu_+ &= \left(\mu + \kappa\right) \quad & \mu_- &= \left(\mu - \kappa\right) \end{aligned} \tag{4.10}$$

The clockwise and counterclockwise waves will propagate with constants proportional to the square root of μ_+ and μ_-, respectively. This nonreciprocal behavior of the RF waves propagating in ferrite media has been exploited in such microwave devices as circulators, isolators, and phase shifters [31, 32, 34]. Figure 4.5 shows a typical behavior of μ_+ and μ_- as a function of frequency where the nonreciprocal nature of the ferrite is made evident. The clockwise waves undergo a resonance at f_o, characterized by a large peak in the imaginary part of the permeability, indicating high attenuation of the clockwise wave. This attenuation peak does not appear in the counterclockwise case. Also, there will be no propagation of the clockwise polarized wave in the frequency range between $f_o < f < f_o + f_m$, where the value of the real part of the permeability is negative.

Since it is cumbersome to deal with a permeability tensor, it is convenient to define an effective scalar permeability μ_{eff} that can eventually be used to predict the tuning range of ferrite-based FAAs. For the case of plane-wave propagation in an unbounded ferrite medium, the value of μ_{eff} will depend on the orientation of the bias field $\boldsymbol{H_o}$ and of the RF field $\boldsymbol{h}$. Assuming propagation in the $+z$-direction, there will be the following three cases:

1. Parallel bias:

$$\boldsymbol{H_o} = H_o\hat{\boldsymbol{v}} \text{ and } \boldsymbol{h} = h_v e^{j\omega t}\hat{\boldsymbol{v}} : \ \mu_{eff} = 1 \tag{4.11}$$

2. Transverse bias:

$$\boldsymbol{H_o} = H_o\hat{\boldsymbol{v}} \text{ and } \boldsymbol{h} = h_u e^{j\omega t}\hat{\boldsymbol{u}} : \ \mu_{eff} = \frac{\mu^2 - \kappa^2}{\mu} \tag{4.12}$$

3. Longitudinal bias:

$$\boldsymbol{H_o} = H_o\hat{\boldsymbol{z}} \text{ and } \boldsymbol{h} = h_u e^{j\omega t}\hat{\boldsymbol{u}} : \ \mu_{eff} = \mu \pm \kappa \tag{4.13}$$

where $\hat{\boldsymbol{u}}$ and $\hat{\boldsymbol{v}}$ represent any pair of mutually orthogonal unit vectors in the x-y plane. The propagation constant within the ferrite for all three cases is calculated as

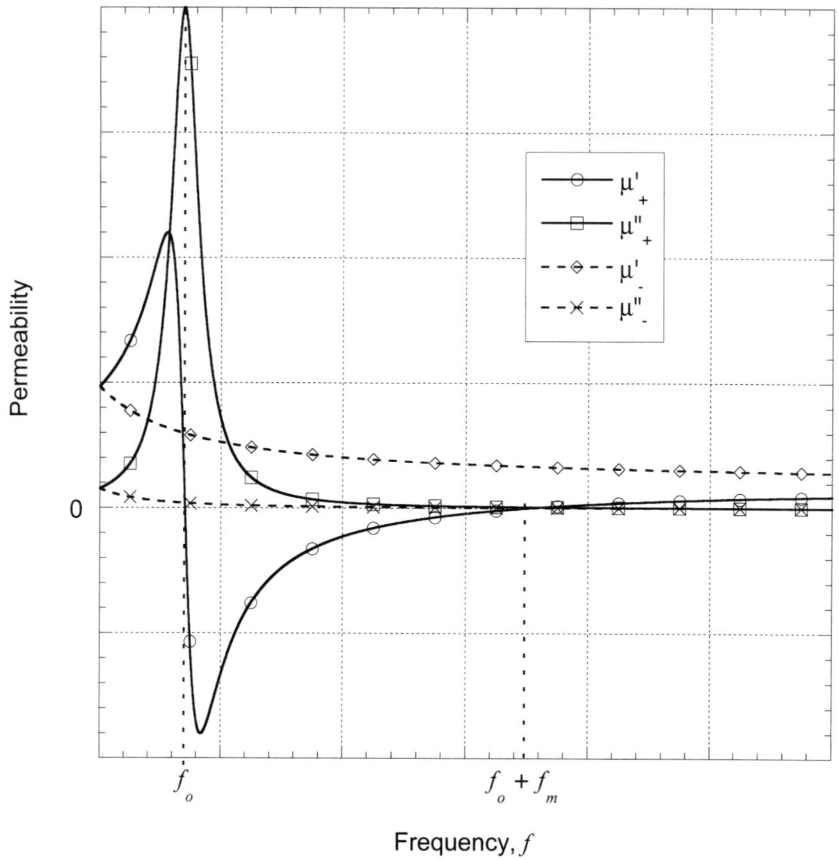

Figure 4.5 Typical behavior of μ_+ and μ_- with frequency.

$$\gamma_{prop} = \alpha + j\beta_e = j\frac{2\pi f}{c}\sqrt{\mu_{eff}\varepsilon_r} \tag{4.14}$$

For case 2, regions where μ_{eff} are negative correspond to a nonpropagating wave and occur in the frequency range where

$$\sqrt{f_o(f_o + f_m)} < f < (f_o + f_m) \tag{4.15}$$

A plot of the normalized frequency (f/f_m) as a function of normalized bias (f_o/f_m) showing the nonpropagating region is illustrated in Figure 4.6.

For case 3, the linearly polarized plane wave can be decomposed into clockwise and counterclockwise circularly polarized components with μ_+ and μ_-, respectively.

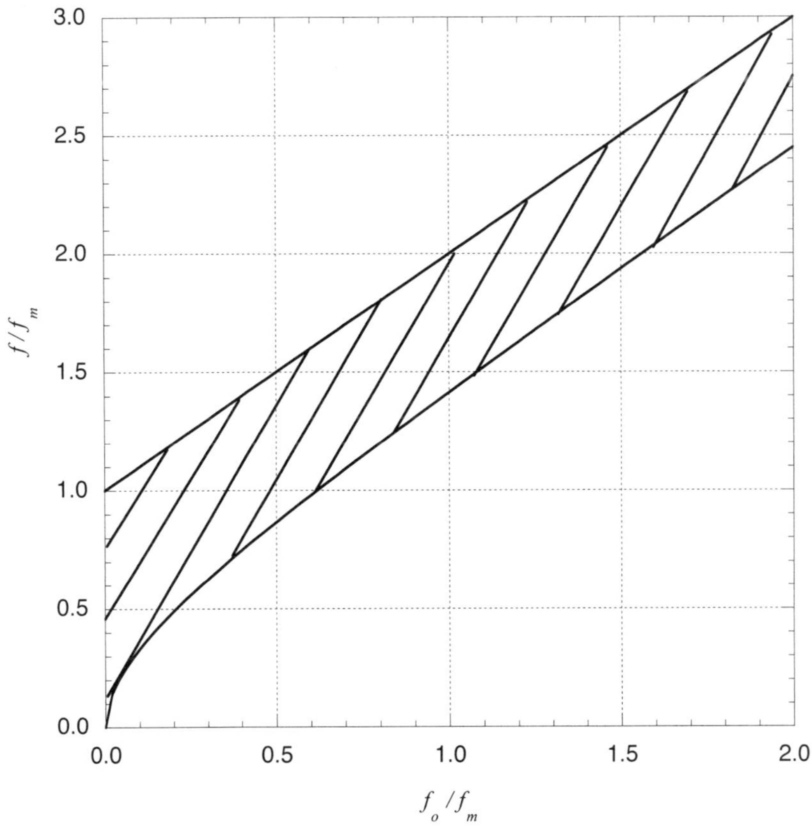

Figure 4.6 Shaded region in the graph of normalized bias versus normalized frequency represents the range where $\mu_{eff} < 0$.

Since the propagation constants of these two components are different, the resultant is a rotation of the polarization with distance, known as Faraday rotation. This phenomenon has been used in the design of phase shifters [35].

The preceding equations also indicate that for the longitudinal and transverse bias (cases 2 and 3), the effective permeability is a strong function of the applied bias field. Figures 4.7–4.9 show plots for μ_{eff} as a function of either transverse and longitudinal bias at different frequencies for the case of a ferrite with a saturation magnetization of $4\pi M_s = 1{,}000$G and a line width of $\Delta H = 100$ Oe. For the transverse bias (Figure 4.7), μ_{eff} shows strong resonance peaks, characterized by high imaginary values, representing regions of high loss. A similar resonance is seen in Figure 4.8 for μ_+ for a clockwise circularly polarized wave propagating in the longitudinally biased ferrite. This resonance peak does not occur, however, for μ_- for a counterclockwise circular polarized wave propagating in the longitudinally biased ferrite (Figure 4.9),

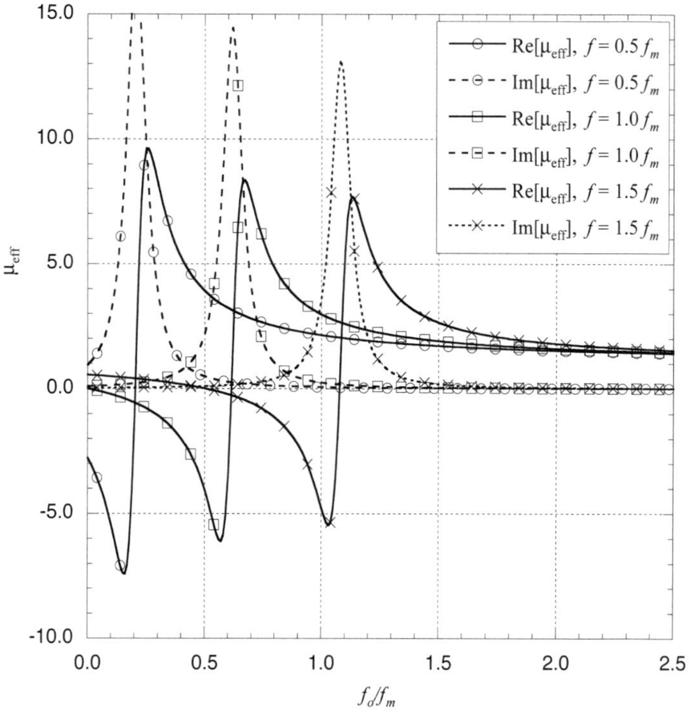

Figure 4.7 μ_{eff} for a transverse polarized ferrite with $4\pi M_s = 1{,}000$G and $\Delta H = 100$ Oe.

once again highlighting the nonreciprocal nature of the ferrite. This behavior parallels the behavior of μ_+ and μ_- with frequency, as already seen in Figure 4.5.

When a lossless ferrite material is saturated, (4.5) and (4.6) can be approximated by [33]

$$\mu \approx \mu_o \,, \quad \kappa \approx \frac{\mu_o \gamma 4\pi M_r}{f} \tag{4.16}$$

where f is the operating frequency and $4\pi M_r$ is the remnant magnetization of the ferrite (i.e., the magnetization within the ferrite after the ferrite has been saturated and the biasing field turned off). Figure 4.10 plots the value of κ (also known as the coupling factor) as a function of frequency for typical values of remnant magnetization. As the frequency increases, the coupling factor decreases. Since this coupling factor is an indication of the maximum amount of tuning that can be achieved, Figure 4.10 indicates that the maximum tuning range of a ferrite-based FAA will decrease with frequency. FAAs based on ferrites include microstrip antennas, cavities, and dielectric resonator antennas. Their performance will be explored in Chapter 6.

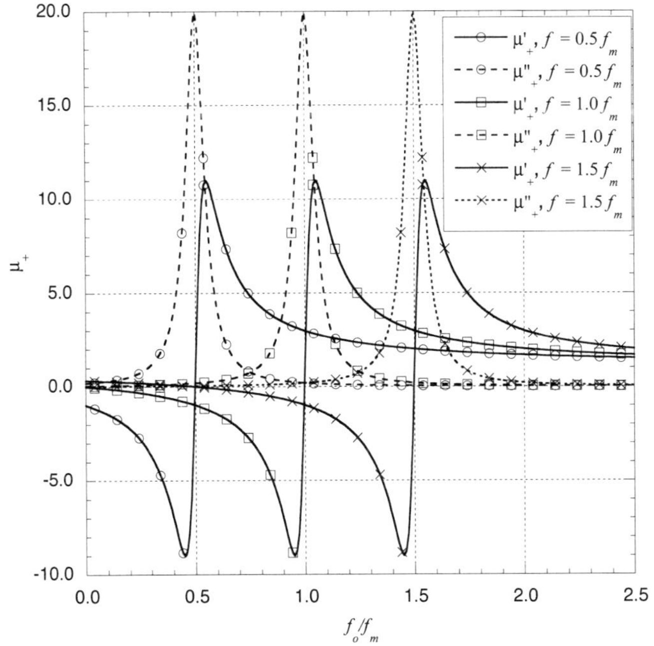

Figure 4.8 μ_+ for a longitudinal polarized ferrite with $4\pi M_s = 1{,}000$G and $\Delta H = 100$ Oe.

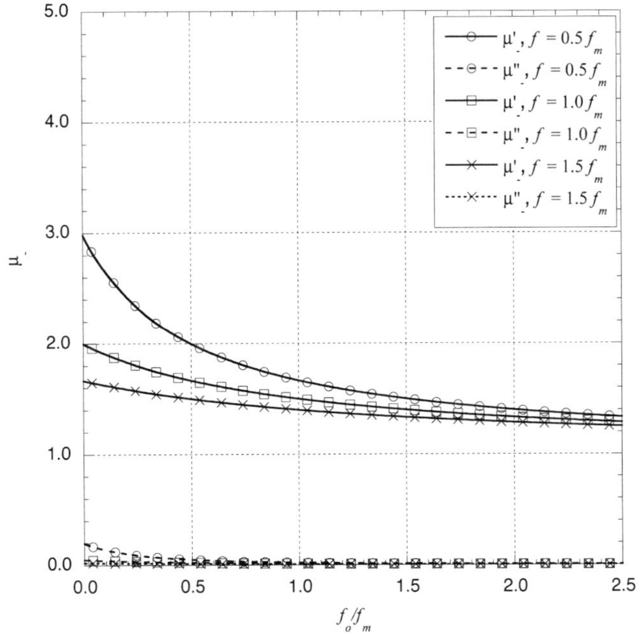

Figure 4.9 μ_- for a longitudinal polarized ferrite with $4\pi M_s = 1{,}000$G and $\Delta H = 100$ Oe.

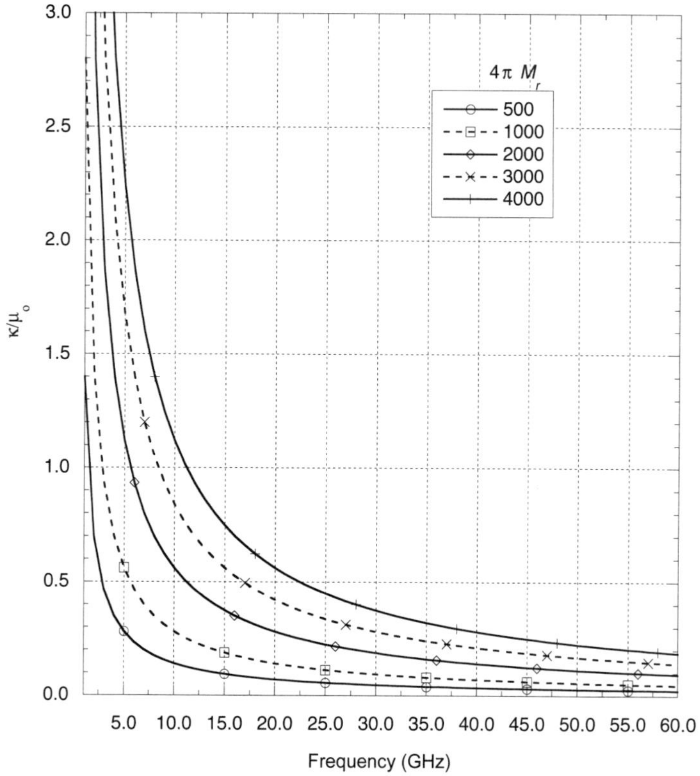

Figure 4.10 Coupling factor versus frequency for different remnant magnetizations.

4.2.2 Ferroelectrics

Ferroelectric substrates offer a behavior analogous to ferrites, where the permittivity of the ferroelectric can be altered by the application of a static electric field. This behavior is the property of certain classes of crystals exhibiting a spontaneous polarization that can be reversed by the application of an electric field. Metal oxides known as perovskites, characterized by the common formula ABO_3 (such as titanates $BaTiO_3$ and niobates $KNbO_3$), have the requisite crystal structure for this behavior [36–40]. The dielectric constant for these materials is typically on the order of several hundred. The polarization of the dipole moments in a normal dielectric is a linear function of the applied electric field. Since the material permittivity is related to the slope of the curve of polarization versus applied E-field [Figure 4.11(a)], then for the normal dielectric, the permittivity remains constant. A material that exhibits paraelectric behavior is one whose polarization is a nonlinear function of the

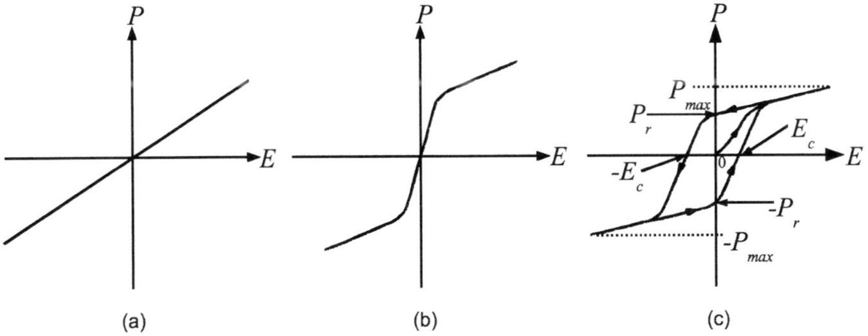

Figure 4.11 Polarization versus applied E-field curves for (a) dielectrics, (b) paraelectrics, and (c) ferroelectrics.

applied electric field, as in Figure 4.11(b). Thus the permittivity of the material is not a constant, but will vary as a function of the applied E-field. A ferroelectric material also exhibits this nonlinear dependency of polarization with the applied E-field; in addition, it exhibits spontaneous polarization (that is, polarization in the absence of an external E-field). The direction of this spontaneous polarization can be reversed by the application of an external E-field. Thus, a plot of the polarization versus applied E-field for a ferroelectric material will exhibit a hysteresis curve (similar to that of a ferrite), as shown in Figure 4.11(c). When a positive electric field is applied to an initially nonpolarized ferroelectric, the polarization will increase nonlinearly to a positive maximum value P_{max}. Unlike a dielectric or paraelectric material, when the E-field is removed from the ferroelectric material, it remains polarized at a value called the remnant polarization P_r. To remove the polarization, a negative E-field of value $-E_c$ (known as the coercive E-field) must be applied. If the magnitude of this negative E-field increases, the ferroelectric will develop a negative polarization which will increase to a maximum value $-P_{max}$. A remnant negative polarization $-P_r$ will then remain if the E-field is removed.

Whether a material exhibits this spontaneous polarization (ferroelectric) behavior or simply the paraelectric behavior depends on temperature. Below the Curie temperature T_c of the material, the ferroelectric exhibits spontaneous polarization (also known as the polar phase). Above T_c, the ferroelectric is in the paraelectric phase. Since, for most ferroelectrics, T_c is quite low (ranging from about 10K to several hundred Kelvin), they have mainly been used in their paraelectric phase for most microwave applications. Even if T_c were much higher, it would not be desirable to operate the ferroelectric in the polar phase at microwave frequencies, due to the hysteresis effect and to the presence of increased losses [39].

The tunability T_f of the ferroelectric is often defined as

$$T_f = \frac{\varepsilon(0) - \varepsilon(V)}{\varepsilon(0)} \tag{4.17}$$

where $\varepsilon(0)$ and $\varepsilon(V)$ are the values of the dielectric constant without and with the applied bias, respectively. Often, it is more convenient to measure the capacitance of a ferroelectric material sandwiched between two parallel plates. The tunability can then be expressed as

$$T_f = \frac{C(0) - C(V)}{C(0)} \tag{4.18}$$

where $C(0)$ and $C(V)$ are the unbiased and biased values of capacitance, respectively.

Biasing a ferroelectric is more convenient than biasing a ferrite, since the external E-field bias is easily achieved by applying a voltage across parallel plates with no current being drawn. Biasing ferrites requires a current-carrying coil or a movable permanent magnet. One drawback with ferroelectrics is the high E-field strength required to achieve the maximum range of tunability (on the order of MV/m). This puts a severe limitation on the thickness of bulk ferroelectric material if reasonable bias voltages are to be used. (A bulk material 1-mm-thick would still require several kilovolts of bias to achieve its maximum tuning range.) At microwave frequencies, values of tunability of up to 20% have been reported for bulk ferroelectrics, with typical loss tangents ranging from $0.0056 \leq \tan\delta \leq 0.044$ at room temperature [39]. The use of bulk ferroelectrics to design an electronically beam scanned antenna composed an array of continuous transverse stubs (CTS) was proposed around the turn of the century [41]. Due to the high bias values required and relatively high losses, the design had limited practicality.

In general, the high values of bias, relatively high losses, and limited tuning range do not make bulk ferroelectric materials attractive candidates for FAAs. However, research on thin-film ferroelectrics (tens of microns thick) has shown more promise. At these thicknesses, reasonable bias levels (tens of volts) are required to achieve the maximum tunability. Tuning ranges of up to 75% with loss tangents between $0.006 \leq \tan\delta \leq 0.06$ at room temperature have been reported [39]. Since the thin-film materials are typically deposited on thicker low-loss substrates, the overall losses of the devices are lower than ones fabricated solely from bulk ferroelectrics. Thin-film ferroelectrics have been used in the design of components such as tunable capacitors (varactors), tunable resonators, phase shifters, and delay lines, as well as in steerable beam antennas, tunable frequency-selective surfaces, tunable impedance surfaces, and phased arrays [42–56]. Examples of FAAs using ferroelectrics will be presented in Chapter 6.

4.2.3 Liquid Crystals

A liquid crystal is a fluid that exhibits a phase state with a certain degree of ordering in the arrangement of its molecules, which lies between the crystalline solid state (at low temperatures) and the ordinary liquid state (at high temperatures) [57–59]. Liquid crystals have been used for many years at optical frequencies in displays for numerous pieces of electronic equipment and more recently for flat-screen televisions. To date, only a limited amount of research has been carried out to examine the use of liquid crystals at microwave or millimeter-wave bands. For microwave applications, the liquid crystals have a behavior similar to that of paraelectrics, where an applied electric field can alter the dielectric constant of the material. In liquid crystals, this change in dielectric constant arises from the change in the orientation of the molecules in response to an applied E-field. The dielectric constants of liquid crystals are much lower than in ferroelectrics, with values typically between $2.5 \leq \varepsilon_r \leq 4$, making them more amenable to integration with printed antennas. A tunability of up to about 30% has been achieved at microwave frequencies, with bias voltages on the order of 15–20V, which is much lower than the bias voltages required for the paraelectric materials discussed in Section 4.2.2. Power consumption is also relatively low. One of the disadvantages of liquid crystals is that they must be kept between about 20°–35°C in order to stay in the liquid crystal phase. If the temperature drops much below 20°C, the material becomes solid, and above 35°C, the material becomes a regular liquid. Liquid crystals also require a relatively long time for tuning over the maximum tunability range. Losses are also relatively high at the lower microwave frequencies, and so liquid crystals are better suited for operation above 20 GHz.

At microwave frequencies, liquid crystals have been used to design phase shifters, tunable capacitors, and tunable frequency-selective surfaces and to control the reflected phase in reflectarray elements [60–78]. Examples of the use of a liquid crystal in the design of FAAs will be presented in Chapter 6.

4.2.4 Bulk Semiconductors

Semiconductors are used in a host of electronic devices including transistors, solar cells, and diodes and in digital and analog integrated circuits. These devices all exploit the fact that the conductivity of semiconductors can be altered by either an external electric field, thermal, or optical energy. This section briefly reviews the properties of intrinsic and extrinsic (doped) semiconductors. The potential use of bulk semiconductors for realizing FAAs is examined in this section, while in Section 4.3 the use of semiconductors in active microwave devices such as PINs, varactors, and transistors will be described.

The physics of semiconductors is well-documented in numerous books (such as [79-81]) and only a short overview is provided in this section. A semiconductor is usually a crystalline or polycrystalline material having an electrical conductivity between that of a conductor and an insulator (generally between $10^3$and 10^{-9} S/cm). Elements in column IV of the periodic table (such as carbon, silicon, and germanium) as well as III-V compounds (such as gallium arsenide and indium phosphide) and II-VI compounds (such as cadmium sulfide and zinc oxide) are considered good semiconductors.

Two complementary models are commonly used to describe the electrical behavior of semiconductors: the covalent bond model and the energy band model. In the covalent bond model, a single-crystal semiconductor is analyzed by examining the material's covalent bond structure at the atomic level. At a temperature near absolute zero, all the electrons are bound in covalent bonds. There is thus no conduction, and the semiconductor behaves as an insulator. As the temperature rises, the added thermal energy breaks some of the bonds and electrons are freed for conduction. The density of these free electrons is known as the intrinsic carrier density (n_i) and is a function of temperature and of the semiconductor band gap energy E_g, which is the minimum energy required to break a covalent bond in the semiconductor crystal. Breaking of this bond creates a free electron with a $-q$ charge ($q = 1.6 \times 10^{-19}$ C), and a bound ion with a $+q$ charge, called a hole. For a pure (intrinsic) semiconductor at thermal equilibrium, the electron charge density (n) and hole charge density (p) are equal and their product is equal to n_i^2. When an electric field is applied, the charged particles will move at a velocity given by

$$\boldsymbol{v}_n = -\mu_n \boldsymbol{E}, \quad \boldsymbol{v}_p = \mu_p \boldsymbol{E} \tag{4.19}$$

where μ_n (μ_p) is the charge mobility (cm^2/Vs) for the electrons (holes). The conductivity σ for an intrinsic semiconductor can then be expressed as:

$$\sigma = q\left(n\mu_n + p\mu_p\right) \quad [\text{S/cm}] \tag{4.20}$$

Since both the charge carrier density and charge mobility are strong functions of temperature, the conductivity will also be temperature-dependent. For intrinsic silicon at room temperature, the conductivity is approximately $\sigma = 3 \times 10^{-5}$ S/cm.

This usually undesirable temperature dependence can be nearly eliminated by the introduction of a small amount of impurities within the semiconductor (a process known as doping). Impurities can be of the donor or acceptor types, resulting in the increase in electron (n) or hole (p) charge densities, respectively. Denoting N_d and N_a as the impurity concentration of donor or acceptor atoms, respectively, then the new charge densities can be approximately given by

$$\begin{aligned} n &\approx \left(N_d - N_a\right), p = n_i^2 / n, \text{ for } N_d > N_a \\ p &\approx \left(N_a - N_d\right), n = n_i^2 / p, \text{ for } N_a > N_d \end{aligned} \tag{4.21}$$

In doped semiconductors, it is usual that $|N_a - N_d| >> 2n_i$ so that either electrons or holes become the majority carriers. A doped semiconductor with $N_d > N_a$ is known as n-type, since its majority carriers are electrons, while if $N_a > N_d$, the semiconductor is known as p-type, with holes being the majority carrier. Since the concentration of impurities (N_a and N_d) is not a function of temperature, the majority carrier density also becomes nearly temperature-insensitive. Although the minority carriers are still highly temperature-sensitive, their concentrations are much lower than the majority carriers, and their contribution to the conductivity becomes small. Thus the conductivity in doped semiconductors also becomes nearly temperature-independent.

The energy band model of semiconductors focuses on the energy states of the electrons. The regular arrangement of atoms in a solid results in the electrons occupying one of a series of energy bands. Each band is delineated by a minimum and maximum energy level, and it may be separated from adjacent bands by a forbidden band gap. The electrical behavior of the material depends on the number of filled electron states in an energy band as well as the size of the band gap. The energy states in the energy band of a conductor are only partially filled, so that when a small amount of energy is applied, electrons can easily move into the unfilled states and behave almost like free electrons, and they are thus highly conductive. An insulator is characterized by a completely filled energy band and a large forbidden band gap ($E_g > 5$ eV). Since the electron states within the band are completely filled, the electrons are bound and are not free to move. Due to the wide energy band gap, it would take a large amount of imparted energy to excite the electrons into the next energy band. Thus there are no electrons to carry current, and the material is an insulator. The semiconductor has an energy band structure similar to that of an insulator, but the size of the forbidden band gap is much smaller (E_g on the order of 1 eV). The band structure of a semiconductor is shown in Figure 4.12. In an intrinsic semiconductor, the uppermost filled band is known as the valence band. Above a temperature of 0K, the valence band is not quite entirely filled, since a small number of electrons have enough energy to jump across the forbidden band gap into the next allowed band, which is called the conduction band, since the electrons in this band can respond to an electric field to form a current. Doping the semiconductor with impurities results in the creation of new energy states. Adding donor impurities

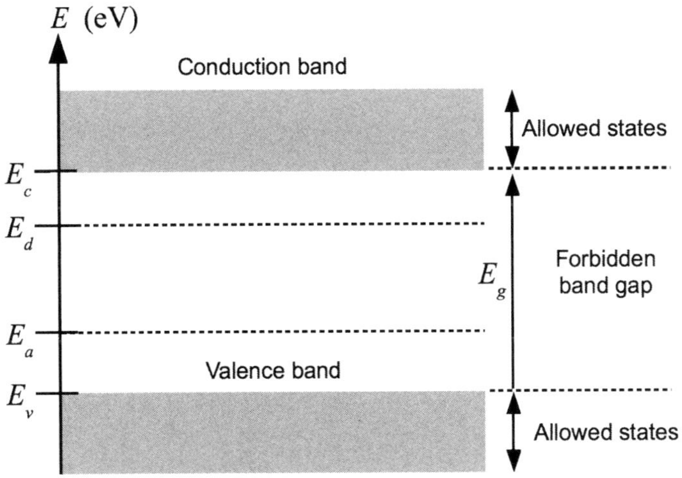

Figure 4.12 Energy band gap diagram of a semiconductor.

creates an energy state E_d slightly below the conduction band, while adding acceptor impurities creates an energy state E_a slightly above the valence band, as seen in Figure 4.12. It requires little energy to move electrons from E_d to E_c or E_v to E_a.

The energy required to break a covalent bond to free an electron can come from either a thermal source or a light source (photon energy). Photogeneration is the process by which photon energy breaks bonds to free electrons (and holes) to act as current carriers. The photon energy must be high enough to overcome the band gap energy of the semiconductor. For the case of silicon (with E_g = 1.124 eV), this requires photons with energy in the far infrared portion of the spectrum (since $E = h\,c/\lambda$). The incident photon energy on the semiconductor thus creates a layer of photoconductivity, whose conductivity and depth are functions of the photon energy (wavelength) as well as of the properties of the semiconductor (e.g., carrier diffusion length, and surface recombination lifetimes). Although the photoconductivity profile is a nonlinear function of penetration depth, a step-function approximation having the maximum photoconductivity value ($\Delta\sigma_m$) over an effective penetration depth (d_e) has been derived, which simplifies analysis of the photoconductive behavior [81]. Figure 4.13 shows a sketch of a typical photoconductivity profile as a function of penetration depth for a continuously illuminated semiconductor sample, along with the equivalent step profile. The values for $\Delta\sigma_m$ and d_e are determined using

$$\Delta\sigma_m = \frac{\Delta\sigma_o}{1+\alpha L_a}\left[\frac{1}{\alpha L_a}\left(\frac{\alpha L_a^2 + v_s\tau}{L_a + v_s\tau}\right)\right]^{\alpha L_a/(1-\alpha L_a)} \tag{4.22}$$

$$d_e = \frac{1}{\alpha}\left[\frac{L_a(1+\alpha L_a)+v_s\tau}{L_a + v_s\tau}\right]\left[\frac{1}{\alpha L_a}\left(\frac{\alpha L_a^2 + v_s\tau}{L_a + v_s\tau}\right)\right]^{\alpha L_a/(1-\alpha L_a)} \tag{4.23}$$

where

$$\Delta\sigma_o = \frac{q}{hc_o}(\mu_N + \mu_p)(1-R)\alpha S\lambda_p\tau\frac{P}{A} \tag{4.24}$$

$$L_a = \sqrt{\left[\frac{(2\mu_n\mu_p\tau k_B T/q)}{(\mu_n + \mu_p)}\right]} \tag{4.25}$$

and α is the radiation absorption coefficient; h *is* Planck's constant (h = 6.626 x 10^{-34} m^2 kg/s); c_o is the speed of light in free space; R is the surface reflectivity; S is the relative spectral response of semiconduction which exhibits a peak response at an optical wavelength of λ_p; P is the incident optical power; A is illuminated area of the semiconductor; v_s is the surface recombination velocity; τ is the excess carrier lifetime; k_B is Boltzmann's constant (k_B = 1.381 x 10^{-23} J/K); and T

is the absolute temperature. (The carrier lifetime and mobilities were assumed to be independent of excess carrier density.) The penetration depth can be approximated by

$$d_e \approx \begin{cases} \dfrac{1}{\alpha}, \text{ for } \alpha L_a \ll 1 \\ L_a, \text{ for } \alpha L_a \gg 1 \end{cases} \tag{4.26}$$

The overall conductivity of the semiconductor, called the plasma conductivity (σ_p), is given by

$$\sigma_p = \sigma_d + \Delta\sigma_m \tag{4.27}$$

where σ_d is the dark conductivity of the semiconductor (i.e., with no illumination).

An example from [81] has a He-Ne laser operating at 633 nm illuminating a silicon semiconductor with an optical power density of P/A = 30 mW/cm^2. For intrinsic silicon: ε_r = 11.7, σ_d = 2.5 x 10^{-4} (S/cm), τ = 10^{-6} s, L_a = 47 μm, $(\mu_n + \mu_p)$ = 2,100 cm^2/V s, R = 0.3, λ_p = 860 nm, v_s = 100 cm/s, α = 3,600 cm^{-1}, and S = 0.70 at 633 nm. Based on (4.22) and (4.23), the resultant values for photoconductivity and penetration depth are $\Delta\sigma_m$ = 4.14 10^{-3} S/cm and d_e = 49.8 μm. Such a change in value of conductivity of semiconductors can potentially be used to design antennas whose frequencies are optically tunable as will be seen in Chapter 6. A summary of the salient attributes of the tunable substrates investigated in this section is presented in Table 4.3.

4.3 ELECTRONIC SWITCHES AND TUNABLE DEVICES

This final section examines methods for the frequency tuning of antennas using discrete electronic components. Due to advances in packaging and miniaturization,

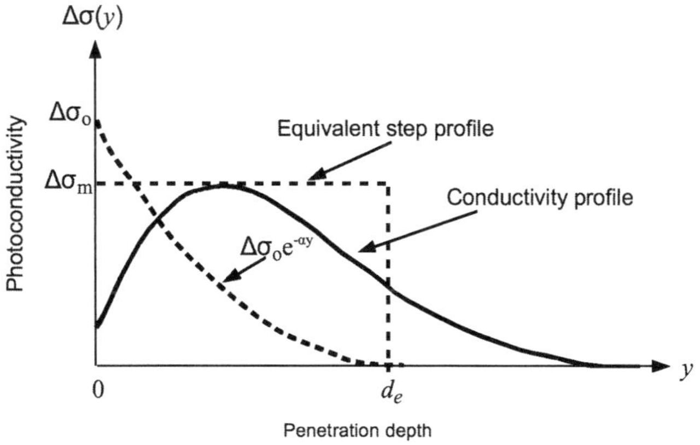

Figure 4.13 Photoconductivity profile in a semiconductor.

Table 4.3

Summary of Tunable Substrates

Tunable Substrate	Bias Method	Attributes
Ferrites	DC magnetic field (current) alters permeability	• Low-loss ferrites available • Medium values of dielectric constant (~10 < ε_r < 25) • Can be used for continuous frequency tuning • Can be integrated with printed antennas • Wide tuning ranges achievable >100% • Tuning range decreases with RF frequency • Switching speed < 5 ms • Temperature-sensitive • Higher DC power consumption
Ferroelectrics	DC E-field (voltage) alters permittivity	• Bulk, thick- and thin-film ferroelectrics available • High dielectric constant values (ε_r > 100) • Requires high bias values (kV/cm) • Can be integrated with printed antennas • Can be used for continuous frequency tuning • Temperature-sensitive • High losses • Tuning speeds < 1 μs (bulk), < 10 ns (thick-film), and < 1 ns (thin-film) • Negligible DC power consumption • Tuning ranges of up to 20% achieved
Liquid crystals	AC E-field (voltage) alters permittivity	• Low bias values required (< 40V) • Low dielectric constant values (ε_r ~ 3) • Negligible DC power consumption • Lossy at lower RF frequencies • Temperature-sensitive • Switching speed < 10 ms • Can be integrated with printed antennas • Can be used for continuous frequency tuning • Tuning ranges of up to 20% achieved
Semiconductors	Optical power alters conductivity	• Can be integrated with printed antennas • Moderate DC power consumption (< 10 mW) • Switching speeds from 10 fs to 10 ms • Typically used for discrete tuning

there are many commercially available devices that are small enough to be integrated into antennas where the packaging allows for convenient application of the required biasing signals. The various devices can be divided into electronic switches used for discrete frequency tuning and tunable devices used for continuous tuning. The switches considered in this section are PIN diode, MEMS switches, FET switches, and optical switches. The varactor diode is examined as an example of an electronically tunable device.

Electronic switches are characterized by several parameters, including the ones listed in Table 4.4. Each of the various electronic switches have their individual strengths and weaknesses, with no one switch demonstrating superior performance in all parameters. The choice of switch will thus strongly depend on the application. This section provides an overview of the different switches, describing their operation, highlighting their strengths and weaknesses, and presenting some simplified models and typical performance behavior.

Table 4.4

Characteristic Parameters of Electronic Switches

Parameter	Description
Actuation voltage	Minimum bias voltage required to effect switching.
Cold/hot switching	Refers to whether switching occurs in the absence/presence of the signal power.
Operational lifetime	Measured by the number of cycles that the switch is able to operate with both DC voltage and RF power applied.
Operating environment	Specifies the temperature range, humidity level, and radiation level the switch must withstand while operating and continuing to meet the required specifications.
Power consumption	DC power required to set the switch in a given state.
RF insertion loss	The loss a signal suffers from traversing the switch.
RF isolation	The signal leakage appearing at the output when the switch is in the Off state.
RF power handling	A measure of how much power a switch can pass. It is usually specified using the 1-dB compression point (which is the input power at which the output power deviates by 1 dB from the linear response). Power handling is limited by two factors: the heat dissipation limit of the transmission line and contacts and forced bias, where additional electrostatic forces generated from the rectified RF signal cause an unwanted actuation of the switch.
Shelf life	The storage time period during which the switch must retain integrity and functionality.
Switching speed	Time required for the switch to respond at the output when the control line input voltage changes. Switching speed includes driver propagation delays and the transition time and is measured from the 50% point to the 90% (10%) value of the RF voltage envelop for the On (Off) state.
Switching transients	Exponentially decaying voltage spikes at the input or output of an RF switch that result when the control voltage changes. Also called video leakage or video feed through.
Transition time	Time required for the RF voltage envelope to go from 10% to 90% of its maximum value for the On time or from 90% to 10% of its value for the Off time.

(From [82])

4.3.1 PIN Diodes

4.3.1.1 Circuit Model

A PIN diode is a circuit device consisting of three semiconductor layers: a lightly doped layer of high resistivity that is almost intrinsic (i), sandwiched by a heavily doped p-type layer on one side and a heavily doped n-type layer on the other side, as shown in cross-section in Figure 4.14(a). Metal contacts are added at the anode and cathode terminals. The PIN diode exhibits nonlinear behavior in terms of its current versus voltage (I-V) characteristics. A typical I-V curve for a silicon PIN diode is shown in Figure 4.14(b). When the voltage applied across the diode is greater than zero, the diode is said to be forward-biased (or in the On state), and with $V = (V_p - V_n) > 1$V, the diode essentially behaves like a short circuit. When the voltage applied across the diode is negative, the diode is said to be reversed-biased (or in the Off state), and it behaves essentially like an open circuit for voltages between 0 and V_b (the breakdown voltage), after which the diode begins to conduct again. This nonlinear behavior can be explained by solid state theory [83–85]. Briefly, when the PIN diode is at zero bias ($V = 0$), the diffusion of holes and electrons from the doped regions to the intrinsic region (high to low concentration) causes space-charge regions of thickness inversely proportional to the impurity concentration to form in the p and n layers adjacent to the i layer. The p layer develops a negative space charge region while the n layer develops a positive space charge. A purely intrinsic layer would be depleted of charge carriers while a lightly doped i layer would contain a small amount of charges. When a reverse bias is applied to the PIN diode ($V < 0$), the space charge regions in the p and n layers increase in width, and an electric field develops across the i layer, which opposes the diffusion and depletes the i layer of charge carriers,

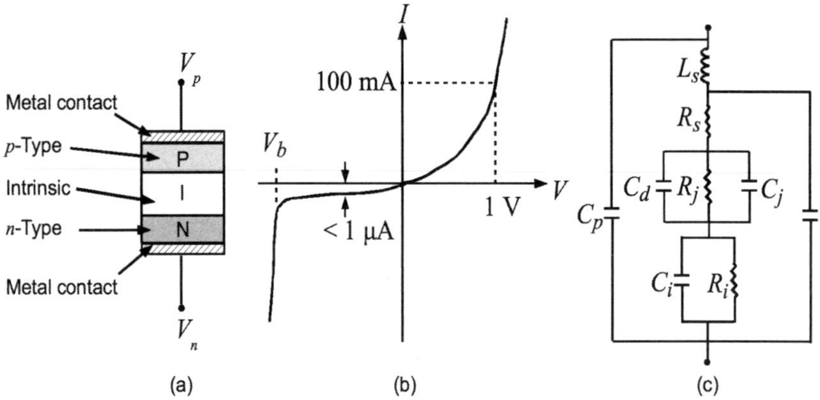

Figure 4.14 PIN diode and equivalent circuit model: (a) PIN diode cross-section, (b) typical I-V characteristics, and (c) circuit model. (After [84].)

thus acting like a highly resistive path or ideally an open circuit. When a forward bias is applied ($V > 0$), the voltage across the i layer enhances diffusion and the i layer is flooded with charge carriers thus behaving like a good conductor and a nearly ideal short circuit.

PIN diodes can be packaged in numerous configurations for various applications. They are often packaged in small plastic rectangular cases from which protrude two flat metal tabs, connected to the anode and cathode terminals of the diode. These packages can have maximum dimensions of less than 2 mm and are suitable for surface-mount applications. Also available are cylindrical packages where the terminals are located at the top and bottom of the cylinder. These diodes can be embedded in substrates, and are useful for applications requiring shunt configurations [86–89].

A thorough circuit model that accounts for capacitance and inductance effects of a typical packaged PIN diode is shown in Figure 4.14(c) [84]. Table 4.5 defines the various circuit elements of the PIN diode model along with some typical values under different bias conditions.

Table 4.5

Circuit Elements for the PIN Diode Model With Typical Values ([84, p. 278])

Circuit Element	Description	Zero Bias ($V = 0$V)	Reverse Bias ($V = -50$V)	Forward Bias ($V > 1$V)
L_s	Total series inductance, mainly from the connection leads.	0.3 nH	0.3 nH	0.3 nH
C_p	Stray capacitances shunted across the wafer by the package.	0.3 pF	0.3 pF	0.3 pF
C_f	Fringing capacitance shunted across the wafer from the leads.	0.02 pF	0.02 pF	0.02 pF
C_j	Charge storage at the boundaries of the depletion region.	0.8 pF	0.17 pF	>> 10 pF
R_j	Reciprocal of conductance caused by carriers generated within the depletion region.	$>10^9\Omega$	$>10^9\Omega$	$< 0.1\Omega$
C_d	Diffusion capacitance representing charge storage as current flows through the depletion region.	Negligible	Negligible	> 3 pF
R_i	Resistance of the intrinsic layer exclusive of the swept-out region.	2,500Ω	0Ω	0.5Ω
C_i	Capacitance of the intrinsic layer exclusive of the swept-out region.	0.25 pF	∞	> 0.25 pF
R_s	Sum of the resistance of the p and n layers and any resistance of the contact layers.	0.3Ω	0.3Ω	0.3Ω

Based on the typical values shown in Table 4.5 for the various circuit elements in the PIN diode model, the model of Figure 4.14(c) can be replaced by two separate but simpler models, one for each of the forward and reverse bias states. These simplified models are shown in Figure 4.15.

To determine the insertion loss, isolation, and return loss of the PIN diode as well as for the other switches to be discussed, it is useful to first review the *s*-parameters of devices connected in either a series or a shunt configuration, as shown in Figure 4.16. For each configuration, the device can be represented either by an impedance $Z_d = R + jX$, or an admittance $Y_d = G + jB$. In Figure 4.16, the characteristic impedance and characteristic admittance of the transmission lines are denoted by Z_o and Y_o, respectively. Since $Y_d = 1/Z_d$, the reactance G and susceptance B can be expressed in terms of resistance R and reactance X as

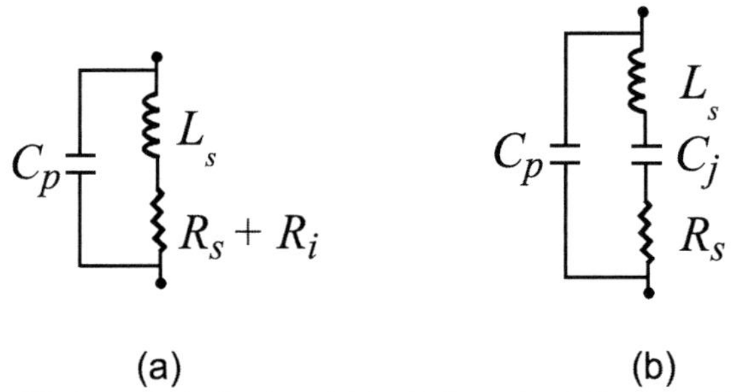

Figure 4.15 Simplified PIN diode circuit model for (a) forward and (b) reverse bias conditions. (After [83, p. 414].)

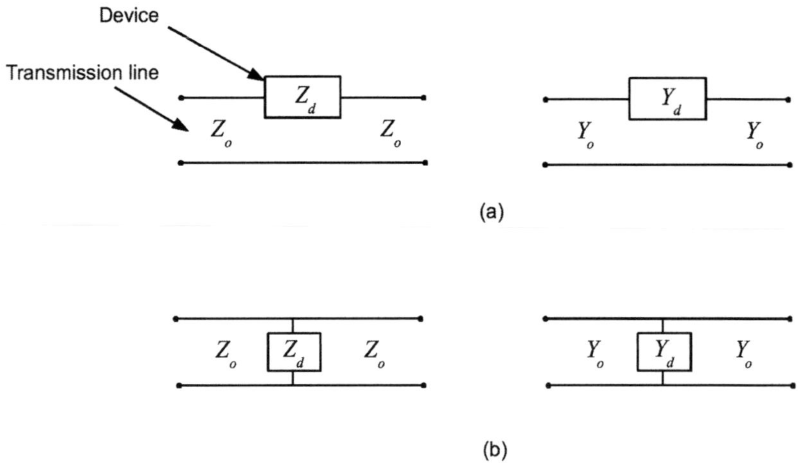

Figure 4.16 (a) Series- and (b) shunt-mounted devices.

$$G = \frac{R}{R^2 + X^2} \tag{4.28}$$

$$B = \frac{-X}{R^2 + X^2} \tag{4.29}$$

For the series configuration, the s-parameters in terms of impedances are [31, p. 178–179]

$$S_{11} = S_{22} = \frac{Z_d}{2Z_o + Z_d} \tag{4.30}$$

$$S_{21} = S_{12} = \frac{2Z_o}{2Z_o + Z_d} \tag{4.31}$$

and in terms of admittances

$$S_{11} = S_{22} = \frac{Y_o}{2Y_d + Y_o} \tag{4.32}$$

$$S_{21} = S_{12} = \frac{2Y_d}{2Y_d + Y_o} \tag{4.33}$$

For the shunt configuration, the s-parameters in terms of impedances are

$$S_{11} = S_{22} = \frac{-Z_o}{2Z_d + Z_o} \tag{4.34}$$

$$S_{21} = S_{12} = \frac{2Z_d}{2Z_d + Z_o} \tag{4.35}$$

and in terms of admittances

$$S_{11} = S_{22} = \frac{-Y_d}{2Y_o + Y_d} \tag{4.36}$$

$$S_{21} = S_{12} = \frac{2Y_o}{2Y_o + Y_d} \tag{4.37}$$

Both sets (impedance and admittance) of s-parameters are shown since, depending on the configuration, it is more convenient to use one set over the other to determine return loss, insertion loss, and isolation.

The RL is defined as

$$\mathrm{RL} = -10 \log |S_{11}|^2 \tag{4.38}$$

For the series configuration [Figure 4.16(a)], the RL is more easily expressed using the admittance form of S_{11}, so that

$$\text{RL} = 10\log\left[\frac{(2G+Y_o)^2+4B^2}{Y_o^2}\right] \tag{4.39}$$

while for the shunt configuration [Figure 4.16(b)], the impedance form is more suitable, giving

$$\text{RL} = 10\log\left[\frac{(2R+Z_o)^2+4X^2}{Z_o^2}\right] \tag{4.40}$$

The attenuation α across the device (which may represent either insertion loss or isolation) is defined as

$$\alpha = -10\log|S_{21}|^2 \tag{4.41}$$

For the series configuration, the attenuation is better expressed using impedances, thus

$$\alpha = 10\log\left[\frac{(R+2Z_o)^2+X^2}{4Z_o^2}\right] \tag{4.42}$$

while for the shunt configuration, admittances are more suitable

$$\alpha = 10\log\left[\frac{(G+2Y_o)^2+B^2}{4Y_o^2}\right] \tag{4.43}$$

The equations (4.39) to (4.43) can be applied to PIN-diode switches as well as to the other switches and to varactors that will be investigated in this chapter.

Looking specifically now at the PIN diode, for the forward-biased case, based on the simplified circuit model of Figure 4.15(a), the diode impedance Z_d at a frequency $\omega = 2\pi f$ is given by

$$R = \frac{R_f}{(1-\omega^2 L_s C_p)^2 + (\omega R_f C_p)^2} \tag{4.44}$$

and

$$X = \frac{\left[\omega L_s(1-\omega^2 L_s C_p) - \omega R_f^2 C_p\right]}{(1-\omega^2 L_s C_p)^2 + (\omega R_f C_p)^2} \tag{4.45}$$

while the admittance Y_d is

$$G = \frac{R_f}{R_f^2 + (\omega L_s)^2} \tag{4.46}$$

$$B = \omega C_p - \frac{\omega L_s}{R_f^2 + (\omega L_s)^2} \tag{4.47}$$

where $R_f = R_s + R_i$.

For the reverse bias condition, Z_d is derived using the simplified reverse bias model in Figure 4.15(b)

$$R = \frac{R_s}{\left[1 - \omega^2 L_s C_p + \frac{C_p}{C_j}\right]^2 + (\omega R_s C_p)^2} \tag{4.48}$$

and

$$X = \frac{\left[\omega L_s - \frac{1}{\omega C_j}\right]\left[1 - \omega^2 L_s C_p + \frac{C_p}{C_j}\right] - \omega R_s^2 C_p}{\left[1 - \omega^2 L_s C_p + \frac{C_p}{C_j}\right]^2 + (\omega R_s C_p)^2} \tag{4.49}$$

while the admittance Y_d is

$$G = \frac{R_s}{R_s^2 + \left[\omega L_s - \frac{1}{\omega C_j}\right]^2} \tag{4.50}$$

$$B = \omega C_p - \frac{\omega L_s - \frac{1}{\omega C_j}}{R_s^2 + \left[\omega L_s - \frac{1}{\omega C_j}\right]^2} \tag{4.51}$$

For the PIN diode mounted in a series configuration, for the forward bias (On) state, the RL, obtained by substituting (4.46) and (4.47) into (4.39), is

$$\mathrm{RL} = 10 \log \left[Z_o^2 \left[\left[\frac{2R_f}{R_f^2 + (\omega L_s)^2} + \frac{1}{Z_o} \right]^2 + 4\left[\omega C_p - \frac{\omega L_s}{R_f^2 + (\omega L_s)^2} \right]^2 \right] \right] \tag{4.52a}$$

For small values of R_f and L_s, and at low frequencies, the RL can be approximated by

$$\mathrm{RL} \approx 20 \log \left[\frac{2Z_o}{R_f} + 1 \right] \tag{4.52b}$$

The insertion loss (IL) is determined by substituting (4.44) and (4.45) into (4.42). For small values of L_s and C_p and at low frequencies, the IL can be approximated by the simplified expression

$$\text{IL} \approx 20 \log \left[1+\frac{R_f}{2Z_o}\right] \tag{4.53}$$

The isolation (IS) is determined when the diode is in the reverse (Off) state by substituting (4.48) and (4.49) into (4.42). For small values of R_s, L_s, and C_p, then $R \sim 0$ and $X \sim 1/\omega C_j$, so that the isolation can be approximated by

$$\text{IS} \approx 10 \log \left[1+\left[\frac{1}{4\pi C_j Z_o f}\right]^2\right] \tag{4.54}$$

For the shunt-mounted PIN diode, shown in Figure 4.16(b), with a reverse bias (Off-state) applied, the RL is determined by substituting (4.48) and (4.49) into (4.40). For small values of R_s, L_s, and C_p, then $R \sim 0$ and $X \sim (-1/\omega C_j)$ so that the RL can be approximated by

$$\text{RL} \approx 10 \log \left[1+\left[\frac{1}{\pi C_j Z_o f}\right]^2\right] \tag{4.55}$$

The IL is determined by substituting (4.50) and (4.51) into (4.43). For small values of R_s, L_s, and C_p, then $G \sim 0$ and $B \sim -\omega C_j$ so that the IL is approximately

$$\text{IL} \approx 10 \log \left[1+\left(\pi f C_j Z_o\right)^2\right] \tag{4.56}$$

For the shunt-mounted PIN diode with a forward bias applied, the IS is determined by substituting (4.46) and (4.47) into (4.43). For small values of L_s and C_p, then $G \sim 1/R_f$ and $B \sim 0_s$ leading to the approximate equation for IS

$$\text{IS} \approx 20 \log \left[\frac{Z_o}{2R_f}+1\right] \tag{4.57}$$

With the following typical values for the PIN diode circuit elements: $L_s = 0.3$ nH, $R_s = 0.5\Omega$, $C_p = 0.08$ pF, $C_j = 0.2$ pF, and $R_i = 0.3\Omega$, and using (4.42) and (4.52a), the PIN diode in series with a transmission line with $Z_o = 50\ \Omega$ would have a response as shown in Figure 4.17, plotted over a frequency range from 1–10 GHz. Also plotted are the RL, IL, and IS computed using the approximations in (4.52b), (4.53), and (4.54), respectively. The approximations for the RL and IL are independent of frequency and diverge from the more accurate expression as frequency increases. The approximation for IS tracks the more accurate expression over frequency but

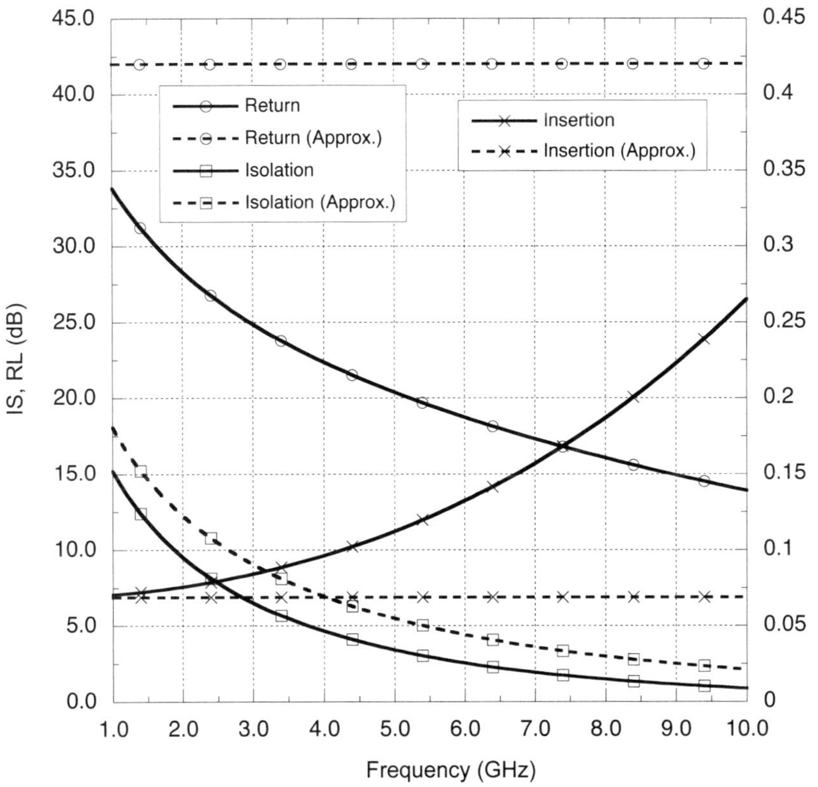

Figure 4.17 Example IL, RL, and IS curves for a series-mounted PIN diode (L_s = 0.3 nH, R_s = 0.5Ω, C_p = 0.08 pF, C_j = 0.2 pF, and R_i = 0.3Ω).

somewhat overestimates isolation for the particular parameter values selected for this example. At 1 GHz, the diode exhibits an insertion loss of 0.07 dB and an isolation of 15.2 dB. As the frequency increases, these values degrade. Used in an FAA, the degradation in the PIN-diode performance with frequency would result in an accompanying decrease in radiation efficiency and perhaps also in pattern distortion.

In the shunt configuration, the PIN diode would exhibit the response shown in Figure 4.18, based on (4.40), (4.42), and (4.43) and on the approximations in (4.55) to (4.57). The approximations diverge with increased frequency from the more accurate expressions. At 1 GHz, the IL would be 0.009 dB and the IS would be 22.0 dB. Thus for this example, the same diode offers better performance when used in a shunt configuration compared to a series configuration. In the series configuration, the switch IS is determined by the diode capacitance C_j, while in the shunt configuration, it is determined by resistance R_s. It is thus important to choose PIN diodes with appropriate values for the desired configuration.

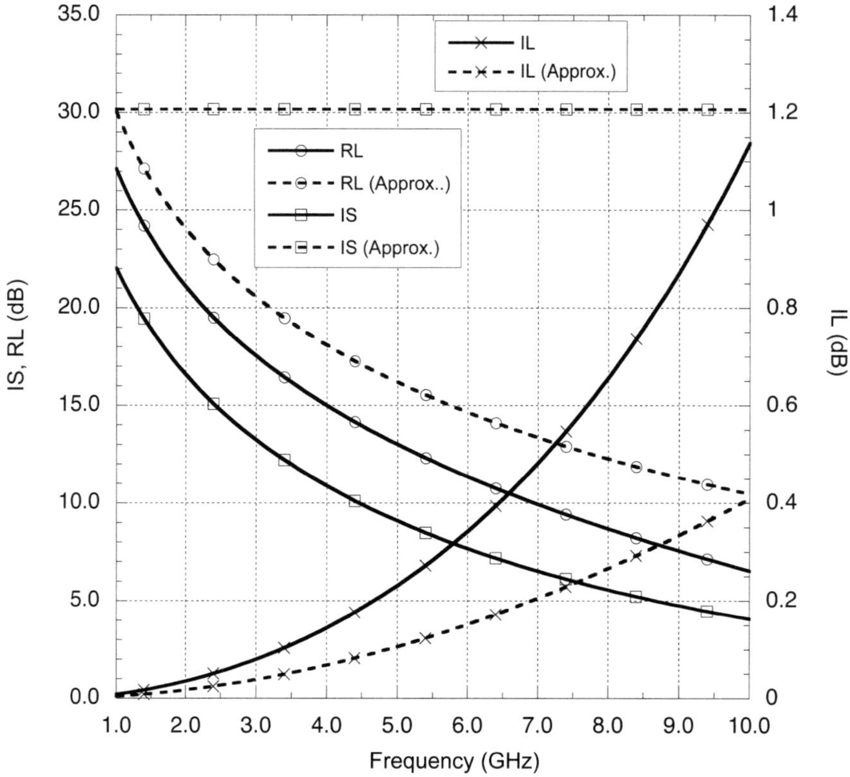

Figure 4.18 Example IL, RL, and IS curves for a shunt-mounted PIN diode (L_s = 0.3 nH, R_s = 0.5Ω, C_p = 0.08 pF, C_j = 0.2 pF, and R_i = 0.3Ω).

4.3.1.2 Low-Frequency Limitation

One important limitation of the PIN-diode switch is its lower frequency limit [90]. The PIN diode behaves as a current-controlled resistor when forward-biased. The resistance R_s in the intrinsic (i) region of the diode can be determined using

$$R_s = \frac{W^2}{(\mu_p + \mu_n) I_f \tau} \tag{4.58}$$

where W is the width of the i-region, μ_p and μ_n are the hole and electron mobilities, respectively, I_f is the forward current, and τ is the minority carrier lifetime (typically 5 ns $\leq \tau \leq$ 300 ns). The lowest operating frequency of the PIN diode is determined by the cutoff frequency (f_c), given by

$$f_c = \frac{1}{2\pi\,\tau} \tag{4.59}$$

At frequencies less than about $0.1f_c$, the PIN diode acts like a PN-junction diode and rectifies the RF signal, making it unsuitable for use. Between $0.1f_c$ and $10f_c$ the PIN diode can be considered to behave as a current-controlled capacitor. At frequencies above $10f_c$, the diode behaves as a current-controlled resistor, which is the behavior required for the switching function. PIN diodes are therefore typically not used below a few megahertz for switching purposes.

4.3.1.3 Biasing

When providing the DC bias to the PIN diodes, it is important to isolate the DC and RF signals from one another. This is typically done by using DC blocks (capacitors) and RF chokes (inductors). Examples of series-mounted and shunt-mounted PIN diodes with these added components are shown in Figure 4.19. The magnitude of the impedance Z_B of the DC block can be estimated using

$$|Z_B| = \frac{1}{2\pi f C_B} \tag{4.60}$$

where C_B is the value of the capacitor. At DC, $f = 0$ and the impedance of the capacitor appears infinite (independent of the value for C_B), thus preventing any DC signal from flowing along the RF path. The value C_B is chosen large enough so that its impedance at the RF operating frequency f appears very small and thus has minimal effect on the RF signal. For example, a circuit operating at 2 GHz with a capacitance of 100 pF would have an impedance of 0.8Ω. Choosing larger values of C_B would, in theory, lower this impedance further; however, the physically larger capacitor may not be as easy to integrate into the RF circuit and may result in RF discontinuities due to packaging or integration geometries.

Along a similar principle, the RF choke is designed to allow DC signals to pass while blocking RF signals. For narrowband lower frequency applications, a lumped inductor can serve as a suitable RF choke. The magnitude of its impedance can be approximated by

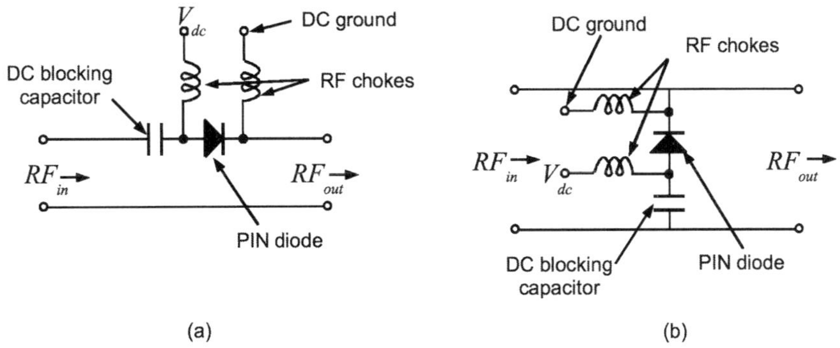

Figure 4.19 Biasing for (a) series- and (b) shunt-mounted PIN diodes. (After [92, 93].)

$$|Z_C| = 2\pi f L_C \tag{4.61}$$

where L_C is the inductance of the choke. In this case, the magnitude of impedance of the choke at DC is zero, thus allowing the DC signal to pass, regardless of the value for L_C. The inductance needs to be large enough to present a high impedance at the desired RF frequency. Again, taking the example of a system at 2 GHz, an RF choke with an inductance of L_c = 100 nH would result in an impedance magnitude of 1.26 kΩ. At higher frequencies, it is difficult to find a lumped inductor component that is purely inductive. In such cases, the DC bias signals can be fed through a quarterwave open-ended stub, which blocks the RF signals from reaching the DC bias lines. Integrating the PIN diode along with the required blocks and chokes into an antenna to make it frequency-agile with minimal deterioration of the antenna's electrical performance is one of the main challenges associated with this tuning technique.

4.3.1.4 RF Power Handling

Since PIN diodes draw current in the forward bias state, they will have an associated power consumption. Some typical values are 100 mA of current for a 1-V forward bias. The maximum RF power that the diode switch can support is limited by the maximum power dissipation allowable for the diode. For a series PIN diode, the dissipated power (P_d) in the diode is given by [91]

$$P_d = \frac{4R_s Z_o}{(2Z_o + R_s)^2}(\Gamma + 1)^2 P_a \tag{4.62}$$

where Γ is the magnitude of the reflection coefficient of the diode, and P_a is the available power from the RF signal generator. The available power can be expressed in terms of the signal generator voltage V_g using

$$P_a = \frac{V_g^2}{4Z_o} \tag{4.63}$$

For a shunt PIN-diode switch, the dissipated power is determined by

$$P_d = \frac{4R_s Z_o}{(Z_o + 2R_s)^2}(\Gamma + 1)^2 P_a \tag{4.64}$$

The maximum RF signal power that the diode can accommodate will differ whether it is configured as a series or a shunt switch. For example, for a diode with R_s = 1Ω and maximum dissipated power of P_{dmax} = 1W, then for the series configuration the maximum available power would be 51W, while for the shunt configuration the maximum available power would be only 13.5W (where (4.62) and (4.64) were used, with Z_o = 50Ω and Γ = 0).

4.3.1.5 Switching Speed

The switching speed of a PIN diode depends on both the driver circuit and on the PIN-diode properties including carrier lifetime and width of the intrinsic region. Typical switching speeds for the PIN diode are on the order of hundreds of nanseconds or less. Numerous examples of PIN-diode loaded FAAs will be reviewed in Chapter 8, and the effects of this integration will be examined.

4.3.2 MEMS Switches

MEMS RF switches make use of micromachined beams or membranes that are deflected to open or close an RF circuit in either a series or shunt configuration. MEMS switches were first demonstrated in 1979 as electrostatically actuated cantilever switches for low-frequency signals [94]. MEMS switches for RF circuits were first developed in the early to mid 1990s [95–99]. The micromachined beams can be designed for either direct (ohmic) or capacitive-coupled contact with the RF circuit. The beams are typically attached at one end (cantilever) or attached at both ends (fixed-fixed beam or membrane), as sketched in Figure 4.20. Actuation of the beams can be accomplished using various techniques including electrostatic, electromagnetic, piezoelectric, or thermomechanical. Electrostatic actuation is most commonly used due to the relative ease of generating electric fields at the microscopic level, fast response times, and low power consumption [82]. Recent articles provide good overviews of the state-of-the art of RF MEMS [100–103], and in-depth analysis can be found in the books by Rebiez [104] and Santos [82] along with numerous articles published over the last two decades. RF MEMS switches are now also becoming commercially available [105, 106].

Summarizing these various references, RF MEMS switches have several attractive features over PIN diodes or other solid-state switches including lower insertion loss, higher isolation, low DC power consumption, and relatively high power handling. The MEMS switches do have certain disadvantages, however, including the requirement for high activation voltages, higher cost, lower reliability, and limited commercial availability. For ohmic contact switches, one of the main sources limiting reliability is the increased resistance that eventually occurs due to damage, pitting, and hardening of the contact area arising from the impact force between the beam and the contact. Microwelding can also be a serious problem for switches handling higher RF power (100 to 1000 mW), where the power dissipated in the ohmic resistance is enough to cause the beam to weld to the contact, resulting in a failure of the switch to open. For capacitive switches, the main source of failure is due to stiction (the inability of the switch to release once the pull-down voltage is removed) between the dielectric layer and the metal layer. This results from the relatively large contact area where the charging effects of the dielectric layer either cause the switch to stick in the down state position or result in an increase in the pull-down voltage, making the

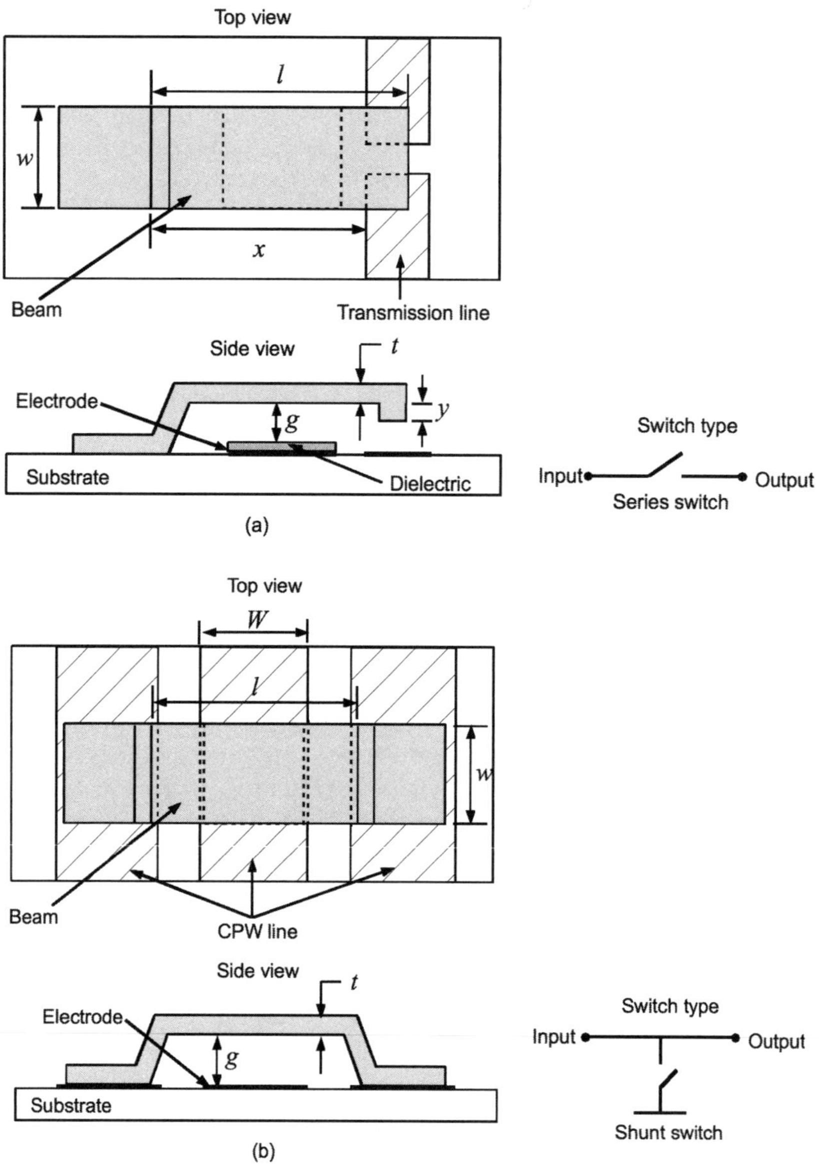

Figure 4.20 Sketches of representative MEMS RF switch configurations: (a) cantilever design, and (b) fixed-fixed beam design.

switch unusable. Also, due to the MEMS's sensitivity to moisture and dirt, the device needs to be hermetically sealed in a nitrogen environment, which increases the cost of these devices.

Electrostatic (voltage-driven) actuators have very low power consumption (microwatt-range) compared to the milliwatt-range for electromagnets, but require higher activation voltages (typically 20–70V). Piezoelectric actuators rely on the deformation of structures caused by the motion of internal charges as a result of an applied electric field [82]. They typically make use of exotic materials, such as lead-zirconate-titanate (PZT), but may require somewhat lower activation voltages. Electromagnetic actuators are current-driven and thus require lower actuation voltages, but they consume higher DC power and it is difficult to realize the required multiturn magnetic windings on a small scale.

4.3.2.1 MEMS Actuation Equations

The required activation voltage of MEMS RF switches is a function of the following parameters: beam (or membrane) material, beam thickness t, gap height g, and beam geometry [107]. The effective spring constant k_c for the capacitive fixed-fixed beam MEMS switch shown in Figure 4.20(a) is given by [102, 104, 108]

$$k_c = 32\, w\, E \left(\frac{27}{49}\right)\left(\frac{t}{l}\right)^3 + 8\, \sigma (1-\nu)\, w \frac{3\, t}{5\, l} \tag{4.65}$$

For the ohmic cantiliver beam MEMS switch in Figure 4.16(b), the effective spring constant k_o is

$$k_o = 2\, w\, E \left(\frac{t}{l}\right)^3 \frac{1-\frac{x}{l}}{3-4\left(\frac{x}{l}\right)^3+\left(\frac{x}{l}\right)^4} \tag{4.66}$$

For (4.65) and (4.66), E is Young's modulus for the beam material, ν is Poisson's ratio, and σ is residual tensile stress in the beam. The voltage V_p required to pull down the fixed-fixed capacitive beam is determined using

$$V_p = \sqrt{\frac{8k_c\left(g+\frac{t_d}{\varepsilon_d}\right)^3}{27\varepsilon_o W w}} \tag{4.67}$$

where t_d and ε_d are the thickness and dielectric constant, respectively, of the dielectric spacer shown in Figure 4.20(a). The voltage V_h required to maintain or hold down the beam after it has been pulled down is given by

$$V_h = \frac{t_d}{\varepsilon_d}\sqrt{\frac{2k_c g}{\varepsilon_o \varepsilon_d W w}} \tag{4.68}$$

For the ohmic cantilever beam, the pull-down (V_p) and hold (V_h) voltages are given by

$$V_p = \sqrt{\frac{8k_o g^3}{27\varepsilon_o W w}} \tag{4.69}$$

and

$$V_h = y\sqrt{\frac{2k_o(g-y)}{\varepsilon_o W w}} \tag{4.70}$$

Typical pull-down voltages range from about 30 to 70V. The switching time t_s can be determined using

$$t_s = \frac{3.67V_p}{V_a \omega_o} \tag{4.71}$$

where V_a is the applied voltage and ω_o is the resonant frequency of the beam of mass m and spring constant k, given by

$$\omega_o = \sqrt{\frac{k}{m}} \tag{4.72}$$

By applying a voltage of V_a greater than the pull-down voltage V_p, switching times can be decreased. Typical RF MEMS switches have switching times of between 2 to 50 μs.

4.3.2.2 MEMS Circuit Model

In terms of an RF circuit model, both the cantilever and membrane configurations can be considered as a series RLC circuit, having negligible resistance R and inductance L. The capacitance C_u in the Off (up) state can be approximated using

$$C_u \approx 1.4\frac{\varepsilon_o w W}{g + \frac{t_d}{\varepsilon_r}} \tag{4.73}$$

where the factor of 1.4 is used to account for the fringing fields of the parallel plate capacitor. Typical values for C_u are on the order of 50 fF or lower. The quality factor is given by

$$Q = \frac{1}{2\pi f R C_u} \tag{4.74}$$

where f is the operating frequency. Typical values for Q-factor range from 20 to 50 for frequencies up to the Ka-band range. In the On (down) state, the fringing fields are negligible and the capacitance C_d is approximated using

$$C_d \approx 0.65 \frac{\varepsilon_o \varepsilon_d wW}{t_d} \tag{4.75}$$

which is somewhat less than the parallel plate capacitance to account for the surface roughness of the dielectric. Typical values for C_d are on the order of 1.2 pF. The capacitance ratio C_u/C_d is on the order of 30. A high value of capacitance ratio is required for achieving a high IS between the up and down states of the switch.

The isolation for a RF MEMS series switch when in the Off state is determined using the expression for attenuation α for a series impedance from (4.42)

$$\alpha = 10 \log \left[\frac{(R_u + 2Z_o)^2 + X_u^2}{4Z_o^2} \right] \tag{4.76}$$

where Z_o is the characteristic impedance of the transmission line, R_u is the up-state resistance, and

$$X_u = 2\pi f L - \frac{1}{2\pi f C_u} \tag{4.77}$$

Since the switch resistance in the Off state is much smaller than the characteristic impedance of the transmission line, and for a negligible switch inductance, the Off-state IS can be approximated by

$$\text{IS} \approx 10 \log \left[1 + \left(\frac{1}{4\pi C_u Z_o f} \right)^2 \right] \tag{4.78}$$

The IL of the switch in the On state is given by

$$\alpha = 10 \log \left[\frac{(R_d + 2Z_o)^2 + X_d^2}{4Z_o^2} \right] \tag{4.79}$$

where R_d is the contact resistance of the switch, and X_d is given by

$$X_d = 2\pi f L - \frac{1}{2\pi f C_d} \tag{4.80}$$

Since, for the On state, C_d is approximately infinite, and assuming L is again negligible, the IL can be approximated by

$$\text{IL} \approx 20 \log \left[1 + \frac{R_d}{2Z_o} \right] \tag{4.81}$$

The RL of the On state is determined by using (4.39)

$$\mathrm{RL} = 10 \log \left[\frac{(2G_d + Y_o)^2 + 4B_d^{\,2}}{Y_o^2} \right] \tag{4.82}$$

where $G_d = \dfrac{R_d}{R_d^2 + X_d^2}$ and $B_d = \dfrac{-X_d}{R_d^2 + X_d^2}$

Making the same assumptions as for the IL leads to $G_d \sim 1/R_d$ and $B_d \sim 0$ so that the RL can be approximated by

$$RL \approx 20 \log \left[\frac{2Z_o}{R_d} + 1 \right] \tag{4.83}$$

The performance of a typical RF MEMS series switch is shown in Figure 4.21, based on equations (4.78) to (4.82). In the On state, the IL is only a few tenths of a decibel, and the RL is better than 20 dB for up to 50 GHz. The switch IS (in the Off state) is below 30 dB up to 60 GHz. This performance is significantly better than PIN-diode switches and does not degrade until much higher frequencies.

For shunt RF MEMS switches, the Off-state IL is given by the attenuation α based on (4.43)

$$\alpha = 10 \log \left[\frac{(G_u + 2Y_o)^2 + B_u^2}{4Y_o^2} \right] \tag{4.84}$$

where $G_u = \dfrac{R_u}{R_u^2 + X_u^2}$ and $B_u = \dfrac{-X_u}{R_u^2 + X_u^2}$.

Assuming the switch resistance is much smaller than its reactance ($R_u << X_u$) and that the switch inductance is negligible, the Off-state IL can be approximated by

$$\mathrm{IL} \approx 10 \log \left[1 + (\pi Z_o C_u f)^2 \right] \tag{4.85}$$

The RL is determined by applying (4.33)

$$\mathrm{RL} = 10 \log \left[\frac{(2R_u + Z_o)^2 + X_u^2}{Z_o^2} \right] \tag{4.86}$$

which, for the same approximations made for the IL, becomes

$$\mathrm{RL} \approx 10 \log \left[1 + \left(\frac{1}{2\pi Z_o C_u f} \right)^2 \right] \tag{4.87}$$

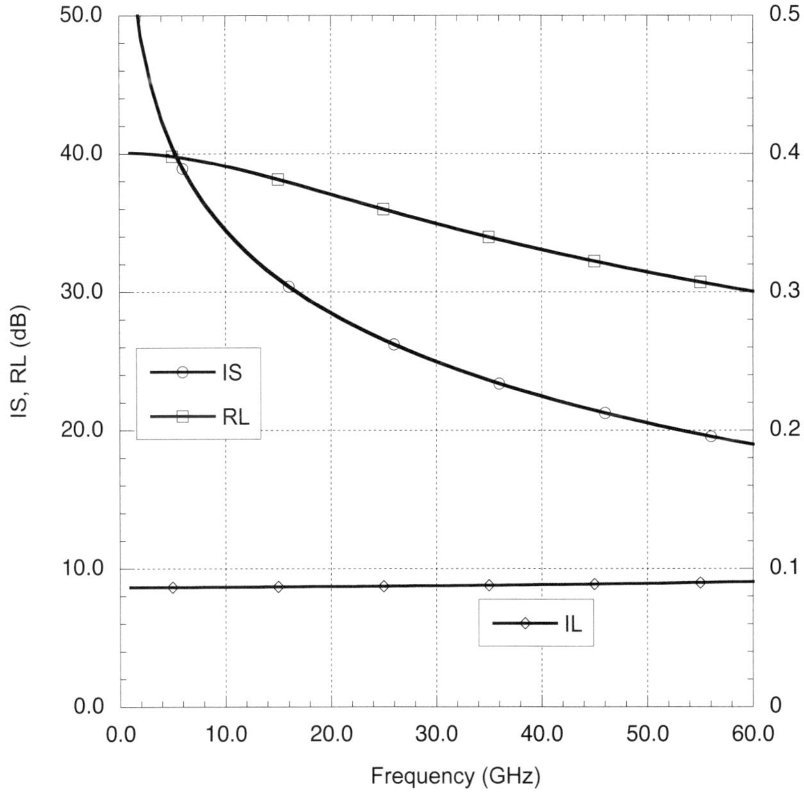

Figure 4.21 Performance of a MEMS series switch (R_s = 1Ω, L = 8 pH, and C_u = 3 fF).

For shunt RF MEMS switches, the down-state IS is determined using

$$\text{IS} = 10 \log \left[\frac{\left(G_d + 2Y_o\right)^2 + B_d^2}{4Y_o^2} \right] \tag{4.88}$$

At the On-state resonant frequency f_d, given by

$$f_d = \frac{1}{2\pi\sqrt{LC_d}} \tag{4.89}$$

and with $B_d = X_d = 0$, the On-state IS at resonance reduces to

$$\text{IS} = 20 \log \left(1 + \frac{Z_o}{2R_d} \right) \tag{4.90}$$

The performance of a typical RF MEMS shunt switch is shown in Figure 4.22, based on (4.85) to (4.90). By optimizing the inductance of the switch, the location of the down-state resonant frequency f_d (where maximum isolation occurs) can be adjusted to be near the desired operating frequency where, in this example, an isolation of 40 dB is achieved. The (Off-state) IL of the shunt switch can be somewhat poorer than the (On-state) IL of the series switch.

A figure of merit used for the switch that is an indication of low-loss performance is the cutoff frequency f_c where

$$f_c = \frac{1}{2\pi C_u R_s} \tag{4.91}$$

The cutoff frequency of RF MEMS series switches is typically 30–80 THz, while for GaAs PIN diodes it is typically 1–2 THz and 0.2–0.5 THz for FET switches (discussed in Section 4.3.3).

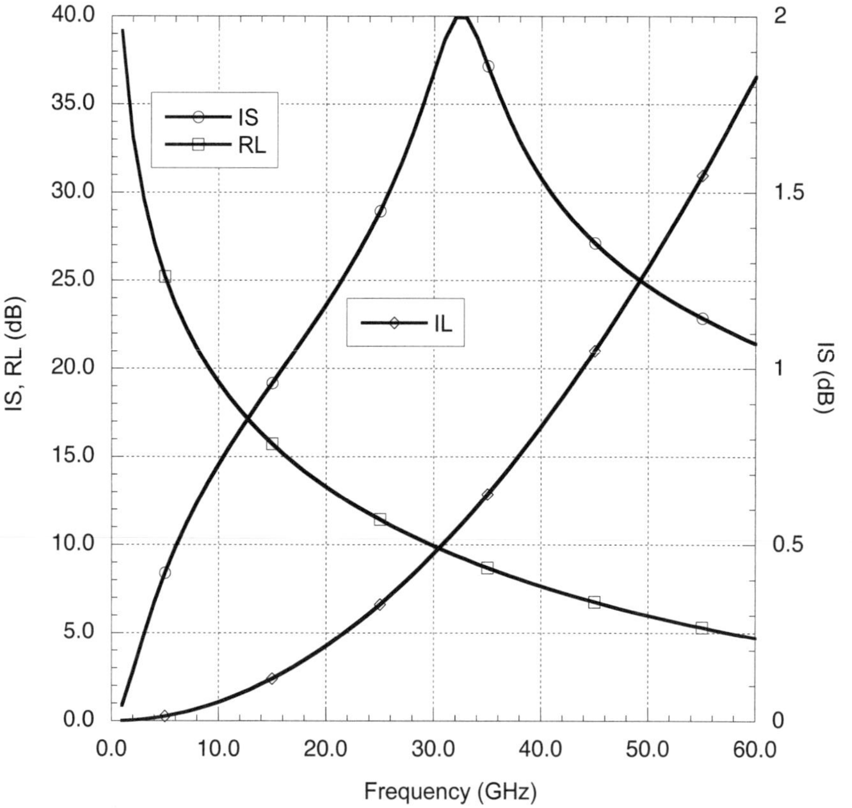

Figure 4.22 Performance of a MEMS shunt switch ($R_s = 0.25\Omega$, $L = 8$ pH, $C_d = 3$ pF, and $C_u = 70$ fF).

4.3.2.3 RF Power Handling

The maximum RF power handling of capacitive MEMS switches refers to the power at which the MEMS devices fail to operate properly. There are two types of failures: RF latching and RF self-actuation. For RF latching, the applied power provides enough force on the membrane to hold the switch down when it should have released (i.e., when the DC bias voltage is removed). RF self-actuation occurs when the RF power is high enough to pull down the membrane to its actuated state without application of DC bias. (Self-actuation powers of 4W are typical while self-latching powers are around 500 mW.) The power (P_{act}) for self-actuation can be approximated by $P_{act} = V_p^2/Z_o$.

4.3.2.4 Biasing, DC Power Consumption, and Switching Speed

Unlike PIN diodes, MEMS switches do not require DC blocks and RF chokes to prevent the DC and RF signals from mixing. In MEMS switches, the DC and RF signals are already inherently isolated. This reduces the complexity of the bias circuitry for MEMS devices to that of a simple pair of electrodes in series with a high-value resistor across which the DC voltage bias is applied. Since the capacitive-type MEMS switch draws no current, and the ohmic-type switch draws negligible current in the On state (due to the presence of the high-value bias resistor), the MEMS switch draws negligible DC power. Due to the physical movement of the cantilever beams, switching speeds are typically slower than for PIN diodes, on the order of microseconds.

MEMS switches have been integrated with various antennas, including microstrip patches, PIFAs, and slots, for the design of FAAs with discrete tuning, examples of which will be presented in Chapter 8.

4.3.3 FET Switches

FETs can also be used as microwave switches [109–111]. Figure 4.23 shows the cross-section of a typical FET in the On and Off states. The FET is a three-terminal device, labeled source (S), drain (D), and gate (G), where the bias voltage applied to the gate controls the flow of current through the channel. When the bias voltages (V_s, V_d, and V_g) to all three terminals are equal (and usually set to 0V), the depletion region (the shaded region under the gate in Figure 4.23) is small, and current flows freely through the channel, which corresponds to the On state of the switch. (V_g could also be set to be somewhat greater than 0V to further reduce the size of the depletion region.) For the Off state, the voltage on the gate is lowered to a value known as the pinchoff voltage (V_p), where the depletion layer increases until it fills the width of the channel, preventing current from flowing. The value for V_p is typically on the order of -5V. The FET switch draws practically no current in either state, giving it an advantage over the PIN-diode switch in terms of a negligible DC power consumption.

4.3.3.1 Circuit Model

The behavior of the FET switch can be modeled using lumped resistors and capacitors, as shown in the circuit model in Figure 4.24. Since the value of R_{sd} in the On state is much lower than the other components, the model can be simplified to a single resistor (R_{sd}). Similarly, in the Off state, the model can be simplified to a series RC circuit. These simplified models are shown in the second row of Figure 4.24.

Following the same derivations as for the previous PIN and MEMS series and shunt switches, the On-state attenuation for the series FET switch (with $R = R_{sd}$, and $X = 0$) corresponds to the IL, approximated by

$$\mathrm{IL} \approx 20\,\log\left[\frac{R_{sd} + 2Z_o}{2Z_o}\right] \tag{4.92}$$

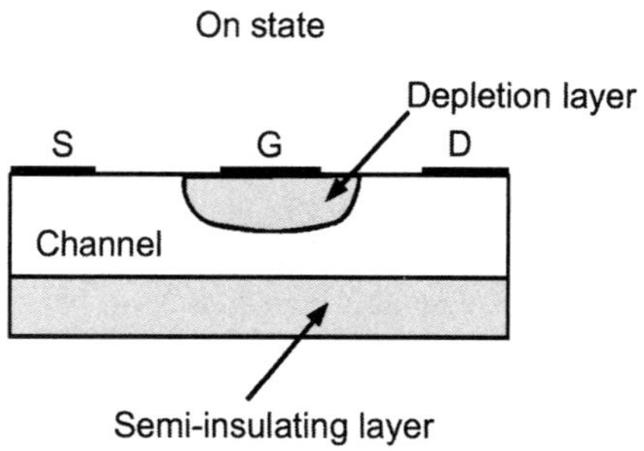

Figure 4.23 Cross-section of a FET in the On and Off states. (After [111].)

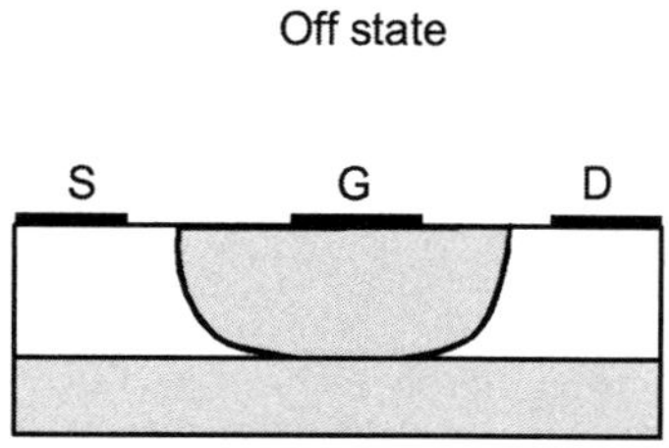

Figure 4.24 FET equivalent circuit models for the On and Off states. (After [111].)

The RL is approximated by

$$\text{RL} \approx 20 \log\left[\frac{2Z_o}{R_{sd}} + 1\right] \tag{4.93}$$

In the Off state, with $R=R_{sd}$ and $X = \dfrac{1}{2\pi f C_{sd}}$, the attenuation corresponds to the IS. Since R_{sd} is usually much smaller than Z_o, the IS can be approximated by

$$\text{IS} \approx 10\log\left[1 + \left[\frac{1}{4\pi Z_o C_{sd} f}\right]^2\right] \tag{4.94}$$

For the shunt FET switch, the attenuation in the On state corresponds to IS, leading to

$$\text{IS} = 20 \log\left[1 + \frac{Z_o}{2R_{sd}}\right] \tag{4.95}$$

In the Off state, assuming that the switch resistance is much smaller than its reactance ($R_u << X_u$), the IL can be approximated by

$$\text{IL} \approx 10 \log\left[1 + \left(\pi Z_o C_{sd} f\right)^2\right] \tag{4.96}$$

and the RL by

$$\text{RL} \approx 10 \log\left[1 + \left[\frac{1}{2\pi Z_o C_{sd} f}\right]^2\right] \tag{4.97}$$

Examples of the performance of FET switches in series and shunt configurations are shown in Figures 4.25 and 4.26, respectively.

4.3.3.2 Biasing, DC Power Consumption, and Switching Speed

Figure 4.27 shows the biasing circuit for the FET switch connected in either a series or shunt configuration. Instead of an RF choke, a high-impedance resistor can be used to prevent any RF signals from entering the gate control line. As mentioned previously, the FET switch draws negligible current (on the order of < 100 μA) and thus consumes virtually no DC power, giving it an advantage over PIN diodes. Switching speeds for FETs are similar to PINS, being on the order of hundreds of nanoseconds or less.

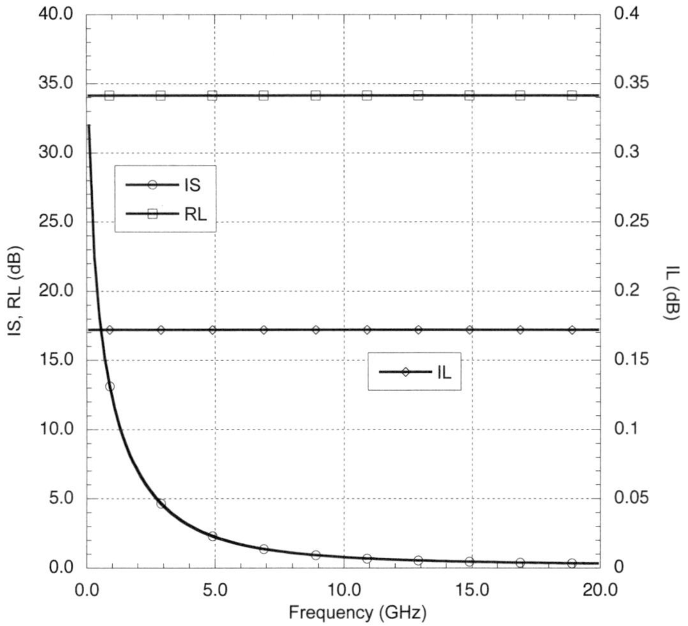

Figure 4.25 FET series switch ($R_{sd} = 2\Omega$, and $C_{sd} = 0.4$ pF).

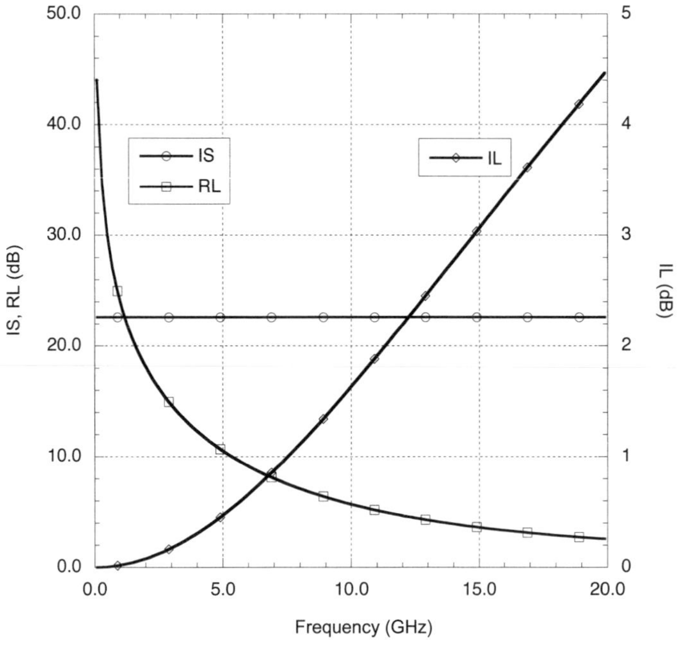

Figure 4.26 FET shunt switch ($R_{sd} = 2\Omega$, and $C_{sd} = 0.4$ pF).

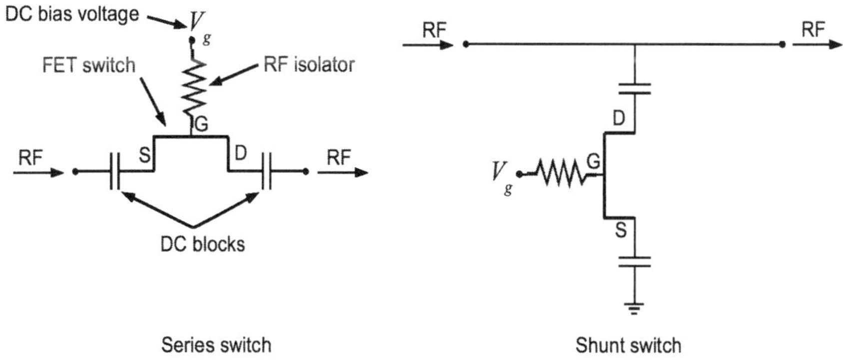

Figure 4.27 Series and shunt configurations of the FET switch.

4.3.3.3 RF Power Handling

In the On state, the maximum RF power P_{max} that the FET can accommodate depends on maximum current I_{max} passing through the drain-source channel, where $P_{max} = 0.5\, I^2_{max} Z_o$; this is usually several watts. In the Off state, the maximum power depends on breakdown voltage. For the switch in a shunt configuration, the maximum power it can accommodate is $P_{max} = 0.5(V_b - V_p)^2/Z_o$ where V_b and V_p are the breakdown and pinchoff voltages, respectively. This value is typically significantly lower than for the On state (~0.5W) [112].

4.3.4 Optoelectronic Switches

As described in Section 4.2.4, the conductivity of semiconductors can be optically controlled. This effect also applies to semiconductor devices such as PIN and FET switches, where switching can be performed through optical bias rather than voltage bias. The semiconductor devices can be controlled by directly illuminating the device or can be controlled indirectly, by illuminating a photovoltaic cell which, in turn, generates a DC voltage that activates the switch [113–115]. Although the indirect control requires additional devices, it results in better performance. Varactor diodes (described in Section 4.3.5) can also be optically controlled. Optical control of these devices has all the advantages associated with optical links such as low loss, light weight, noise immunity, and RF circuit IS. Examples of FAAs designed with optoelectronic switches, including microstrip patch antennas, monopoles, and dipoles, will be described in Section 8.4.

4.3.5 Varactors

A varactor (also known as a varicap) is a circuit element usually based on a p-n junction semiconductor device, whose capacitance can be controlled by the application of a bias voltage [116, 117]. A typical cross-section of the semiconductor is shown in Figure 4.28. The variable junction capacitance C_j is a function of the width w_d of the depletion region. The size of the depletion layer width, in turn, is controlled by the applied voltage $V_{pn} = V_p - V_n$. The depletion layer width increases as V_{pn} becomes more negative and decreases as V_{pn} becomes more positive and is governed by the expression

$$w_d = w_o\left[1 - \frac{V_{pn}}{\phi}\right]^{\gamma} \tag{4.98}$$

where w_o is the width of the depletion layer when $V_{pn} = 0$ and ϕ is the junction potential when $V_{pn} = 0$, which depends on the semiconductor material ($\phi = 0.7$V for silicon, 1.2V for gallium arsenide, and 0.3V for germanium). The exponent γin (4.98) depends on the doping profile of the p-n junction. For a linear graded profile, where the carrier concentration is a linear function of distance across the junction, $\gamma = 1/3$. For an abrupt junction, where the carrier concentration is constant on both sides of the junction, $\gamma = 1/2$. For a hyperabrupt junction, where the carrier concentration in the n-layer is rapidly reduced as the distance increases from the junction, then $1 < \gamma < 2$. Hyperabrupt junction diodes typically have a larger variation in capacitance with applied voltage than abrupt junction diodes but suffer from lower Q-factors (higher losses).

4.3.5.1 Circuit Model

Circuit models for the varactor diode in reverse bias are shown in Figure 4.29. In Figure 4.29(a), L_s represents the series lead inductance; the capacitance effects of the

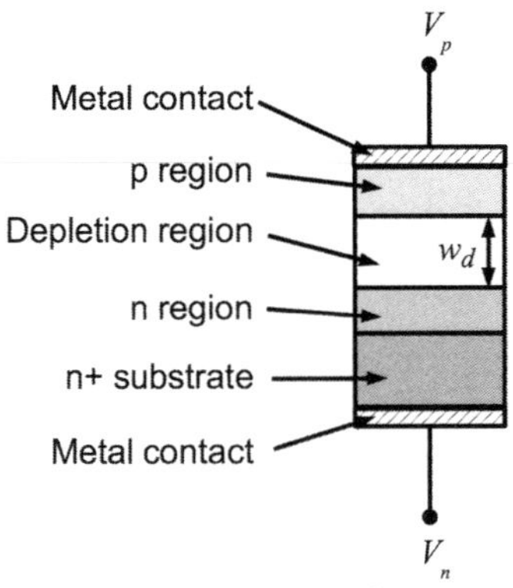

Figure 4.28 Typical cross-section of a varactor diode. (After [83, 84].)

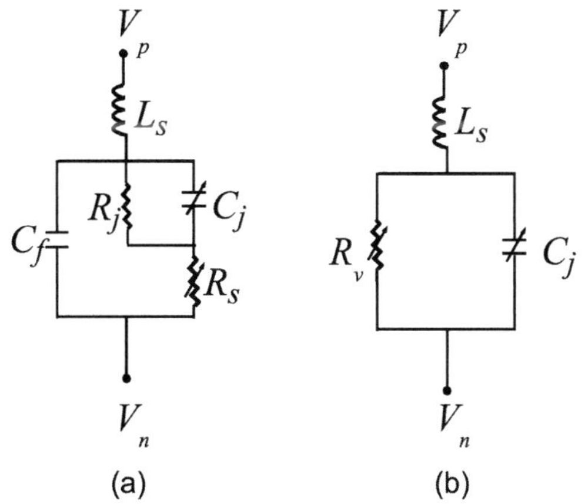

Figure 4.29 Circuit models for a varactor diode in reverse bias: (a) circuit model and (b) simplified circuit model. (After [83, 84].)

package are represented by C_f; R_j is the leakage resistance across the diode junction; R_s represents the variable series resistance from the contacts, leads, and bulk material; and C_j is the variable junction capacitance. In reverse bias, the junction resistance R_j is typically high ($R_j > 10$ MΩ) and can be ignored in the circuit model. The simplified model of Figure 4.29(b) is often sufficient to predict the varactor behavior at lower frequencies, while at higher frequencies the parasitic effects should be included to properly model the behavior. [A series RC circuit could also be used in place of the parallel RC circuit shown in Figure 4.29(b).] The circuit model for forward bias is more complicated (since it includes the diffusion capacitance of the injected carriers and the effect of these carriers on conductance of semiconductors), but the model is not required for the varactor, which is always operated with a reverse bias.

Since the capacitance C_j of the junction is inversely proportional to the junction width, it can be expressed as

$$C_j = \frac{C_o}{\left[1 - \frac{V_{pn}}{\phi}\right]^{\gamma}} \tag{4.99}$$

where C_o is the capacitance when $V_{pn} = 0$. Plots of the change junction capacitance C_j with applied voltage V_{pn} for a GaAs varactor ($\phi = 1.2$V) with selected values of γ are shown in Figure 4.30. Higher values of γ result in a wider and more rapid variation of C_j with applied voltage.

An important parameter for the varactor is its capacitance ratio C_x, commonly expressed as

$$C_x = \frac{C_o}{C_{\min}} \tag{4.100}$$

where C_{min} is the minimum attainable value of capacitance (obtained at the largest magnitude of applied voltage V_{pn}). This minimum capacitance of the varactor will be limited by the reverse-bias breakdown voltage of the device. Typical values for C_x range from about 2 to 3 for abrupt junctions and up to 10 or more for hyperabrupt junctions.

As with the switches described in the previous section, the varactor can be inserted either in a series or a shunt configuration within a circuit, similar as shown in Figure 4.16 for the PIN diode. The bias circuitry for the varactor is also similar to that of the PIN diode, as shown in Figure 4.19. Since the varactor is operated in reverse bias there is little DC current flowing through the device, and the RF choke could be replaced by a high-impedance resistor.

Using the simplified circuit model of Figure 4.29(b), the varactor impedance ($Z_v = R + jX$) and admittance ($Y_v = G + jB$) can be approximated by

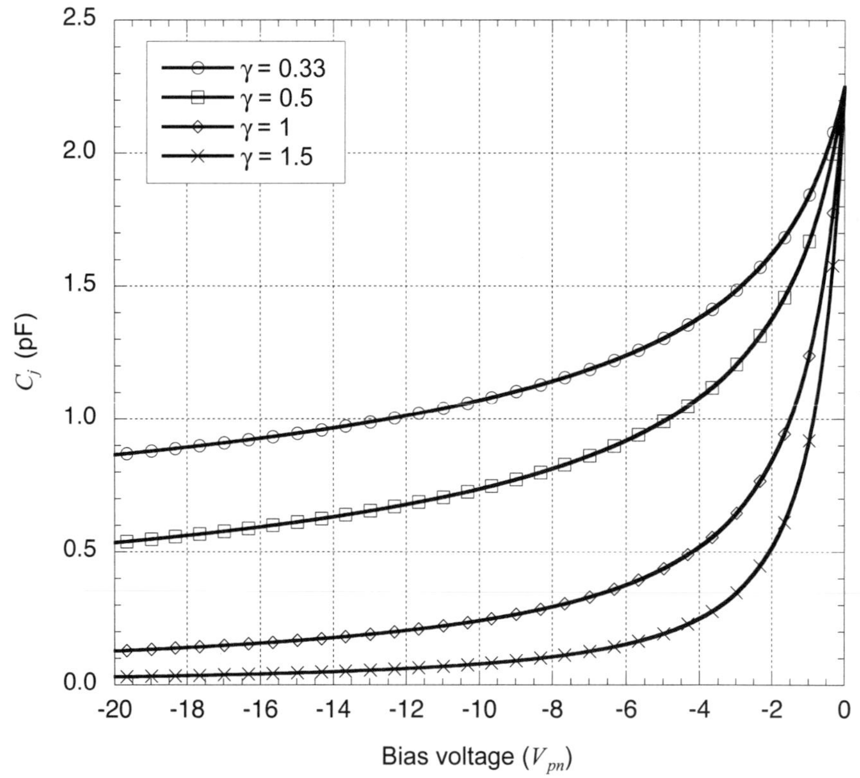

Figure 4.30 C_j versus V_{pn} for a GaAs varactor ($\phi = 1.2$V) with $C_o = 2.25$ pF.

$$R \approx \frac{R_v}{(2\pi f R_v C_j)^2 + 1} \tag{4.101}$$

$$X \approx 2\pi f L_s - \frac{(2\pi f R_v^2 C_j)}{(2\pi f R_v C_j)^2 + 1} \tag{4.102}$$

$$G \approx \frac{R_v}{R_v^2(1 - 4\pi^2 f^2 L_s C_j)^2 + (2\pi f L_s)^2} \tag{4.103}$$

$$B \approx \frac{2\pi f R_v^2 C_j(1 - 4\pi^2 f^2 L_s C_j) - 2\pi f L_s}{R_v^2(1 - 4\pi^2 f^2 L_s C_j)^2 + (2\pi f L_s)^2} \tag{4.104}$$

where both R_v and C_j are functions of the applied bias voltage. For the series configuration, the RL and IL of the varactor can be estimated using (4.39) and (4.42), respectively. For the shunt configuration, (4.40) and (4.43) can be used for the RL and IL, respectively.

Commercial manufactures of varactors often provide complex circuit models of their devices that capture the behavior of the various parasitic effects that manifest themselves at higher frequencies. These models can be run using circuit simulators to accurately predict the behavior of the devices. In many instances, the response of these circuit models can be closely emulated by the simple model in Figure 4.29(b), when appropriate values of R_v are selected for the various bias voltage values. Values of R_v can be obtained by curve-fitting the response of the simple model to that of the complex model. Table 4.6 list values for R_v and C_j obtained by curve-fitting to a commercially available circuit model [118]. These values can then be used in (4.101) to (4.104) to obtain values for the varactor impedance or admittance, which in turn can be used to predict the RL and IL for the shunt or series configuration. The effect of altering the voltage bias V_{pn} on the return and insertion losses can thus be estimated. Using selected values from Table 4.6 in the simplified circuit model of Figure 4.29(b), and with $L_s = 0.7$ nH, the RL and IL response for the series and shunt configurations can be estimated, as shown in Figures 4.31 and 4.32, respectively. In both configurations, the effect of altering the bias voltage is to shift the maximum RL and minimum IL response over a significant range of frequencies. This tuning of the response of the varactor with frequency will be exploited in the design of continuous tuning FAAs. Numerous examples of varactor-loaded antennas will be presented in Chapter 7.

Table 4.6
Component Values of the Simplified Varactor Circuit Model of Figure 4.29(b)

V_{pn} (V)	R_v (Ω)	C_j (pF)
0	75	2.22
-1	100	1.60
-2	140	1.25
-4	225	0.84
-6	290	0.63
-18	1,200	0.245
-19	1,400	0.235

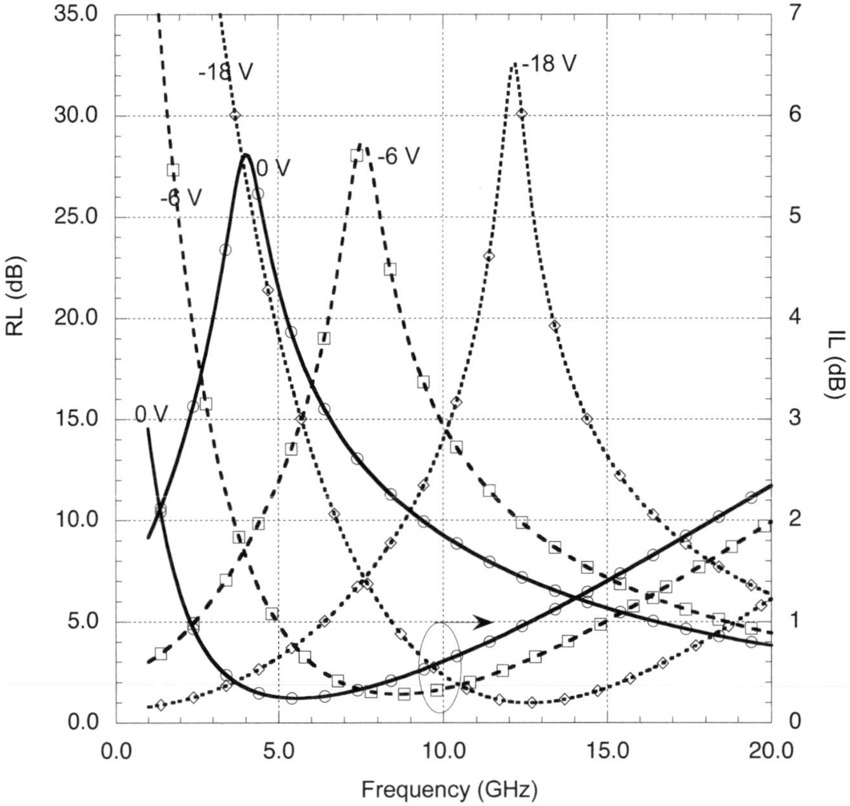

Figure 4.31. Effect of bias voltage on a varactor in series.

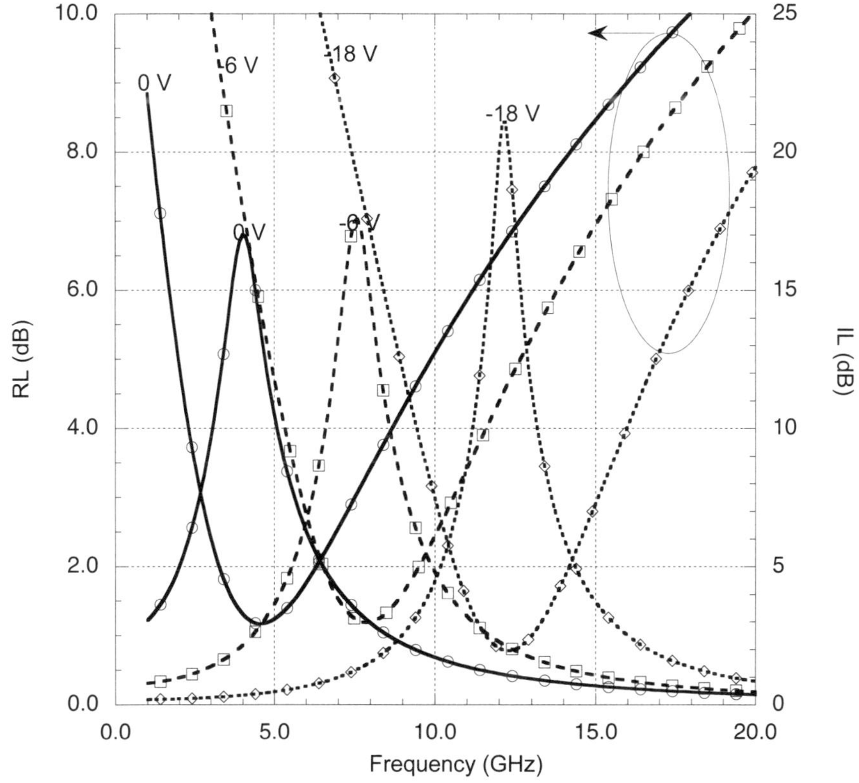

Figure 4.32. Effect of bias voltage on a varactor in shunt.

4.3.6 Summary

This section provided an overview of the various electronic devices that could potentially be incorporated with antennas to achieve frequency agility. A summary of the attributes of these devices is provided in Table 4.7. Numerous examples of their use in FAAs will be provided in Chapters 7 and 8.

Table 4.7

Summary of Active Devices Attributes

Electronic Device	Attributes
PIN-diode switch	• Provides discrete tuning • Requires low bias voltage • Fast switching speeds • Relatively inexpensive • Readily available commercially • Moderately low IL • Moderate DC power consumption • Moderate levels of IS • May require complex biasing network • Low-to-moderate power handling
FET switch	• Provides discrete tuning • Low DC power consumption • Bias levels < ~5V • Fast switching speeds • Low-to-moderate power handling • Relatively inexpensive • Readily available commercially • May require moderately complex biasing network
MEMS switch	• Provides discrete tuning • High IS • Low IL (<0.1–0.2 dB) • Low power consumption • High linearity • Moderate power handling capability (1–10W) • Requires high activiation voltages (20–70V) • Requires hermetic packaging • Relatively expensive • Lower reliability • Limited commercial availability
Optoelectronic switch	• Similar attributes to PIN or FET switches • Optical control of these devices offers better noise immunity and improved RF circuit isolation
Varactor diode	• Provides continuous tuning • Easy-to-integrate • Capacitance ratios of up to ~10 • Moderately high bias levels < ~30V • Low DC power consumption • Moderate power handling capability • May require complex biasing network • Moderate IL

References

[1] Funakubo, H., ed., *Actuators for Control, in Precision Machinery and Robotics, Vol. 2,* New York, NY: Gordon and Breach Science Publishers, 1991.

[2] Uchino, K., *Piezoelectric Actuators and Ultrasonic Motors,* Norwell, MA: Kluwer Academic Publishers, 1997.

[3] Bodlea, I., and S. A. Nasar, *Electric Actuators and Generators,* Cambridge, UK: Cambridge University Press, 1997.

[4] Busch-Vishniac, I. J., *Electromechanical Sensors and Actuators,* (Mechanical Engineering Series), New York, NY: Springer-Verlag, 1998.

[5] Janocha, H., ed., *Actuators: Basics and Applications,* Berlin, Germany: Springer-Verlag, 2004.

[6] Pons, J. L., *Emerging Actuator Technologies: A Micromechatronic Approach,* Chichester, West Sussex, England: John Wiley & Sons, 2005.

[7] Gomis-Bellmunt, O., and L. F. Campanile, *Design Rules for Actuators in Active Mechanical Systems,* London: Springer-Verlag, 2010.

[8] Haertling, G. H., "Chemically Reduced PLZT Ceramics for Ultra-High Displacement Actuators," *Ferroelectrics,* Vol. 154, No. 1, April 1994, pp. 101-106.

[9] Haertling, G. H., "Rainbow Actuators and Sensors: A New Smart Technology," *Proceedings of SPIE,* Vol. 3040, No. 81, 1997, pp. 81-92.

[10] Brauer, J. R., *Magnetic Actuators and Sensors,* New York, NY: John Wiley & Sons, 2006.

[11] Tabib-Azar, M., *Microactuators: Electrical, Magnetic, Thermal, Optical, Mechanical, Chemical and Smart Structures,* Norwell, MA: Kluwer Academic Publishers, 1998.

[12] Kohl, M., *Shape Memory Microactuators,* Berlin, Germany: Springer-Verlag, 2004.

[13] Fukuda, T., and W. Menz (eds), *Micro Mechanical Systems: Principles and Technology, Vol. 6 of Handbook of Sensors and Actuators,* (S. Meddelhoek, Series Editor), Amsterdam, Netherlands: Elsevier, 1998.

[14] http://www.newscaletech.com/press_releases/pr080219.html

[15] http://www.namiki.net/product/kbo/dc/servo.html

[16] http://www.micro-drives.com

[17] http://www.faceinternational.com

[18] Krivits, I. L., and G. V. Krejnin, *Pneumatic Activating Systems for Automatic Equipment: Structure and Design,* Boca Raton, FL: CRC Press, 2006.

[19] Barber, A., *Pneumatic Handbook,* Eighth ed., Oxford, UK: Elsevier Science Ltd, 1997.

[20] Majumdar, S. R., *Pneumatic Systems: Principle and Maintenance,* New Delhi, India: Tata McGraw-Hill, 2006.

[21] Ilango , S., and V. Soundararajan, *Introduction to Hydraulics and Pneumatics,* New Delhi, India: Prentice-Hall of India, 2007.

[22] Parr, A., *Hydraulics and Pneumatics: A Technician's and Engineer's Guide*, Oxford, UK: Elsevier Press, 2006.

[23] Polder, D., "On the Theory of Electromagnetic Resonance," *Philosophical Magazine,* Vol. 40, 1949, pp. 99-115.

[24] Fox, A. G., S. E. Miller, and M. T. Weiss, "Behavior and Applications of Ferrites in the Microwave Region," *Bell System Technical Journal,* January 1955, pp. 5-103.

[25] Waldron, R. A., *Ferrites: An Introduction for Microwave Engineers,* London, UK: D. Van Nostrand Co., 1961.

[26] Lax, B., and K. Button, *Microwave Ferrites and Ferrimagnetics,* New York, NY: McGraw, 1962.

[27] Dionne, F. G., "A Review of Ferrites for Microwave Applications," *Proceedings of the IEEE,* Vol. 63, No. 5, May 1975, pp. 777-789.

[28] Green, J. J., and F. Sandy, "Microwave Characterization of Partially Magnetized Ferrites," *IEEE Transactions on Microwave Theory & Techniques,* Vol. 22, No. 6, June 1974, pp. 641-644.

[29] Clarricoats, P. J. B., *Microwave Ferrites,* New York, NY: Wiley & Sons, 1961.

[30] Helszajn, J., *Principles of Microwave Ferrites Engineering,* New York, NY: Wiley-Interscience, 1969.

[31] Collin, R. E., *Foundations for Microwave Engineering,* New York, NY: McGraw-Hill, 1966, pp. 286-309.

[32] Pozar, D., *Microwave Engineering,* Boston, MA: Addison-Wesley, 1990, pp. 529-578.

[33] Hord, W. E., C. R. Boyd, and F. J. Rosenbaum, "Application of Reciprocal Latching Ferrite Phase Shifters to Lightweight Electronic Scanned Phased Arrays," *IEEE Proceedings,* Vol. 56, No. 11, November 1968, pp.1931-1939.

[34] Hord, W. E., "Ferrite Control Components." In *Handbook of Microwave and Optical Components, Vol. 1*, Chapter 5, K. Chang, (ed.), New York, NY: John Wiley, 1989.

[35] Reggia, F., and E. Spencer, "A New Technique in Ferrite Phase Shifting for Beam Scanning of Microwave Antennas," *Proceedings of the IRE,* Vol. 45, No. 11, November 1957, pp. 1510-1517.

[36] Burfoot, J. C., *Ferroelectrics: An Introduction to the Physical Principles,* London, UK: D. Van Nostrand Company Ltd., 1967

[37] Uchino, K., *Ferroelectric Devices,* New York, NY: Marel Dekker, 2000.

[38] Martienssen, W., H. Warlimont, (eds.), "Ferroelectrics and Antiferroelectrics." In *Springer Handbook of Condensed Matter and Materials Data,* pp. 903-915, Berlin, Germany: Springer, 2005.

[39] Gevorgian, S. S., and E. L. Kolberg, "Do We Really Need Ferroelectrics in Paraelectric Phase Only in Electrically Controlled Microwave Devices," *IEEE Transactions on Microwave Theory and Techniques,* Vol. 49, No. 11, November 2001, pp. 2117-2124.

[40] Gevorgian, S., "Agile Microwave Devices," *IEEE Microwave Magazine,* August 2009, pp. 93-98.

[41] Iskander M. F., et al., "Design of a Low-Cost 2-D Beam-Steering Antenna Using Ferroelectric Material and the CTS Technology," *IEEE Transactions on Microwave Theory and Techniques,* Vol. 49, No. 5, May 2001, pp. 1000-1003.

[42] Miranda, F. A., et al., "Design and Development of Ferroelectric Tunable Microwave Components for Ku- and K-Band Satellite Systems," *IEEE Transactions on Microwave Theory and Techniques,* Vol. 48, No. 7, July 2000, pp. 1181-1189.

[43] Van Keuls, F. W., et al., "Current Status of Thin Film (Ba,Sr)TiO3 Tunable Microwave Components for RF Communications," NASA/TM, 2000-210233, 2000.

[44] Subramanyam, G., et al., "Performance of a K-Band Voltage-Controlled Lange Coupler Using a

Feroelectric Tunable Microstrip Configuration," *IEEE Microwave and Guided Wave Letters,* Vol. 10, No. 4, April 2000, pp. 136-138.

[45] Romanofsky, R., et al., "K-Band Phased Array Antennas Based on Ba0.60Sr0.40TiO3 Thin-Film Phase Shifters," *IEEE Transactions on Microwave Theory and Techniques,* Vol. 48, No. 12, December 2000, pp. 2504-2510.

[46] Zhao, Z., et al, "Ferroelectric Phase Shifters at 20 and 30 GHz," *IEEE Transactions on Microwave Theory and Techniques,* Vol. 55, No. 2, February 2007, pp. 430-437.

[47] Vendik, O. G., "Insertion Loss in Reflection-Type Microwave Phase Shifter Based on Ferroelectric Tunable Capacitor," *IEEE Transactions on Microwave Theory and Techniques,* Vol. 55, No. 2, February 2007, pp. 425-429.

[48] Velu, G., et al., "A 360° BST Phase Shifter with Moderate Bias Voltage at 30 GHz," *IEEE Transactions on Microwave Theory and Techniques,* Vol. 55, No. 2, February 2007, pp. 438-444.

[49] Hu, W., et al., "Investigation of Ferroelectric Thick-Film Varactors for Microwave Phase Shifters," *IEEE Transactions on Microwave Theory and Techniques,* Vol. 55, No. 2, February 2007, pp. 418-424.

[50] Romanofsky, R., "Advances in Scanning Reflectarray Antennas Based on Ferroelectric Thin-Film Phase Shifters for Deep-Space Communications," *IEEE Proceedings,* Vol. 95, No. 10, October 2007, pp. 1968-1975.

[51] Tombak, A., "A Ferroelectric-Capacitor-Based Tunable Matching Network for Quad-Band Cellular Power Amplifier," *IEEE Transactions on Microwave Theory and Techniques,* Vol. 55, No. 2, February 2007, pp. 370-375.

[52] Norling, M., et al., "A Low-Noise K-Band VCO Based on Room-Temperature Ferroelectric Varactors," *IEEE Transactions on Microwave Theory and Techniques,* Vol. 55, No. 2, February 2007, pp. 361-369.

[53] Fu, J. S., et al., "Improving Linearity of Ferroelectric-Based Microwave Tunable Circuits," *IEEE Transactions on Microwave Theory and Techniques,* Vol. 55, No. 2, February 2007, pp. 354-360.

[54] Lugo, C., et al., "Frequency and Bandwidth Agile Millimeter-Wave Filter Using Ferroelectric Capacitors and MEMS Cantilevers," *IEEE Transactions on Microwave Theory and Techniques,* Vol. 55, No. 2, February 2007, pp. 376-382.

[55] Subramanyam, G., F. W. Van Keuls, and F. A. Miranda, "A K-Band-Frequency Agile Microstrip Bandpass Filter Using a Thin-Film HTS/Ferroelectric/Dielectric Multilayer Configuration," *IEEE Transactions on Microwave Theory and Techniques,* Vol. 48, No. 4, April 2000, pp. 525-530.

[56] Rao, J. B. L., D. P. Patel, and V. Krichevsky, "Voltage-Controlled Ferroelectric Lens Phased Arrays," *IEEE Transactions on Antennas and Propagation,* Vol. 47, No. 3, March, 1999, pp. 458-468.

[57] de Gennes, P. G., and J. Prost, *The Physics of Liquid Crystals,* Oxford, UK: Clarendon Press, 1993.

[58] Chandrasekhar, S., *Liquid Crystals,* Second ed., Cambridge, UK: Camridge University Press, 1992.

[59] Collings, P. J., and M. Hird, *Introduction to Liquid Crystals Chemistry and Physics,* London, UK: Taylor & Francis, 1997

[60] Weil, C., et al., "Highly-Anisotropic Liquid-Crystal Mixtures for Tunable Microwave Devices," *Electronics Letters,* Vol. 39, No. 24, November 2003. pp. 1732-1733.

[61] Jakoby, R., et al., "Nonlinear Dielectrics for Tunable Microwave Components," *Proc. IEEE International Conference on Microwaves, Radar and Wireless Communications,* Vol. 2, Warsaw,

Poland, May 17-19, 2004, pp. 369-378.

[62] Yeh, J. A., et al., "Microwave Characteristics of Liquid-Crystal Tunable Capacitors," *IEEE Electron Device Letters,* Vol. 26, No. 7, July 2005, pp. 451-453.

[63] Chang, C. A., C. C. Cheng, and J. A. Yeh, "Analysis and Modeling of Liquid-Crystal Tunable Capacitors," *IEEE Transactions on Electron Devices,* Vol. 53, No. 7, July 2006, pp. 1675-1682.

[64] Dolfi, D., et al., "Liquid Crystal Microwave Phase Shifter," *Electronics Letters,* Vol. 29, No. 10, May 1993, pp. 926-928.

[65] Fujikake, H., et al., "Thick Polymer-Stabilized Liquid Crystal Films for Microwave Phase Control," *Journal of Applied Physics,* Vol. 89, No. 10, May 2001, pp. 5295-5298.

[66] Kuki, T., H. Fujikake, and T. Nomoto, "Microwave Variable Delay Line Using a Dual-Frequency Switching-Mode Liquid Crystal," *IEEE Transactions on Microwave Theory and Techniques,* Vol. 50, No. 11, November 2002, pp. 2604-2609.

[67] Martin, N., et al., "Technological Evolution and Performances Improvements of a Tunable Phase-Shifter Using Liquid Crystal," *Microwave and Optical Technology Letters,* Vol. 43, No. 4, November 2004, pp. 338-341.

[68] Mirfatah, A., and J. J. Laurin, "Tunable Hairpin Resonator Based on Liquid Crystal," *Proc. Symposium on Antenna Technology and Applied Electromagnetics ANTEM and the Canadian Radio Science Meeting Digest,* Banff, Canada, February 15-18, 2009.

[69] Patel, J. S., and M. W. Maeda, "Multiwavelength Tunable Liquid-Crystal Etalon Filter," *IEEE Photonics Technology Letters,* Vol. 3, No. 7, July 1991, pp. 643-644.

[70] Tentillier, N., et al., "A Liquid-Crystal, Tunable, Ultra-Thin Fabry-Perot Resonator in Ka Band, *IEEE Antennas and Wireless Propagation Letters,* Vol. 8, 2009, pp. 701-704.

[71] Yang, F., and S. R. Sambles, "Microwave Liquid Crystal Wavelength Selector," *Applied Physics Letters,* Vol. 79, No. 2, November 2001, pp. 3717-3719.

[72] Yang, F., and S. R. Sambles, "Microwave Liquid Crystal Variable Phase Grating," *Applied Physics Letters,* Vol. 79, No. 2, September 2004, pp. 2041-2043.

[73] Hu, W., et al., "Liquid Crystal Tunable mm Wave Frequency Selective Surface," *IEEE Microwave and Wireless Component Letters,* Vol. 17, No. 9, September 2007, pp. 667-669.

[74] Moessinger, A., et al., "Electronically Reconfigurable Reflectarrays with Nematic Liquid Crystals," *Electronics Letters,* Vol. 42, No. 16, August 2006, pp. 899-900.

[75] Gaebler, A., et al., "Liquid Crystal – Reconfigurable Antenna Concept for Space Applications at Microwave and Millimterwaves," *International Journal of Antennas and Propagation,* 2009, Article ID 876989.

[76] Hu, W., et al., "Design and Measurement of Reconfigurable Millimeter Wave Reflectarray Cells with Nematic Liquid Crystal," *IEEE Transactions on Antennas and Propagation,* Vol. 56, No. 10, October 2008, pp. 3112-3117.

[77] Hu, W., et al., "Tunable Liquid Crystal Reflectarray Patch Element," *Electronics Letters,* Vol. 42, No. 9, April 2006, pp. 509-511.

[78] Tanaka, M., and S. Sato, "Focusing Properties of Liquid Crystal Lens Cells with Stack-Layered Structure in the Millimeter-Wave Region," *IEEE Microwave and Wireless Components Letters,* Vol. 12, No. 5, May 2002, pp. 163-165.

[79] Jaeger, R. C., *Microelectronic Circuit Design,* New York, NY: McGraw-Hill, 1997.

[80] Muller, R. S., and T. I. Kamins, *Device Electronics for Integrated Circuits,* Second ed., New York,

NY: John Wiley & Sons, 1986.

[81] Platte, W., Optically CW-Induced Losses in Semiconductor Coplanar Waveguides," *IEEE Transactions on Microwave Theory and Techniques,* Vol. 37, No. 1, January 1989, pp. 139-149.

[82] De Los Santos, H. J., *Introduction to Microelectromechanical Microwave Systems,* Second ed., Norwood, MA: Artech House, 2004.

[83] Trew, R. J., "Active Devices." In *Microwave Solid State Circuit Design,* Chapter 7, I. Bahl, and P. Bhartia, (eds.), New York, NY: John Wiley and Sons, Second ed., 2003.

[84] Olson, H. M., "p-i-n Diodes." In *Microwave Semiconductor Devices and Their Circuit Applications,* Chapter 9, H. A. Watson, (ed.), New York, NY: McGraw-Hill, 1969.

[85] Chang, K., *Microwave Solid-State Circuit and Applications,* New York, NY: John Wiley and Sons, 1994, pp. 202-234.

[86] Infineon. www.infineon.com

[87] Skyworks, www.skyworksinc.com

[88] Microsemi, www.microsemi.com

[89] MA/Com, www.macom.com

[90] Agilent Technologies Application Note 5989-7618EN, "Understanding RF Microwave Solid State Switches and Their Applications," May, 2010.

[91] Skyworks Application Note, "Design with PIN Diodes," Skyworks Solutions Inc., 2009.

[92] Doherty, Jr., W. E., and R. D. Joos, "PIN Diode Circuit Designer's Handbook," Microsemi-Watertown Corporation, 1998.

[93] Agilent Technologies Application Note AN922, "Applications of PIN Diodes," November 1999.

[94] Peterson, K. E. ,"Micromechanical Membrane Switches on Silicon," *IBM Journal of Research and Development,* Vol. 23, No. 4, July 1979, pp. 376-385.

[95] Larson, L. E., et al., "Micromachined Microwave Actuator Technology – A New Tuning Approach for Microwave Integrated Circuits," *Proc. IEEE Microwave and Millimeter-Wave Monolithic Circuit Symposium,* Boston, MA, June 10-11, 1991, pp. 27-30.

[96] Goldsmith, C., et al., "Micromechanical Membrane Switches for Microwave Applications," *Proc. IEEE Microwave Theory and Techniques Symposium,* Orlando, FL, May 16-20, 1995, pp. 91-94.

[97] Goldsmith, C., et al., "Characteristics of Micromachined Switches at Microwave Frequencies," *Proc. IEEE Microwave Theory and Techniques Symposium,* San Francisco, CA, June 17-21, 1996, pp. 1141-1144.

[98] Yao, J. J., and M. F. Chang, "A Surface Micromachined Miniature Switch for Telecommunications Applications with Single Frequencies from DC to 4 GHz," *Transducers '95 Proceedings,* Boston, MA, June 1995, pp. 384-387.

[99] Zhou, S., X. Sun, and W. N. Carr, "A Micro Variable Inductor Chip Using MEMS Relays," *Proc. Transducers '97,* Chicago, IL, June 19, 1997, pp. 1137-1140.

[100] Yao, Z. J., et al., "Micromachined Low-Loss Microwave Switches," *IEEE Journal of Microelectromechanical Systems,* Vol. 8, No. 2, June 1999, pp. 129-134.

[101] Rebeiz, G. M., and J. B. Muldavin, "RF MEMS Switches and Switch Circuits," *IEEE Microwave Magazine,* December 2001, pp. 59-71.

[102] Van Caekenberghe, K., "RF MEMS on the Radar," *IEEE Microwave Magazine,* October 2009,

pp. 99-116.

[103] Rebeiz, G. M., et al., "Tuning in to RF MEMS," *IEEE Microwave Magazine,* October 2009, pp. 55-72.

[104] Rebeiz, G. M., *RF MEMS Theory, Design and Technology,* Hoboken, NJ: Wiley, 2003.

[105] RadantMEMS, www.radantmems.com

[106] Memtronics, www.memtronics.com

[107] Kingsley, N., et al., "RF MEMS Sequentially Reconfigurable Sierpinksi Antenna on a Flexible Organic Substrate with Novel DC-Biasing Technique," *Journal of Microelectricmechanical Systems,* Vol. 16, No. 5, October 2007, pp. 1185-1192.

[108] Muldavin, J. B., and G. M. Rebeiz, "High-Isolation CPW MEMS Shunt Switches – Part 1: Modeling," *IEEE Transactions on Microwave Theory and Techniques,* Vol. 48, No. 6, June 2000, pp. 1045-1052.

[109] Atwater, H. A., and R. W. Sudbury, "Use of Switching Q in the Design of FET Microwave Switches, *Proc. IEEE Microwave Theory and Techniques Symposium,* Los Angeles, CA, June 15-19, 1981, pp. 370-372.

[110] McLevige, W. V., and V. Sokolov, "Resonated GaAs FET Devices for Microwave Switching," *IEEE Transactions on Electron Devices,* Vol. 28, No. 2, February 1981, 198-204.

[111] Gopinath, A., and J. B. Rankin, "GaAs FET RF Switches," *IEEE Transactions on Electron Devices,* Vol. 32, No. 7, July 1985, pp. 1272-1278.

[112] Shifrin, M. N., P. I. Kazin, and Y. Ayasli, "Monolithic FET Structures for High-Power Control Component Applications," *IEEE Transactions on Microwave Theory and Techniques,* Vol. 37, No. 12, December 1989, pp. 2134-2141.

[113] VanBlaricum, M. L., "A Brief History of Photonic Antenna Reconfiguration," *Proc. International Topical Meeting on Microwave Photonics MWP* 2000, Oxford, UK, September 11-13, pp. 9-12.

[114] Nagra, A.S., et al., "Indirect Optical Control of Microwave Circuits Using Monolithic Optically Variable Capacitors," *IEEE Transactions on Microwave Theory and Techniques,* Vol. 47, No. 7, July 1999, pp 1365-1362.

[115] Daryoush, A., K. Bontzos, and P. Herczfeld, "Optically Tuned Patch Antenna for Phased Array Applications," *Proc. IEEE International Symposium on Antennas and Propagation,* Philadelphia, PA, June 8-13,1986, pp. 361-364.

[116] Norwood, M. H., and E. Shatz, "Voltage Variable Capacitor Tuning: An Review," *Proceedings of the IEEE,* Vol. 56, No. 5, May 1968, pp. 788-798.

[117] Cory, R., "Nuts and Bolts of Tuning Varactors," *High Frequency Electronics,* February 2009, pp. 42-51.

[118] Desjardins, J., "Reconfigurable Dielectric Resonator Antennas," Master's Thesis, University of Ottawa, 2010.

Chapter 5

FAAs Based on Mechanical Tuning Techniques

Chapter 4 considered various types of actuators that could potentially be used to change one or more physical dimensions of an antenna, resulting in a shift in resonant frequency. This chapter presents how several of these mechanical tuning techniques can be incorporated with various antennas for achieving frequency agility. The major sections in this chapter are organized by antenna type, and contain one or more applicable tuning techniques. The antennas examined include microstrip patches, PIFAs, helical antennas, slot dipoles, and DRAs. The types of actuators used include electrostatic, piezoelectric, magnetostatic, hydraulic, and pneumatic. Information for each example will include tuning range, total spectrum, bias requirements, and antenna performance.

5.1 MICROSTRIP PATCH ANTENNAS

Microstrip patch antennas are examined in Section 3.4 with particular attention paid to patches of rectangular, circular, and annular shapes. Regardless of the shape, the resonant frequency of all microstrip patch antennas is a function of the patch dimensions, the substrate thickness, and the dielectric constant of the substrate. If one or more of these parameters could be dynamically altered, the microstrip patch antenna would become frequency-agile. Using mechanical actuation as the tuning method, one of the more straightforward approaches for achieving this frequency agility is to alter the height of the patch above the ground plane. Figure 5.1 shows a rectangular metal patch placed at a distance h above a metallic ground plane. If the length L and width W of the patch are kept fixed while the height h is increased, the resonant frequency of the patch will decrease. The resonant frequency of the rectangular patch antenna can be estimated by solving (3.16) and is plotted in Figure 5.2 for the sample case of an air dielectric with $L = W = 14$ cm, as h is varied from 1 to 20 mm.

Although application of (3.16) will result in an ever decreasing resonant frequency with increased height, higher-order modes would start to appear as the height of the patch approaches a half wavelength, resulting in unwanted radiation. The input impedance will also change with height, and high mismatch losses would also limit the usable range of values for h. This is seen in Figure 5.3 where the RL of the square patch is plotted as a function of height. The RL was based on the reflection coefficient of the patch connected to a 50Ω transmission line, where the input impedance of the patch was calculated using (3.33) and (3.34) for a probe radius of $a = 1$ mm and a probe inset of $s = 5$ cm from the resonant edge of the patch. For this example then, if a minimum value of 10 dB were required for the RL, the height of the patch would be limited to about $h = 5$ mm, and the resonant frequency range would be limited to between 1,060 and 1,020 MHz, for a tuning range of 3.8%.

The following examples make use of this dependence of resonant frequency on height to design frequency-agile microstrip antennas through various forms of mechanical actuation.

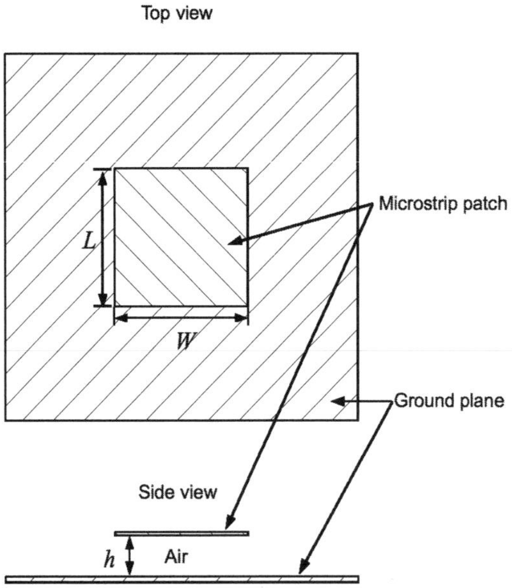

Figure 5.1 Rectangular microstrip patch antenna with an air substrate.

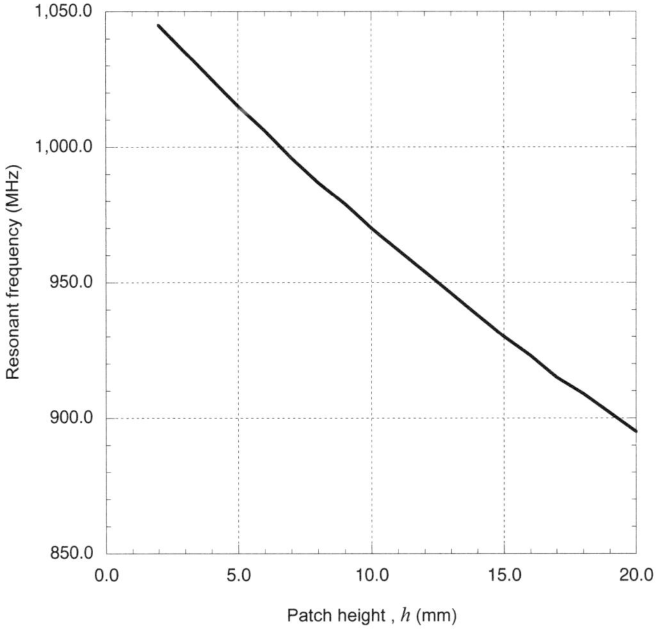

Figure 5.2 Resonant frequency versus patch height for a square microstrip patch antenna ($L = W = 14$ cm, and $\varepsilon_r = 1$).

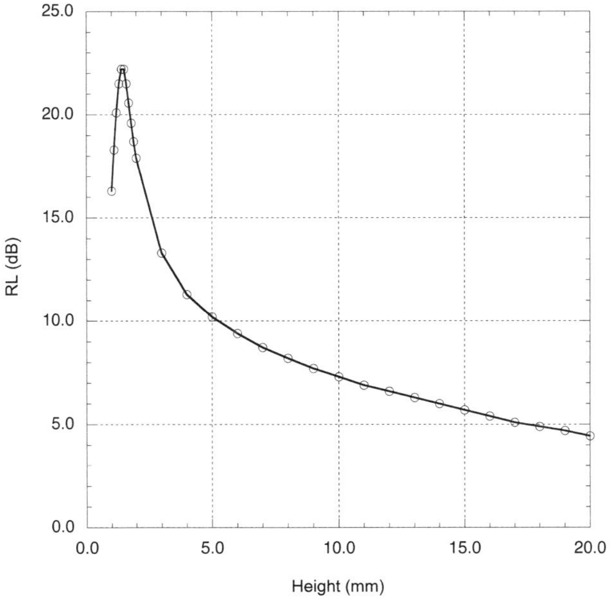

Figure 5.3 RL versus patch height for a probe-fed square microstrip patch antenna ($L = W = 14$ cm, $s = 5$ cm, and $\varepsilon_r = 1$).

5.1.1 Stacked Patches

It was demonstrated early in the 1980s that the resonant frequency of a stacked-patch microstrip antenna could be tuned by altering the distance between the upper and lower patches [1]. For the circular stacked-patch geometry shown in Figure 5.4, the lower patch is fed with a probe (not shown) and the upper patch is excited through electromagnetic coupling. With a gap of $g = 0$ mm between the patches, resonances occured at 765 MHz and at 830 MHz. When the upper patch was raised to form a gap of up to $g = 1$ mm, the upper resonant frequency shifted from 830 MHz to 910 MHz ($TR = 9\%$) in a nearly linear manner, while the lower resonant frequency shifted only slightly upward. The reason for the upward shift in this case is the drop in effective dielectric constant when $g > 0$.

Nearly 20 years later, a stacked rectangular microstrip antenna prototype was developed incorporating high-deflection piezoelectric actuator wafers (called Rainbows) for controlling the gap between patches, as illustrated in Figure 5.5 [2]. Each Rainbow wafer (having a diameter of 50.8 mm and a thickness of 1 mm) was capable of a maximum deflection of 0.5 mm and so 20 actuators were stacked in order to achieve the required travel of 10 mm.

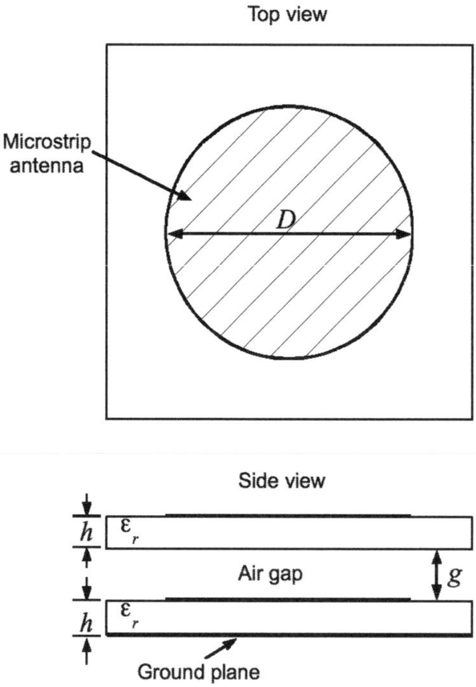

Figure 5.4 Mechanically tuned-stacked circular microstrip patch ($D = 68$ mm, $h = 3.18$ mm, $0 \le g \le 1$ mm, and $\varepsilon_r = 2.3$). (After [1].)

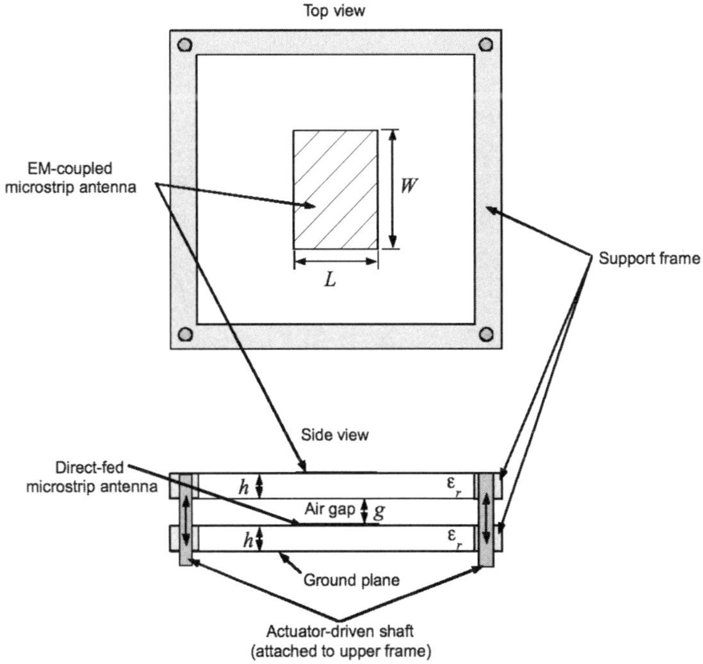

Figure 5.5 Stacked microstrip antenna with piezoelectric actuator control (L = 28 mm, W = 35.2 mm, h = 1.588 mm, $0 \leq g \leq 10$ mm, and ε_r = 2.94). (After [2].)

The upper electromagnetically coupled microstrip antenna was placed in a frame to which drive shafts were attached, controlled by the piezoelectric actuators. The separation distance between the upper electromagnetically coupled patch and the lower coaxially fed microstrip antenna was controlled by varying the actuation voltage of the piezoelectric actuators. To achieve the 10-mm travel required an actuation voltage ranging from -800V to +200V, resulting in a shift in frequency from 2,793 to 2,988 MHz (TR = 6.7%), as plotted in Figure 5.6 (with the corresponding gap separation g labeled directly on the plot). A 10-dB RL was not maintained over the entire range of travel, resulting in a total spectrum[1] of TS = 5%. The gain of the antenna varied from a minimum of 6.7 dBi to a maximum of 8.3 dBi for different gap spacings. Although the use of stacked Rainbow actuators resulted in little degradation in antenna performance, their bulky nature and the high activation voltages required would make them impractical for most lower frequency applications. The piezoelectric actuators also exhibit hysteresis and require a feedback system to detemine their position. The

1 Unless otherwise stated, the tunable impedance bandwidth equals the total spectrum for FAAs with continuous tuning.

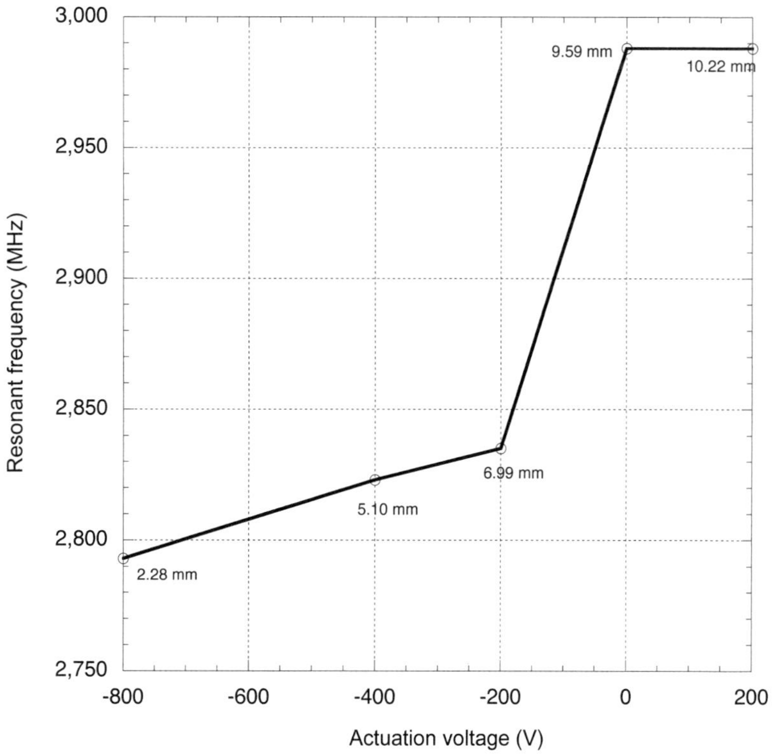

Figure 5.6 Resonant frequency versus actuation voltage for the frequency-agile stacked microstrip antenna controlled using piezoelectric actuators. (From [2].)

use of piezoelectric actuators may be better suited for higher microwave or millimeter-wave frequencies, where the required travel would be much smaller and where a single actuator would suffice.

5.1.2 Electrostatic Actuation

Investigations have also been carried out where microstrip antennas are fabricated directly on piezoelectric substrates [3, 4]. Figure 5.7 shows the dimensions of a square microstrip patch on a quartz substrate, designed to resonate at 8 GHz. The curves in Figure 5.8 show the change in resonant frequency with applied DC electric field for three different crystal plane cuts (labeled as X, Y, and Z) of the quartz substrate. Depending on the crystal plane cut, the application of the electric field either causes a small change in height due to the piezoelectric effect (X-cut) or a small change in the dielectric constant (Y- and Z-cuts). The tuning range represented by these curves for this FAA is small (maximum TR ~ 0.5%) and

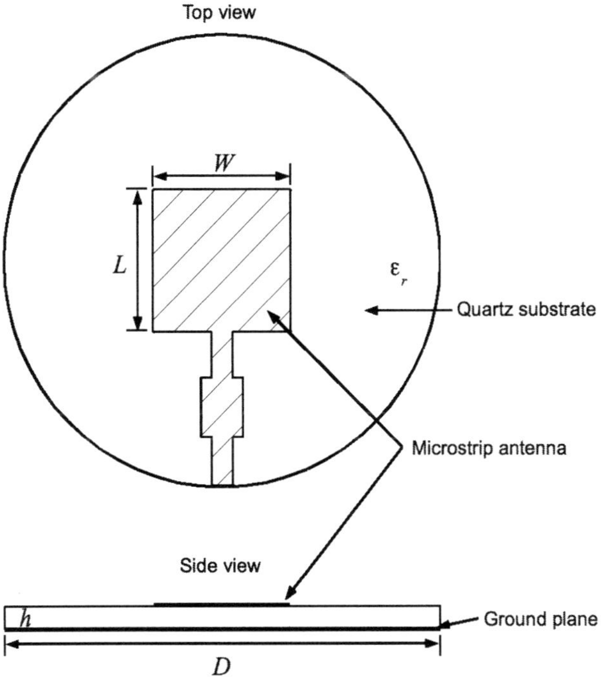

Figure 5.7 Microstrip antenna on quartz substrate ($L = W = 8.65$ mm, $h = 0.5$ mm, $D = 76.2$ mm, and $\varepsilon_r = 4.6$). (After [4].)

would be suitable only for narrowband applications with fine-tuning requirements. The high bias field strength required may also limit practical implementation.

A second example of a frequency-agile microstrip antenna based on electrostatic tuning uses as the tuning mechanism a thin micromachined membrane forming a deformable ground plane layer for the microstrip patch [5, 6]. As sketched in Figure 5.9, the rectangular microstrip antenna was patterned onto the top of a thin glass substrate. Underneath the patch, a thin conducting membrane was fabricated using micromaching techniques. By applying a DC voltage through a pull-down electrode, the membrane was deflected down into the air gap in the silicon wafer. A deflection of 138 microns resulted in a shift in the resonant frequency from 16.8 to 17.82 GHz (TR = 5.9%), as shown in Figure 5.10. A slight narrowing of the impedance bandwidth was noted. The total spectrum for this design was TS = 8%. The voltage required to deflect the membrane was 1,000V for the fabricated prototype. An equation to estimate the activation voltage V as a function of deflection, based on a parallel plate model, is given by

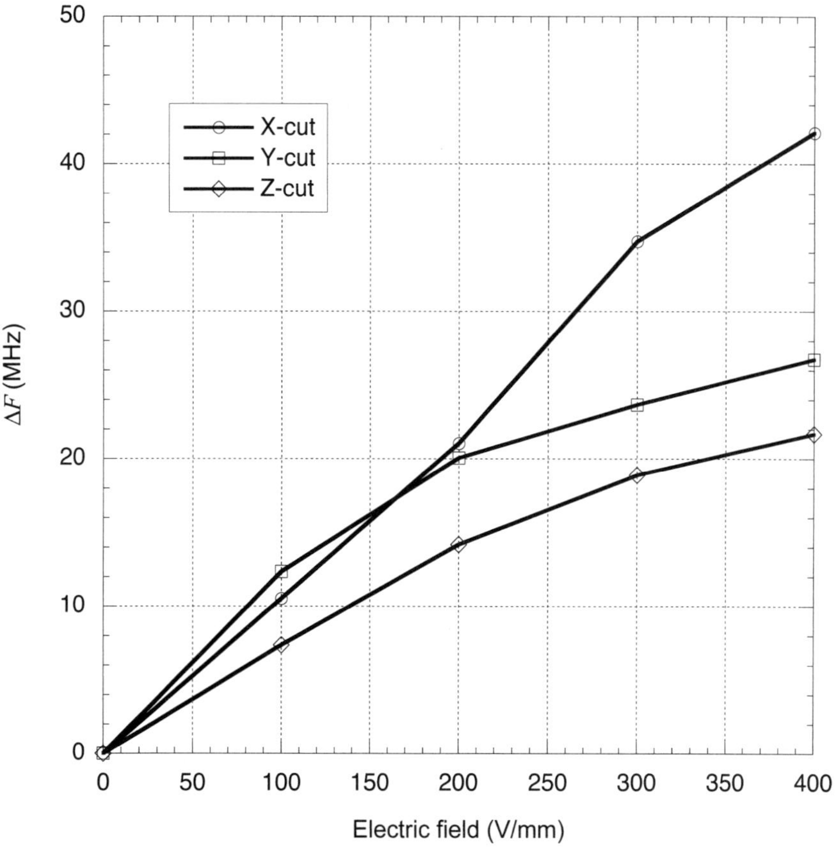

Figure 5.8 Change in resonant frequency with applied E-field for the micostrip antenna on quartz. (From [4].)

$$V = \sqrt{\frac{2kz(d-z)^2}{\varepsilon_{eff}\pi R^2}} \tag{5.1}$$

$$k = \frac{\pi A_p E t^3 + B_p E z^2}{R^2} \tag{5.2}$$

$$A_p = \frac{2(q+2)(q+1)}{3\left[1-\left(\frac{v}{q}\right)^2\right]} \tag{5.3}$$

$$B_p = \frac{32}{\left(q^2-9\right)}\left(\frac{1}{6} - \frac{3-v}{(q-v)(q+3)}\right) \tag{5.4}$$

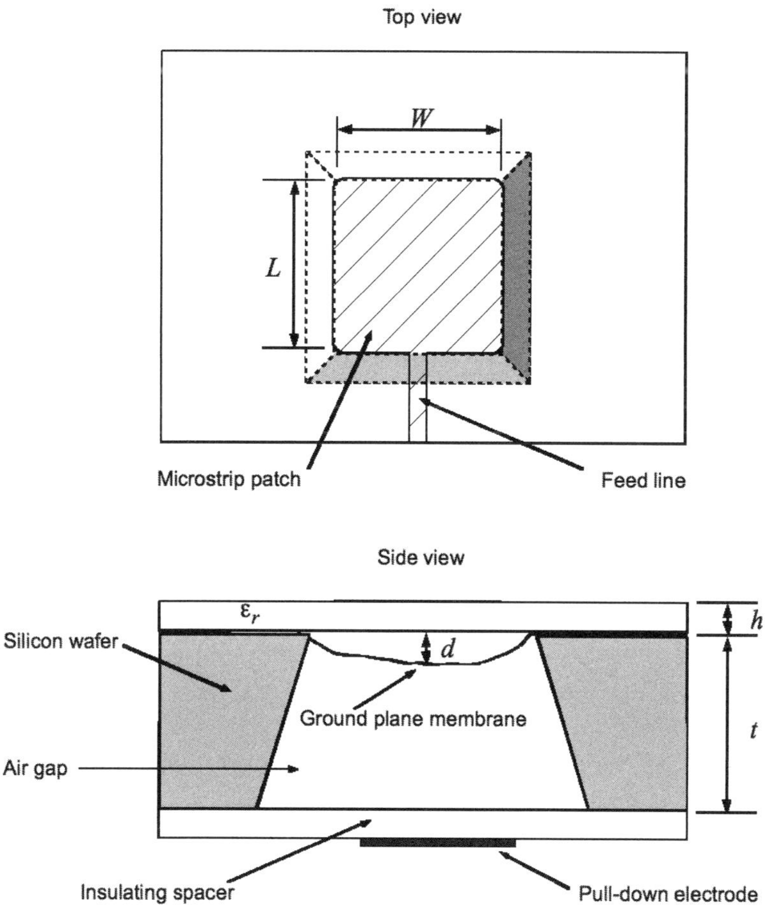

Figure 5.9 Electrostatically controlled ground plane membrane for tuning a microstrip antenna (L = 6 mm, W = 7 mm, h = 0.1 mm, t = 0.4 mm, and ε_r = 6.2). (After [5, 6].)

$$q = \sqrt{\frac{S}{L_c}\left[1 + \frac{3}{2}\left(\frac{H}{t}\right)^2\right]} \tag{5.5}$$

where k is the spring constant of a sinusoidally corrugated membrane [7], z is the membrane vertical deflection, d is the distance between the membrane and the drive voltage plate, R is the membrane radius, ε_{eff} is the effective dielectric constant of the glass-air stacked dielectric, v is Poisson's ratio, H is the corrugation

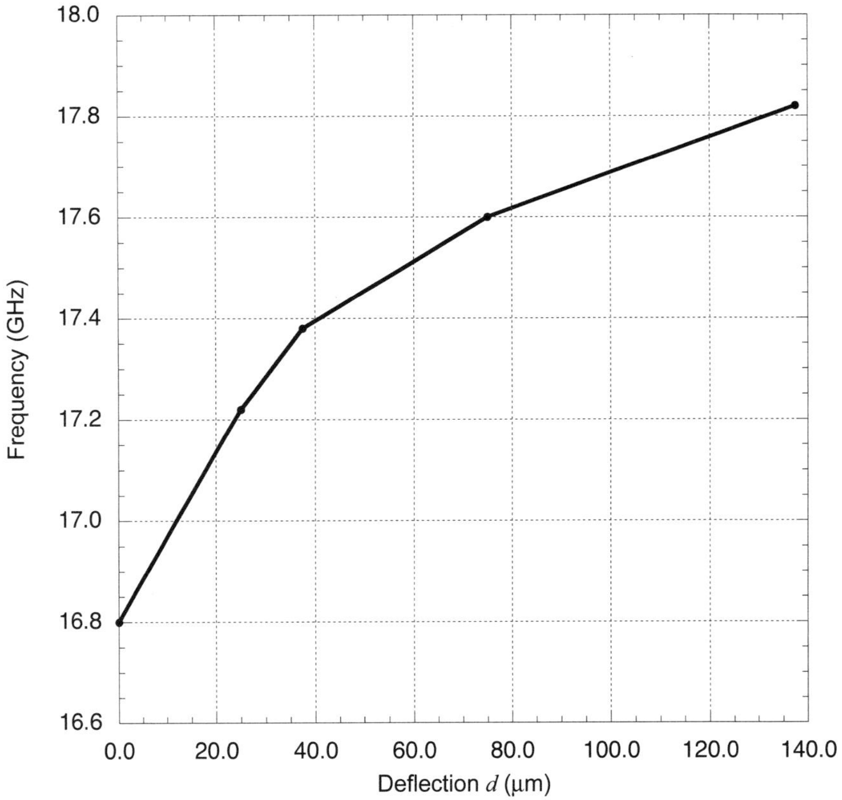

Figure 5.10 Resonant frequency versus gap of the electrostatically actuated microstrip antenna. (From [6].)

height, S is the corrugation arc length, and L_c is the corrugation period. Based on (5.1) to (5.5), the authors estimated that by using a thinner silicon wafer (t = 150 microns instead of 400 microns), the actuation voltage for a deflection of z = 150 microns would be reduced to 120V. The measured gain and radiation efficiency of the antenna were not reported.

A third example of a tunable microstrip antenna using electrostatic actuation is shown in Figure 5.11 [8]. In this case, the microstrip patch was etched on a thin Kapton E polyimde membrane (t = 0.05 mm), which was suspended above the ground plane by a distance approximately equal to the membrane thickness t. When a bias voltage of up to 268V was applied to the microstrip patch through resistive lines (to prevent the RF signal from mixing with the DC bias), the membrane was pulled down with a corresponding shift in the resonant frequency from 18.34 to 17.95 GHz (TR = 2%). The antenna achieved a total spectum of TS = 7%. The measured radiation patterns under various bias voltages showed

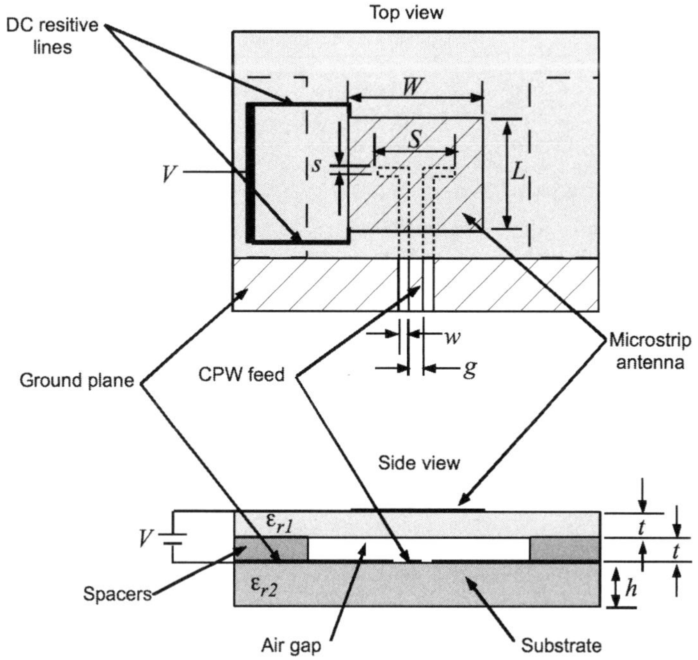

Figure 5.11 Electrostatically controlled microstrip antenna ($L = W = 6$ mm, $t = 0.05$ mm, $\varepsilon_{r1} = 3.1$, $h = 0.76$ mm, ε_{r2}=2.94, $S = 3.5$ mm, $s = 0.05$ mm, $w = 0.1$ mm, and $g = 1.627$ mm). (After [8].)

significant ripples, which were not predicted in the simulations and were attributed to warpage of the antenna. The cross-polarization levels did not deteriorate with applied bias, but neither the antenna gain nor radiation efficiency was reported.

A similar design was reported in [9], using a microstrip patch antenna directly fed by a microstrip line. The patch was etched on a thin (h = 0.5 mm) layer of flexible polyethylene terephthalate (PET) material (ε_r = 3.5) and was suspended above the ground plane at a distance of g = 0.2 mm when unbiased. Applying a DC voltage between the patch and the ground plane generated an electrostatic force that pulled the patch down toward the ground plane. This increased the effective dielectric constant (ε_{eff}) of the patch thereby lowering the resonant frequency. The effective dielectric constant can be estimated using

$$\varepsilon_{eff} = \frac{\varepsilon_r(g+h)}{\varepsilon_r g + h} \tag{5.6}$$

which can then be used in (3.15) and (3.14) to determine the values of ΔL and L_{eff}, respectively. The resonant frequency of the antenna can then be estimated using (3.16).

Three antennas were designed to operate at frequencies of f_1 = 6 GHz, f_2 = 8 GHz, and f_3 = 10 GHz, in the unbiased state. When a DC bias voltage of up to 150V was applied, the three antennas achieved tuning ranges of TR_1 = 3.7%, TR_2 = 3.2%, and TR_3 = 4.1%. The corresponding values for total spectrum were TS_1 = 10%, TS_2 = 12%, and TS_3 = 13%. The measured gain or radiation efficiency of these antennas was not reported.

The final example of an electrostatically controlled microstrip patch antenna is shown in Figure 5.12 [10]. The antenna, etched on a high-resistivity silicon substrate, had metal rolls on its sides that were in their furled position when no DC field was applied [Figure 5.12(a)]. In this state the antenna resonated at 23.7 GHz. When a DC voltage of 150V was applied between the microstrip patch and the ground plane, the electrostatic field caused the metal rolls to unfurl, as shown in Figure 5.12(b). The increased area of the patch lowered the resonant frequency to 12.5 GHz, resulting in a tuning range of TR = 61.8%. The antenna achieved a total spectrum of TS = 64% but with a tunable impedance bandwidth of only BW_{TZ} = 5%. The measured gain or the radiation efficiency of the antenna was not reported.

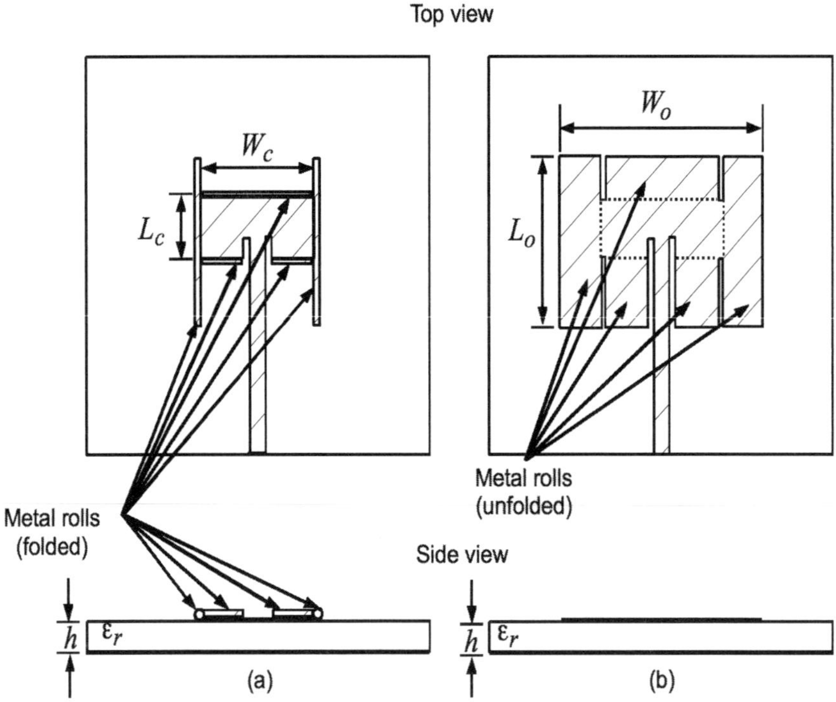

Figure 5.12 Electrostatically actuated foldable microstrip antenna: (a) folded state and (b) unfolded state (L_c = 1.40 mm, W_c = 3.095 mm, L_o = 4.10 mm, W_o = 5.495 mm, h = 0.525 mm, and ε_r =11.2). (After [10].)

5.1.3 Magnetostatic Actuation

In this example involving the micromachining of a microstrip antenna, shown in Figure 5.13, magnetostatic actuation was used to tune the resonant frequency of a microstrip antenna [11, 12]. The rectangular microstrip antenna was only attached to the microstrip feed line and not to the substrate, so that it was free to lift off. The patch was coated with a thin layer of magnetic material (permalloy) and by applying a DC magnetic field, the microstrip antenna could be raised from an initially horizontal position to a vertical one ($0^\circ \leq \theta \leq 90^\circ$), with an accompanying shift in resonant frequency. The major shift in frequency occured for $0^\circ \leq \theta \leq 15^\circ$, where the frequency was altered from 25.6 to 28.1 GHz (TR = 9.3%). The antenna achieved a total spectrum of TS = 12%. Once the patch was raised to above 45°, the radiation pattern changed from being that of a typical microstrip antenna, to one that more resembled a monopole pattern. High cross-polarization levels were reported for this antenna as well as poor radiation efficiency, due to the lossy nature of the permalloy.

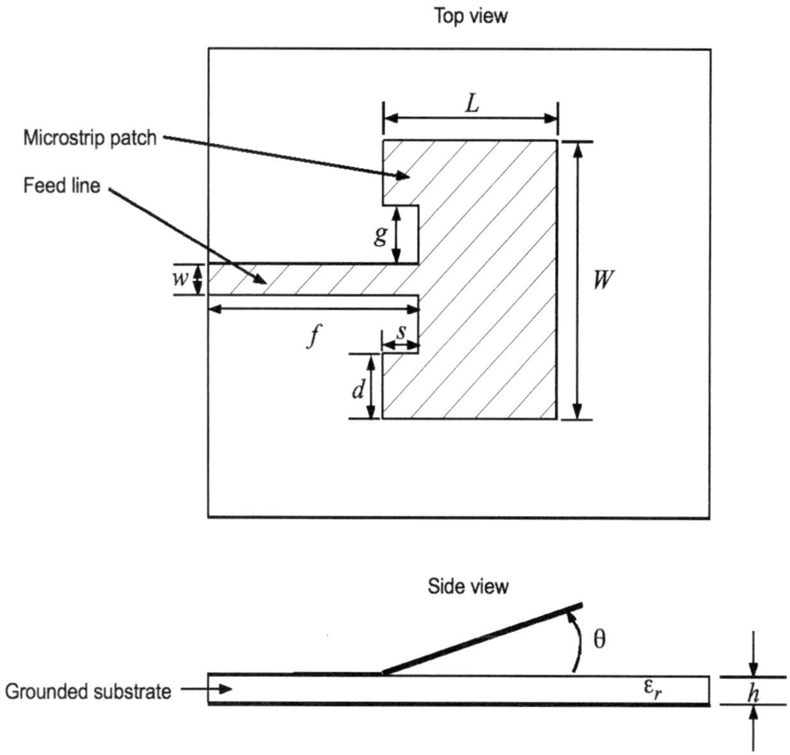

Figure 5.13 Deflection tuned microstrip patch (L = 4.46 mm, W = 5.0 mm, h = 0.6 mm, ε_r = 2.16, s = 1.4 mm, g = 1.0 mm, f = 11.1 mm, w = 2 mm, and $0^\circ \leq \theta \leq 90^\circ$). (After [11].)

The authors also noted that their design was only suitable for a small number of frequency tuning cycles, since stress failure would occur at the joint between the microstrip antenna and the feed line. The required external magnetic bias strength was not reported; however the authors mentioned that high values were needed and that these values would have to increase for larger sized antennas, resulting in bulkier and heavier bias magnetic field sources. This would place a practical limitation on the physical size of an antenna tuned using this magnetostatic approach.

5.1.4 Pneumatic Actuation

The final microstrip antenna example in this section is based on a change in height using pneumatic actuation to tune the frequency of the antenna. Figure 5.14 shows a sketch of the microstrip antenna printed on a thin (t = 0.02 mm) polydimethylsiloxane (PDMS) membrane, suspended at a height h above the

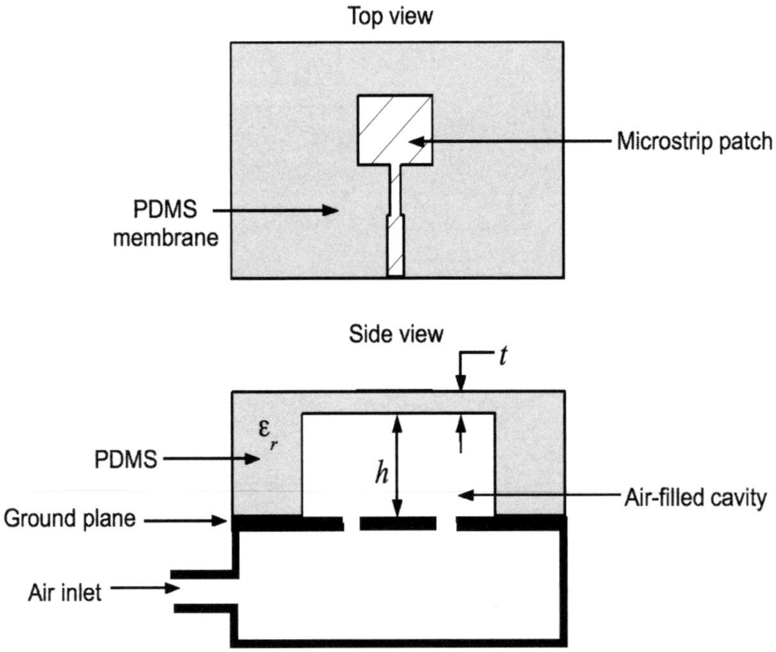

Figure 5.14 Pneumatically tuned microstrip patch (ε_r = 2.68, t = 0.02 mm, 0.2 mm ≤ h ≤ 0.575 mm). (After [13, 14].)

ground plane, thereby forming a small air-filled cavity [13, 14]. Using a syringe pump to force air into or out of this cavity, the membrane could be deformed, resulting in a change in height from 0.09 mm $\leq h \leq$ 0.575 mm using 55 μL of air for the displacement. The resonant frequency of the antenna was tuned from 55.35 down to 51 GHz (TR = 8.2%), as shown in Figure 5.15. The total spectrum achieved was greater than TS = 36%. Meaured gains varied from 4.5 to 5.0 dBi, with radiation efficiencies between 36 to 39%. Much of the losses were attributed to a lossy feed line printed on the PDMS membrane. The authors estimated that tuning speeds of 50–100 ms could be expected by using appropriate actuators. Further investigation is still required to determine the reliability of this technique. With such small displacements, this technique is limited to millimeter-wave frequency applications.

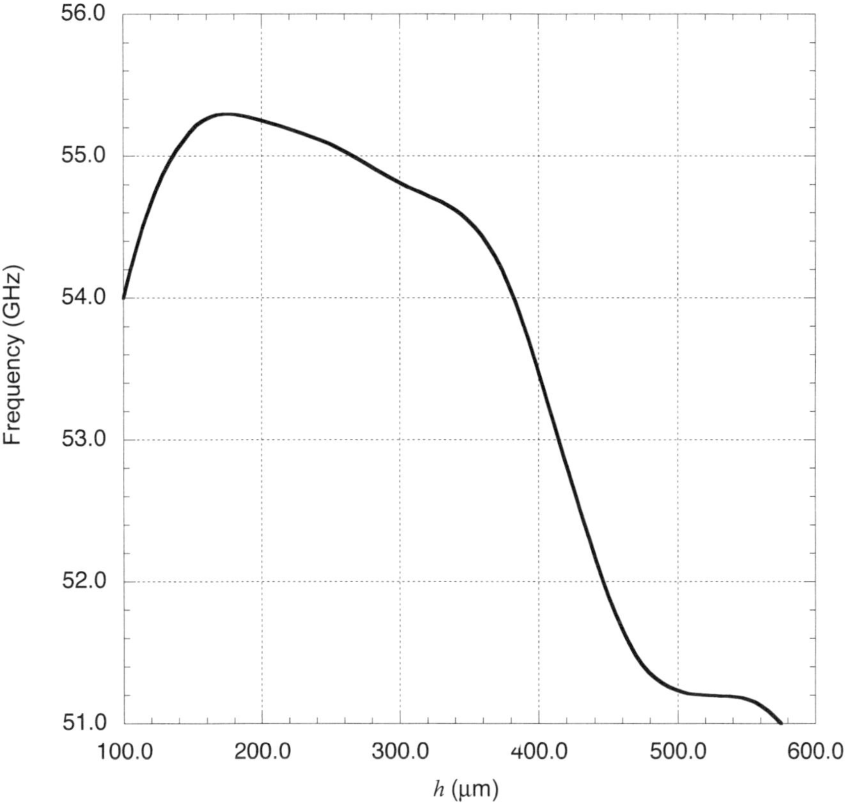

Figure 5.15 Resonant frequency versus air gap height of the pneumatically tuned microstrip patch. (From [13].)

5.1.5 Rotating Stacked Patches

A mechanically tunable stacked patch configuration was proposed in [15], consisting of a fixed lower circular microstrip patch and a rotatable upper hexagonal patch, as shown in Figure 5.16. Two thin slits (12.5 mm x 0.7 mm) were introduced into the circular patch as a means of lowering the resonant frequency by about 20%. The upper hexagonal patch was added to tune the resonant frequency by introducing a capacitance over the slits that could be varied as the patch was rotated. For the prototype shown in Figure 5.16, a rotation of the upper hexagonally shaped patch by up to $\phi = 90^\circ$ resulted in a shift of the resonant frequency from 2.46 to 2.12 GHz (TR = 15%), as plotted in Figure 5.17. The total spectrum for this antenna was TS = 16%, with a tunable bandwidth of

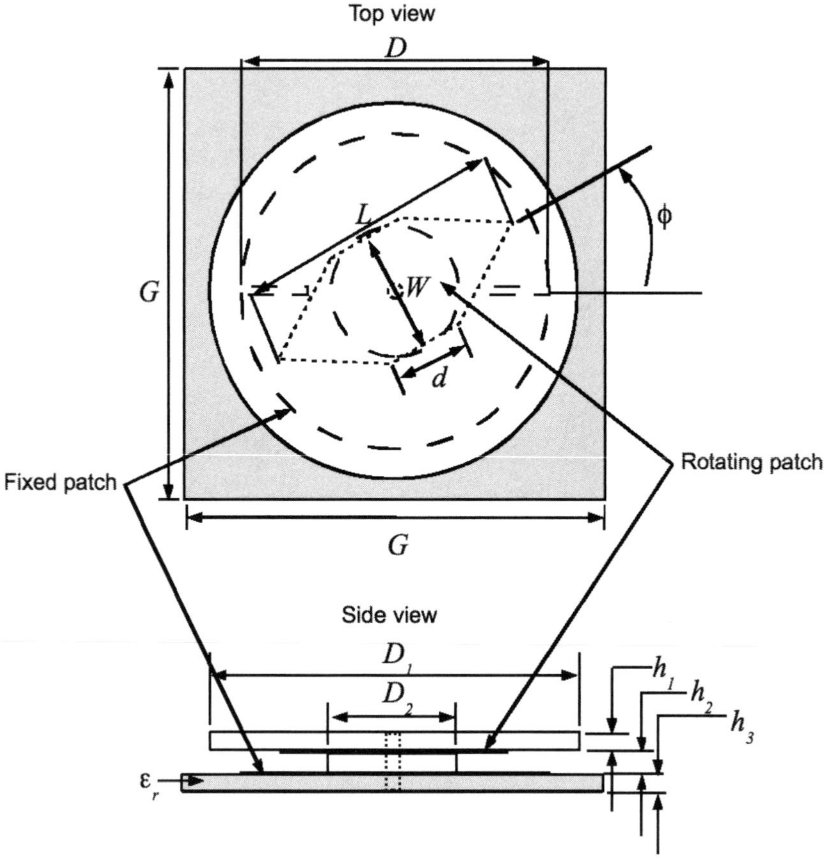

Figure 5.16 Mechanically rotating stacked-patch microstrip antenna ($D = 40$ mm, $L = 36$ mm, $W = 15.08$ mm, $d = 8$ mm, $D_1 = 60$ mm, $h_1 = 0.8$ mm, $D_2 = 25$ mm, $h_2 = 0.1$ mm, $G = 140$ mm, $h_3 = 1.52$ mm, and $\varepsilon_r = 2.5$). (After [15].)

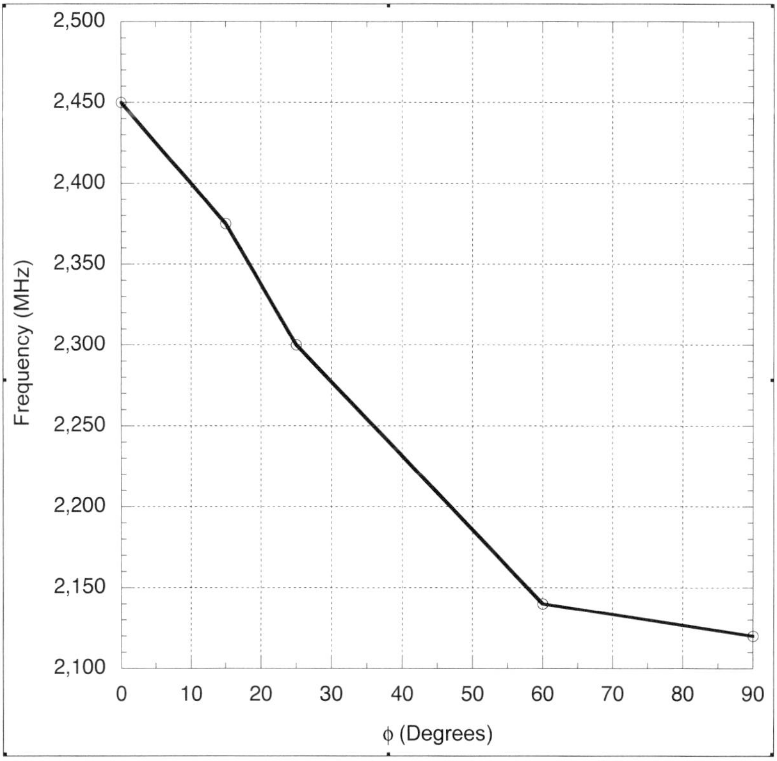

Figure 5.17 Frequency versus top patch angle of the rotating stacked-patch microstrip antenna. (From [15].)

BW_{TZ} ~ 8%. Measured gain or radiation efficiency values were not reported. Although only a passive prototype was presented, a small motor could be incorporated into the design to rotate the upper hexagonal patch in order to make this antenna frequency-agile. Circular polarized versions of the rotating stacked-patch microstrip antenna were also presented in [15].

A second example of a mechanically rotated antenna is shown in Figure 5.18 [16]. This antenna was paired with a wideband printed antenna (not shown), intended for cognitive radio applications. It consists of five microstrip antennas of various shapes and sizes arranged on a rotating disk. Each microstrip antenna was designed to operate over a different frequency band, covering the entire range 2.1–10 GHz without gaps (TS = 131%), as listed in Table 5.1. (The peak measured gain for each antenna is also provided in Table 5.1) The backside of the disk was attached to a stepper motor, which, when activated, rotated the disk to one of five positions, as indicated at the bottom of Figure 5.18. A 12-V power supply was required to power both the stepper motor and the accompanying drive circuitry.

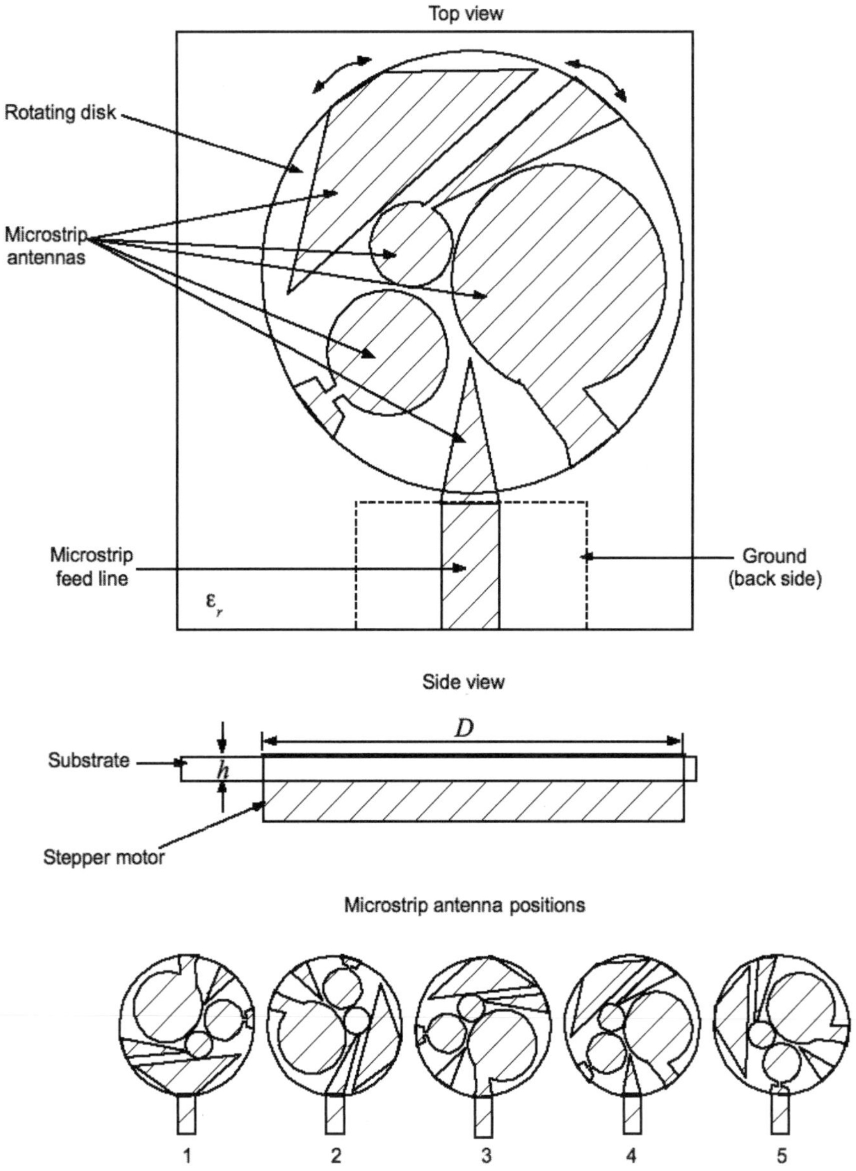

Figure 5.18 Assembly of rotating microstrip antennas (D = 38 mm, h = 1.6 mm, and ε_r = 2.2). (After [16].)

Table 5.1

Operating Frequency of the Rotating Microstrip Antennas

Antenna Position	Operating Frequency (GHz)	Peak Measured Gain (dBi)
1	3.4-5.56	6.62
2	6.3-10	8.45
3	2.1-3.0	7.77
4	5.4-6.2	6.67
5	3.0-3.4	7.40

From [16].

For each of these positions, one of the five antennas was aligned and made physical contact with the fixed microstrip feed line, thus activating this antenna while the others remained inactive. The use of the stepper motor behind the antenna minimized interference with the antenna patterns since it eliminated the need for a biasing network in the plane as the radiating aperture.

5.2 QUADRIFILAR HELIX

The basic principles of a helical antenna are reviewed in Section 3.9, where the type of helical antenna analyzed consists of a single wire and is known as a monofilar helix. Quadrifilar helical antennas (consisting of four interwound wires) have been used for applications such as mobile satellite communications where compact, low-gain, circular polarized antennas are required. Quadrifilar helical antennas exhibit excellent axial ratio and front-to-back ratio performance within a compact design. A frequency-tunable design of a quadrifilar helix was proposed in [17]. The helix was split into two halves, with overlapping arms, as shown in Figure 5.19. By adjusting the separation between the two halves of the split quadrifilar helix antenna, and by rotating one half with respect to the other, tuning of the resonant frequency was demonstrated. Figure 5.20 highlights the variables involved in either a straight translation between the upper and lower sections, or a combination of translation and rotation. (For clarity, only a single set of arms is shown in Figure 5.20.) For a simple translation by an amount ΔL between lower and upper sections [Figure 5.20(a)], the spacing S between the lower and upper arms remains constant. However, if a rotation of relative angle θ is also applied [Figure 5.20(b)], the spacing s between the two arms will also change. Table 5.2 shows the change in resonant frequency, gain, and axial ratio for different cases.

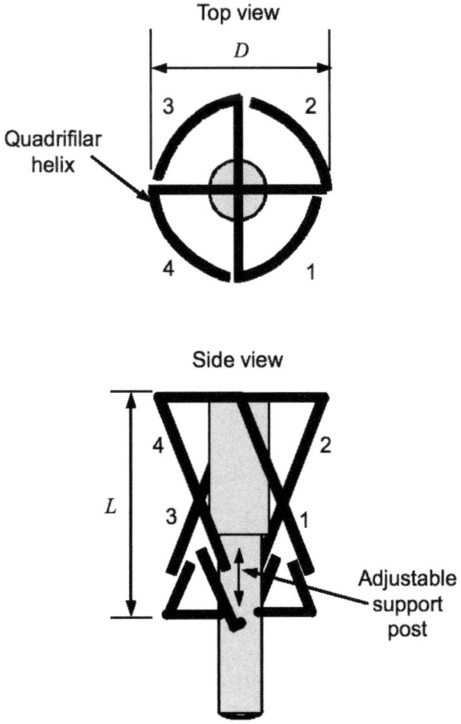

Figure 5.19 Mechanically tuned split quadrifilar helix ($D = 44$ mm, and $42 \le L \le 60$ mm). (After [17].)

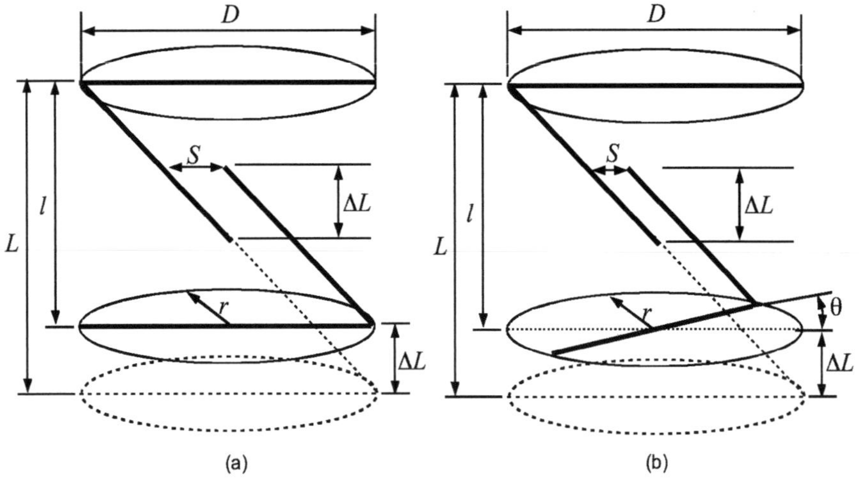

Figure 5.20 Relationship between upper and lower arms of the split quadrifilar helix: (a) translation only; and (b) translation and rotation. (After [17].)

Table 5.2

Operating Frequency of the Tunable Quadrifilar Helix

Antenna Position	l/L	Radial Gap s (mm)	Resonant Frequency (MHz)	Gain (dB)	Axial Ratio (dB)
1	1.0	0	1,200	5.52	2.85
2	0.9	7	1,255	4.76	4.44
3	0.8	6	1,290	4.81	4.0
4	0.8	2.3	1,350	5.2	4.2
5	0.7	5.5	1,365	4.64	4.6

From [17].

For the prototype design, the resonant frequency was tuned from 1,200 to 1,365 MHz (TR = 12.9%) with a total spectrum of TS = 25%. The gain variation was approximately 1 dB (5.5 dB to 4.6 dB) with a degradation in the maximum axial ratio from 2.85 to 4.6 dB. A linear actuator could be incorporated in the design in order to make the antenna electronically tunable.

5.3 SLOT DIPOLE WITH PIEZOELECTRIC TRANSDUCER

Rectangular slot antennas, introduced in Section 3.5, behave similarly to dipoles. A frequency-agile design is illustrated in Figure 5.21 [18]. The antenna consists of

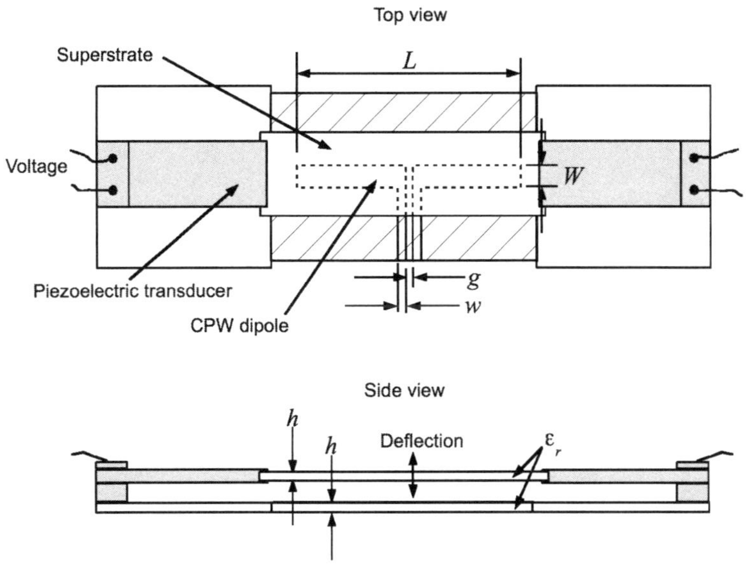

Figure 5.21 Tunable CPW slot dipole using piezoelectric deflection of superstrate (L = 30.02 mm, W = 4.47 mm, w = 1.524 mm, g = 0.254 mm, h = 1.27 mm, and ε_r = 6.15). (After [18].)

a CPW-fed rectangular slot dipole above which a dielectric superstrate is suspended. The two ends of the superstrate are each attached to a piezoelectric transducer. The application of a DC voltage to the transducers results in a deflection of the superstrate, which, in turn, causes a shift in the resonant frequency of the slot dipole. For the prototype shown in Figure 5.21, the resonant frequency was tuned from 4.77 down to 4.35 GHz (TR = 9%) and achieved a total spectrum of TS = 15% when the voltage was increased from -90V to 60V, as graphed in Figure 5.22. The measured radiation patterns did not degrade significantly when tuned, but actual gain values were not reported. This example is another illustration of the relatively high-bias voltages required to actuate piezoelectric transducers, which would restrict their use to applications that could accommodate such large DC supplies.

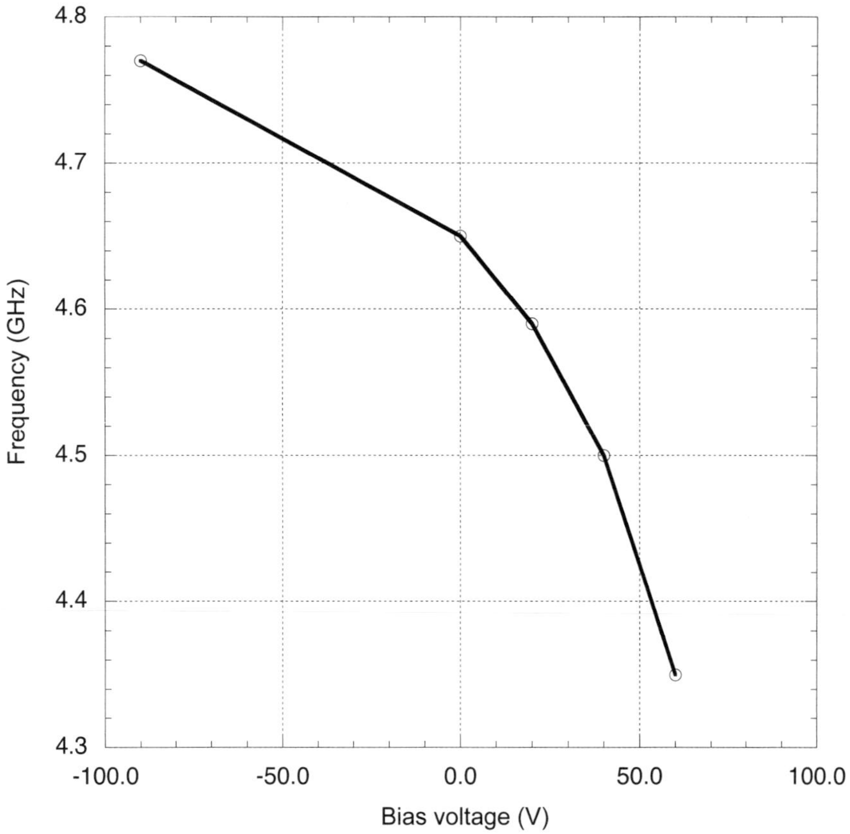

Figure 5.22 Resonant frequency versus bias voltage of the tunable CPW-fed slot dipole. (From [18].)

5.4 PIFAs

PIFAs are reviewed in Section 3.6. An example of a mechanically tunable PIFA is shown in Figure 5.23 [19]. A tuning plate was added between the PIFA and the ground plane, with the height h of the plate being adjustable via a tuning screw. The resonant frequency of the PIFA was controlled over a range of 2.5–3.3 GHz (TR = 28%) and attained a total spectrum of TS > 37% for a height adjustment of between 3.6 mm $\leq h \leq$ 5 mm, as plotted in Figure 5.24. This technique achieved a reasonable tuning range and did not introduce losses in the antenna. The manually adjusted tuning screw could potentially be replaced by a small linear actuator for electronic tuning capability.

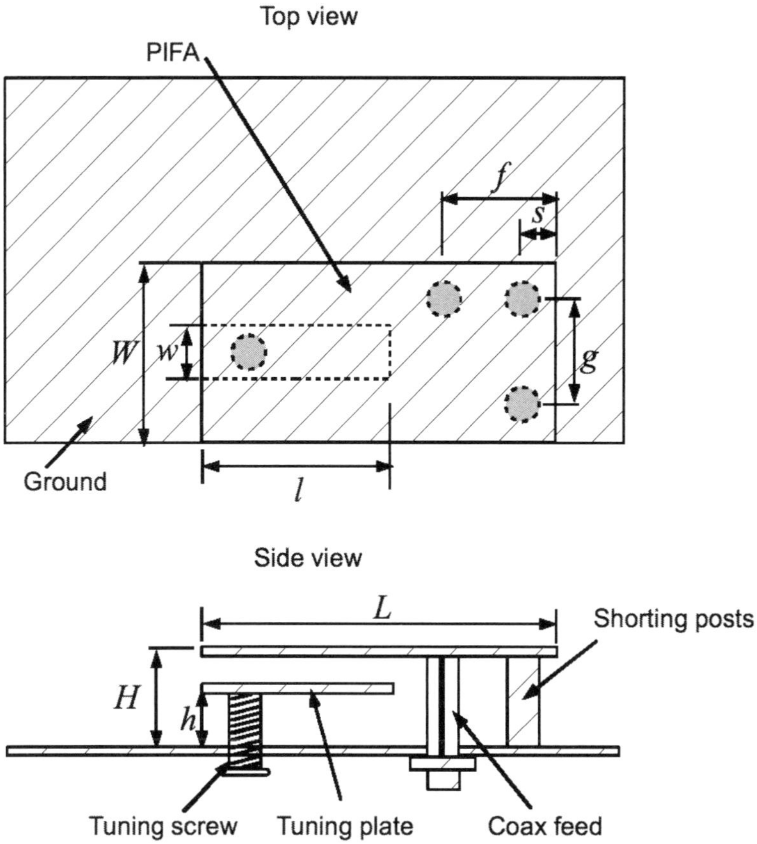

Figure 5.23 Mechanically tuned PIFA (L = 30 mm, W = 5 mm, H = 6 mm, g = 3 mm, s = 1 mm, f = 8.5 mm, l = 10.8 mm, w = 2 mm, and 3.6 mm $\leq h \leq$ 5 mm). (After [19].)

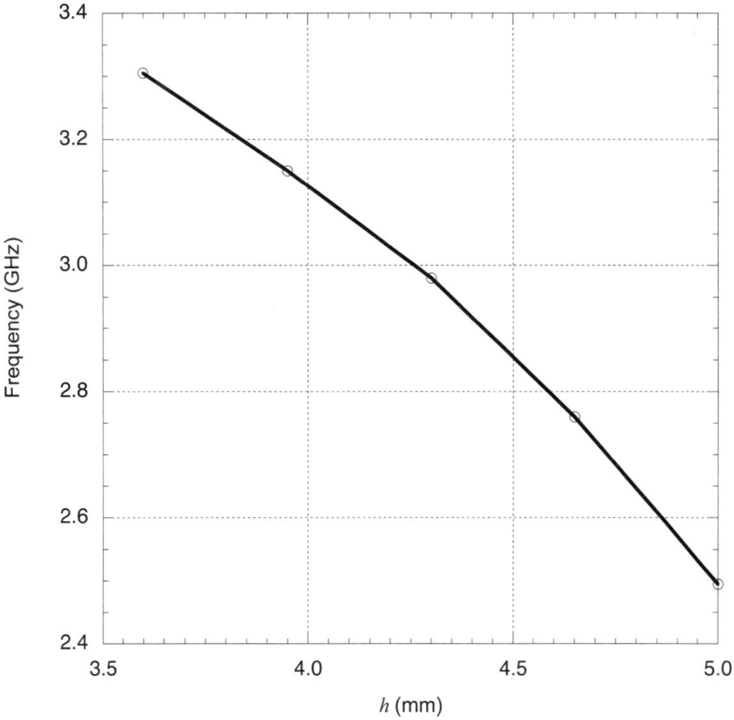

Figure 5.24 Resonant frequency versus tuning plate height of the mechanically tuned PIFA. (From [19].)

5.5 LIQUID-BASED DRAs

Some investigations on liquid-filled tubes have demonstrated that they can radiate as monopoles or as DRAs [20–22]. DRAs of cylindrical or rectangular shape are introduced in Section 3.8. A DRA consisting of a water-filled cylindrical tube was developed in the late 1990s, intended for low-frequency radar applications [20]. The use of liquid dielectrics can help solve some of the problems associated with the solid dielectrics, such as the difficulty in inserting feed probes and the potential for air gaps between the dielectric and ground plane, which can have considerable impact on the input impedance. Water has a dielectric constant ranging from about $88 \geq \varepsilon_r \geq 70$, for a temperature 0°–30°C, and the dielectric losses are fairly low for frequencies below 1 GHz. By adjusting the level of water in the tube, it was demonstrated that the resonant frequency of the DRA could be tuned [23]. Figure 5.25 shows the geometry of a prototype probe-fed water-filled DRA. Tap water at 8°C was used ($\varepsilon_r \sim 87$, $\tan\delta \sim 0.003$), and the DRA was excited in the $HEM_{11\delta}$ mode (see Section 3.8.1).

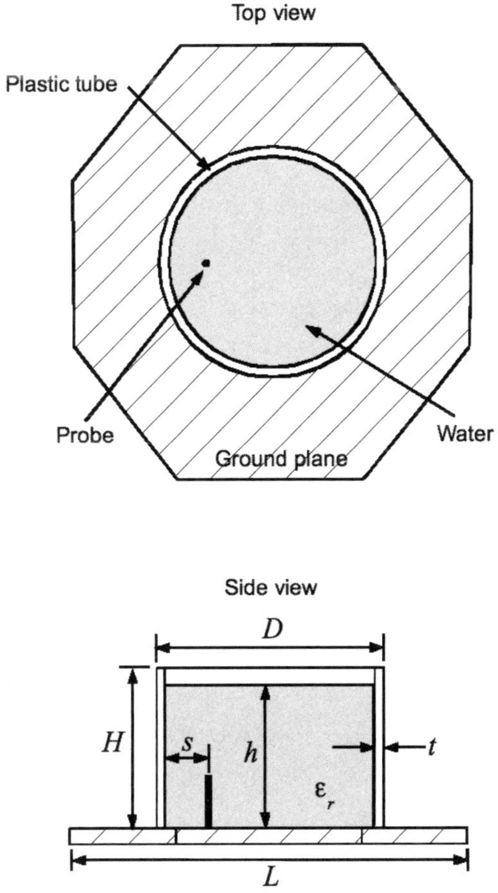

Figure 5.25 Water-filled DRA (L = 800 mm, D = 550 mm, H = 300 mm, 156 mm $\leq h \leq$ 276 mm, s = 70 mm, t = 5 mm, and $\varepsilon_r \sim 87$). (After[23].)

The water level was varied over the range of 276 mm $\geq h \geq$ 95 mm, resulting in a shift in resonant frequency from 50 to 96.8 MHz (TR = 64%), as graphed in Figure 5.26. The theoretical resonant frequency for the $HEM_{11\delta}$ mode of a cylindrical DRA based on (3.88) is also plotted. For the initial probe height of 175 mm, the RL degraded when the water level dropped below h = 156 mm, and a shorter (134 mm) probe was used for the lower water levels. In practice, the authors proposed the use of a folded probe with integrated varactor diodes for electronically tuning the impedance match as the water level changed. It was suggested that a simple pumping arrangement could be used to alter the water level, but no details of such a system were provided.

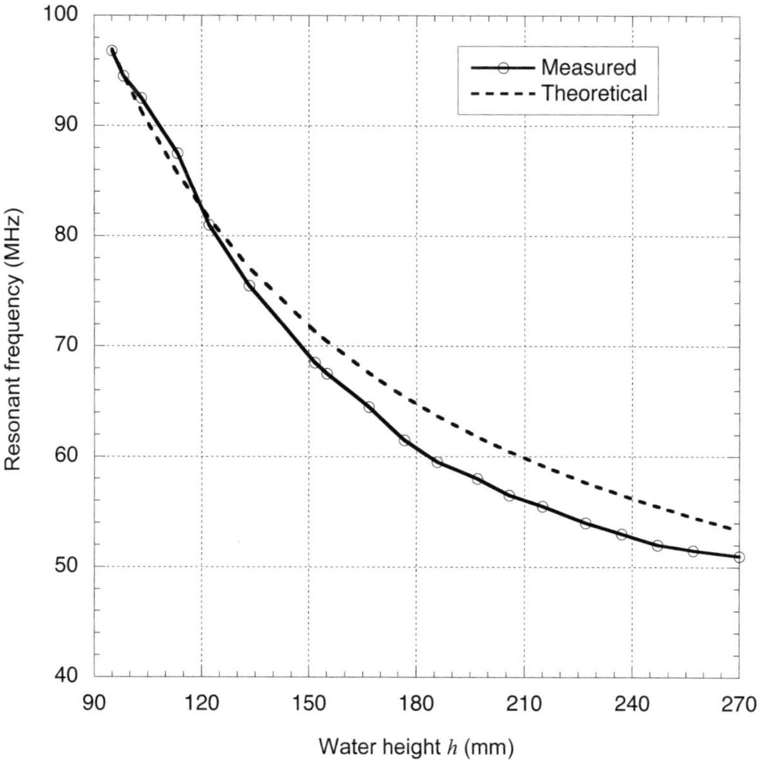

Figure 5.26 Resonant frequency versus height h of water-filled DRA. (From [23].)

Since the dielectric losses of water become unacceptably high at microwave frequencies, some research on nonaqueous liquids has recently been carried out. A cylindrical DRA filled with a colloidal dispersion designed to operate between 2.5 and 4.5 GHz was proposed in [24]. By controlling the volume fraction of the particulates dispersed within the liquid, the desired effective dielectric constant can be determined using

$$\varepsilon_{eff} = \varepsilon_2 + 3V\varepsilon_2 \frac{\varepsilon_1 - \varepsilon_2}{\varepsilon_1 + 2\varepsilon_2 - V(\varepsilon_1 - \varepsilon_2)} \tag{5.7}$$

where V is the volume fraction and $\varepsilon_1 = \varepsilon_{r1}(1 - j\tan\delta_1)$ and $\varepsilon_2 = \varepsilon_{r2}(1 - j\tan\delta_2)$ are the relative complex permittivity of the colloids and the liquid, respectively. In [21], a colloidal dispersion consisting of napthenic oil ($\varepsilon_{r2} = 2.1$, $\tan\delta_2 = 0.001$) with barium strontium titanium oxide (BSTO) particles ($\varepsilon_{r1} \sim 500$, $\tan\delta_1 \sim 0.01$) at volume fraction of $V = 0.5$ was used, with an effective dielectric constant of $\varepsilon_{eff} = 8.24$. A similar configuration as shown in Figure 5.25 was used, except the ground plane was a square of length $L = 100$ mm. The polycarbonate tube had a thickness of t = 1.6 mm, height of H = 30 mm, and diameter of D = 27 mm. A probe of height 8 mm and offset s = 4 mm from the inner side of the tube was used. As graphed in Figure 5.27, the resonant frequency of the DRA was tuned from 2.5 to 4.5 GHz (TR = 46%) for a fluid level variation over the range of $24 \geq h \geq 8$ mm. The resonant frequency for the TM_{011} mode is estimated using

$$f_{011} = \frac{c}{\pi(D-2t)\sqrt{\mathrm{Re}\left[\varepsilon_{eff}\right]}}\sqrt{\left(\chi'_{11}\right)^2 + \frac{\pi(D-2t)}{4h}} \tag{5.8}$$

where $\chi'_{11} = 1.841$ is the first root of the first derivative of the Bessel function of the first kind, and c is the speed of light. The prototype antenna that was manufactured lacked a top cover to contain the liquid, and thus radiation patterns and gain performance were not measured. Again, there was no consideration given to the hydraulic system that would be required to alter the level of the fluid.

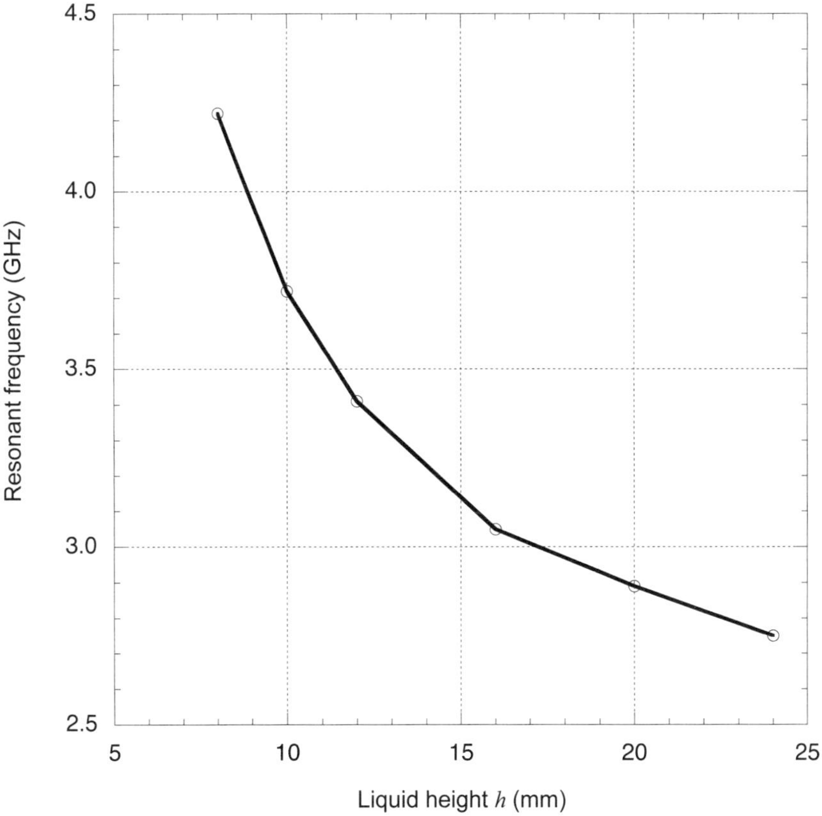

Figure 5.27 Resonant frequency versus height h of water-filled DRA. (From [24].)

5.6 SUMMARY

This chapter explores various antenna designs that incorporate some form of mechanical actuation to achieve frequency tuning. The examples have shown some of the widely diverse range of possible combinations of actuation mechanisms and antenna elements. A summary of the examples is listed in Table 5.3. The use of mechanical actuation for achieving frequency agility is far from its mature state, and there are still many avenues to explore and considerable effort required to fully exploit this tuning technique.

Table 5.3

Summary of FAA Examples Using Mechanical Actuation

Antenna	Actuation	Bias	Frequency Range (GHz)	Tuning Range (TR)	Reference
Microstrip patch	Mechanical translation	-	0.83–0.91	9%	[1]
Microstrip patch	Piezoelectric	-800–200V	2.793–2.988	6.7%	[2]
Microstrip patch	Piezoelectric	0–400 V/mm	8.0–8.05	0.5%	[3, 4]
Microstrip patch	Electrostatic	0–1,000V	16.8–17.8	8%	[5, 6]
Microstrip patch	Electrostatic	0–268V	17.95–18.34	2%	[8]
Microstrip patch	Electrostatic	0–150V	9.5–9.9	4.1%	[9]
Microstrip patch	Electrostatic	0–150V	12.5–23.7	61.8%	[10]
Microstrip patch	Magnetostatic	N/A	25.6–28.1	9.3%	[11, 12]
Microstrip patch	Pneumatic	N/A	51.0–55.35	8.2%	[13, 14]
Microstrip patch	Mechanical rotation	N/A	2.12–2.46	15%	[15]
Microstrip patch	Mechanical rotation	12V	2.1–10.0	131%	[16]
Helix	Mechanical translation and rotation	N/A	1.2–1.365	12.9%	[17]
Printed dipole	Piezoelectric	-90–60V	4.35–4.77	9%	[18]
PIFA	Mechanical translation	N/A	2.5–3.3	28%	[19]
DRA	Hydraulic	N/A	0.050–0.097	64%	[23]
DRA	Hydraulic	N/A	2.5–4.5	46%	[24]

References

[1] Dahele, J. S., and K. F. Lee, "A Tunable Dual-Frequency Stacked Microstrip Antenna," *Proc. IEEE International Symposium on Antennas and Propagation,* Albuquerque, NM, May 24-28, 1982, pp. 308-311.

[2] Bernhard, J. T., E. Kiely, and G. Washington, "A Smart Mechanically Actuated Two-Layer Electromagnetically Coupled Microstrip Antenna with Variable Frequency, Bandwidth, and

Antenna Gain," *IEEE Transactions on Antennas and Propagation,* Vol. 49, No. 4, April 2001, pp. 597-601.

[3] Song, J. T., I. H. Cho, and Y. H. Kim, "Frequency Agile Microstrip Patch Antenna Using Piezoelectric Substrates," *Japanese Journal of Applied Physics,* Vol. 40, May 2001, pp. L515-L517.

[4] Song, J. T., H. I. Kang, and M. H. Yong, "Frequency Agile Microstrip Patch Antenna Using Quartz Substrates," *Japanese Journal of Applied Physics,* Vol. 43, January, 2004, pp. L142-L143.

[5] Al-Dahleh, R., C. Shafai, and L. Shafai, "Frequency-Agile Microstrip Patch Antenna Using A Reconfigurable MEMS Ground Plane," *Microwave and Optical Technology Letters,* Vol. 43, No. 1, October 2004, pp. 64-67.

[6] Al-Dahleh, R., et al., "Low Actuation Voltage Ground Plane Membrane for a Frequency Agile Microstrip Patch Antenna," *Proc. 10th International Symposium on Antenna Technology and Applied Electromagnetics ANTEM and URSI Conference 2004,* Ottawa, Canada, July 20-23, 2004, pp. 21-24.

[7] Jerman, J. H., "The Fabrication and Use of Micromachined Corrugated Silicon Diaphragms," *Sensors and Actuators,* Vol. A21-23, 1990, pp. 988-992.

[8] Goteti, R. V., R. E. Jackson Jr., and R. Ramadoss, "MEMS-Based Frequency Switchable Microstrip Patch Antenna Fabricated Using Printed Circuit Processing Techniques," *IEEE Antennas and Wireless Technology Letters,* Vol. 5, 2006, pp. 228-230.

[9] Kang, H. I., and J. T. Song, "Electrically Tunable Rectangular Microstrip Antenna," *Electronics Letters,* Vol. 46, No. 1, January 2010, pp. 18-19.

[10] Blondy, P., D. Bouyge, A. Crunteanu, and A. Pothier, "A Wide Tuning Range MEMS Switched Patch Antenna," *Proc. IEEE Microwave Theory and Technique Symposium,* San Francisco, CA, June 11-16, 2006, pp. 152-154.

[11] Langer, J. C, et al., "Micromachined Reconfigurable Out-of-Plane Microstrip Patch Antenna Using Plastic Deformation Magnetic Actuation," *IEEE Microwave and Wireless Components Letters,* Vol. 13, No. 3, March 2003, pp. 120-122.

[12] Zou, Z., et al., "Plastic Deformation Magnetic Assembly (PDMA) of Out-Of-Plane Microstructures: Technology Development and Application," *IEEE Journal of Microelectromechanical Systems,* Vol. 10, June 2001, pp. 302-309.

[13] Hage-Ali, S., et al., "A Millimeter-Wave Frequency Tunable Microstrip Antenna on Ultraflexible PDMS Substrate," *Proc. IEEE International Symposium on Antennas and Propagation Symposium,* Toronto, Canada, July 11-17, 2010.

[14] Hage-Ali, S., et al., "A Millimeter-Wave Inflatable Frequency-Agile Elastomeric Antenna," *IEEE Antennas and Wireless Propagation Letters,* Vol. 9, 2010, pp. 1131-1134.

[15] Bokhari, S. A., et al., "A Small Microstrip Patch Antenna with a Convenient Tuning Option," *IEEE Transactions on Antennas and Propagation,* Vol. 44, No. 11, November 1996, pp. 1521-1528.

[16] Tawk, Y., et al., "Implementation of a Cognitive Radio Front-End Using Rotatable Controlled Reconfigurable Antennas," *IEEE Transactions on Antennas and Propagation,* Vol. 59, No. 5, May 2011, pp. 1773-1778.

[17] Amin, M., R. Cahill, and V. F. Fusco, "Mechanically Tunable Multiband Compact Quadrifilar Helix Antenna With Dual Mode Operation," *IEEE Transactions on Antennas and Propagation,* Vol. 56, No. 6, June 2008, pp. 1528-1532.

[18] Kim, S. G., and K. Chang, "Frequency Tunable CPW-Fed Slot Dipole Using Piezoelectric Transducer," *Electronics Letters,* Vol. 39, No. 13, June 2003, pp. 991-992.

[19] Chiu, C. Y., K. M. Shum, and C. H. Chan, "A Tunable Via-Patch Loaded PIFA With Size Reduction," *IEEE Transactions on Antennas and Propagation,* Vol. 55, No. 1, January 2007, pp. 65-71.

[20] Kingsley, S. P., and S. G. O'Keefe, "Beam Steering and Monopulse Processing of Probe-Fed Dielectric Resonator Antenna," *IEE Proceedings on Radar, Sonar and Naviagtion,* Vol. 146, No. 3, 1999, pp. 121-125.

[21] Fayad, H., and P. Record, "Broadband Liquid Antenna," *Electronics Letters,* Vol. 42, No. 3, February 2006, pp. 133-134.

[22] Fayad, H., and P. Record, "Experimental Investigation on New Steerable Dielectric Resonator Antenna," *Electronics Letters,* Vol. 43, No. 19, September 2007, pp. 1009-1010.

[23] O'Keefe, S. G., and S. P. Kingsley, "Tunability of Liquid Dielectric Resonator Antennas," *IEEE Antennas and Wireless Propagation Letters,* Vol. 6, 2007, pp. 533-536.

[24] Huff, G. H., et al., "A Frequency Reconfigurable Dielectric Resonator Antenna Using Colloidal Dispersions," *IEEE Antennas and Wireless Propagation Letters,* Vol. 9, 2010, pp. 288-290.

Chapter 6

FAAs Using Tunable Substrates

This chapter examines how the tunable substrates introduced in Chapter 4 have been incorporated into various antennas to make them frequency-agile. Ferrites, ferroelectrics, liquid crystal substrates, and semiconductors combined with antenna elements such as microstrip patches, cavities, DRAs, and coplanar slot antennas have demonstrated continuous frequency-tuning capabilities. Details of how these designs have been implemented and their resultant performance are presented. The selected examples are grouped by the type of tunable substrate used to achieve frequency agility.

6.1 FAAs DESIGNED USING FERRITES

The theory of ferrite behavior at microwave frequencies is presented in Section 4.2.1, which shows that the application of a DC magnetic bias field results in a change in the effective permeability of the ferrite. In this section, FAAs that make use of this property are examined. Examples include microstrip patch antennas printed on ferrite substrates, ferrite-filled cavity-backed slot antennas, and DRAs fabricated with ferrite materials.

6.1.1 Microstrip Patch Antennas

The equations used in Section 3.4 to predict the resonant frequency of microstrip patch antennas of various shapes were based on the assumption that these antennas were fabricated using a substrate characterized by a dielectric constant of $\varepsilon_r \geq 1$ and a relative permeability of $\mu_r = 1$. However, printing the microstrip antennas on a ferromagnetic substrate ($\mu_r > 1$) will have an impact on the resonant frequency. To a first-order approximation, the resonant frequency of microstrip patch antennas printed on ferromagnetic materials will be inversely proportional to $(\varepsilon_r \mu_r)^{1/2}$. Consequently, altering the value of μ_r will lead to a shift in the resonant frequency of the microstrip antenna.

Rectangular and circular microstrip patches printed on ferromagnetic substrates were first proposed in the early 1980s as a method for lowering the resonant frequency compared to a nonmagnetic substrate, since any increase in μ_r results in a decrease in the resonant frequency [1-4]. The ferrite remained unbiased, and designs were carried out in the UHF band (200–660 MHz).

A study on the effects of a DC magnetic bias on circular disk resonators printed on ferrite substrates was also carried out in the early 1980s [5]. Although the study focused primarily on the circular disk as a resonator element, it was recognized that the application of a DC magnetic bias oriented perpendicular to the ferrite substrate would cause the degenerate modes in the disk to split, resulting in clockwise and counterclockwise circular polarized radiation. The shift in resonant frequency as a function of DC bias was also investigated. For a disk of radius $a = 10.1$ mm printed on a ferrite substrate with $4\pi M_s = 650$G, $\varepsilon_r = 9.35$, and a thickness of $h = 1$ mm, the resonant frequency was tuned from 2.9 to 3.8 GHz (TR = 27%) for an applied bias of up to 900 Oe for clockwise polarization, and from 2.9 down to 2.3 GHz (TR = 23%) for an applied bias of up to 600 Oe for counterclockwise polarization. This study only examined the effects on RL and quality factors and did not investigate the effects on radiation patterns (such as gain, radiation efficiency, and cross-polarization levels).

The use of a DC magnetic field to tune the resonant frequency of a rectangular microstrip patch printed on a ferrite substrate was then proposed by Pozar and Sanchez [6]. Figure 6.1 shows the geometry of the prototype rectangular patch, which was resonant along length L in the x-direction. Permanent magnets were used to bias the ferrite substrate in directions parallel to each of the three axes (shown as H_x, H_y, and H_z in Figure 6.1). The ferrite substrate was made from YIG material (having $4\pi M_s = 1{,}720$G and $\varepsilon_r = 15$) with a thickness of $h = 1.27$ mm. The resonant frequency of the unbiased patch was $f_o = 4.6$ GHz. When a DC magnetic bias was applied using a permanent magnetic placed below the ground plane, the resonant frequency was altered to $f_x = 5.5$ GHz, $f_y = 2.8$ GHz, and $f_z = 5.5$ GHz, where the subscripts correspond to the H_x, H_y, and H_z bias, respectively. The resulting tuning ranges were TR = 21.8% in the x- and z-directions and TR = 48.6% in the y-direction. (The authors noted that the H_z field was not as strong as H_x and H_y since only one pole piece of the magnet was in contact with the ground plane, while two pole pieces were in contact for the other two directions.) Polarization purity and the impedance match were both found to

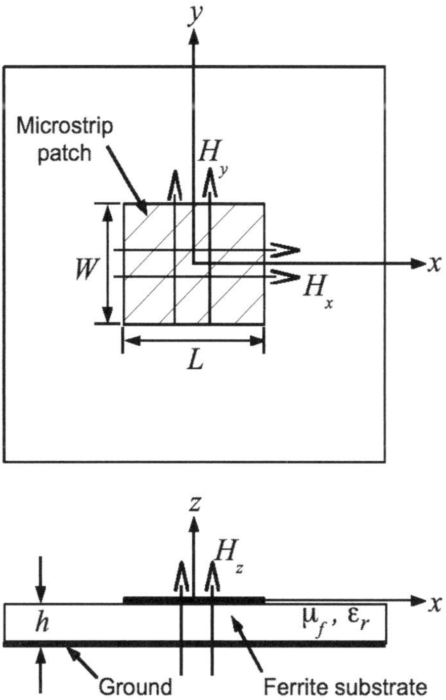

Figure 6.1 Microstrip antenna on a ferrite substrate showing the magnetic bias directions (L = 14 mm, W = 18 mm, h = 1.27 mm, ε_r = 15, and $4\pi M_s$ = 1,720G). (After [6].)

degrade with bias. The radiation efficiency of the antenna was determined to be approximately 70%, causing a reduction of about 1.5 dB in the gain of the patch, to a value of 4.5 dBi. This was attributed mainly to the surface waves in the ferrite due to the relatively high value of the dielectric constant.

In [7], it was demonstrated that the polarization of a square microstrip antenna printed on a ferrite substrate and excited with a single point feed could be switched from linear for the unbiased case to circular for the case of a z-directed DC bias (H_z). It was further shown that when a two-point feed was used for generating either right-hand circular polarization (RHCP) or left-hand circular polarization (LHCP), the resonance frequencies of the RHCP or LHCP were different under biased conditions (where the ferrite was operated in saturation). The shift in resonance frequency was estimated using a cavity model for a circular patch (to approximate the square patch) on the ferrite substrate. The resonant frequency f_r of a circular patch of radius a was found from the solution to

$$J_1'(ka) \pm \frac{\kappa}{ka\mu} J_1(ka) = 0 \tag{6.1}$$

where $k = 2\pi f_r\sqrt{\mu_{eff}\varepsilon_r\varepsilon_o}$ and $\mu_{eff} = \dfrac{\mu^2 - \kappa^2}{\mu}$ [i.e., the transverse bias case, (4.12) in Section 4.2.1], with μ and κ defined in (4.4) and (4.5), respectively. $J_1(x)$ is the Bessel function of the first kind, and + is for LHCP while – is for RHCP. For the unmagnetized ferrite, $\kappa = 0$ and $ka \sim 1.841$. When the magnetic bias is applied, the +/- roots are different, leading to different resonant frequencies for RHCP and LHCP. It was also earlier pointed out (in Section 4.2.1) that a band gap region will be formed for the transverse bias case, corresponding to the region where $\mu_{eff} < 0$, which occurs for frequencies in the range $\sqrt{f_o(f_o + f_m)} < f < (f_o + f_m)$, where f_o and f_m were defined in (4.6) and (4.7), respectively. In this range, the azimuth modes in the ferrite material are in the cutoff state since the direction of propagation for these modes is perpendicular to the bias field, similar to the cutoff effect for an extraordinary plane wave propagating in a ferrite medium in a direction perpendicular to the bias field.

A microstrip antenna printed on a ferrite substrate and biased using an electromagnet, as shown in Figure 6.2, was introduced in [8]. In this configuration,

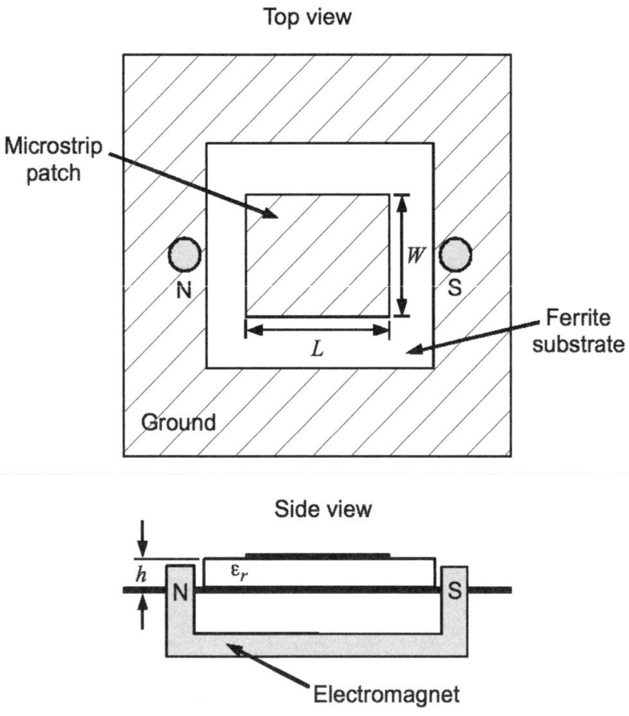

Figure 6.2 Rectangular microstrip patch antenna on a ferrite substrate (L = 40 mm, W = 11.1 mm, h = 2 mm, ε_r = 14.78, and $4\pi M_s$ = 3,000G). (After [8].)

the bias field could be accurately controlled by adjusting the current on the electromagnet, instead of having to physically move a permanent magnet. As shown in Figure 6.3, the unbiased microstrip patch was resonant at 200 MHz, and a frequency shift up to 1,000 MHz was achieved with a bias field of 6.25 kOe, representing a tuning range of 133%.

An example of an annular (or ring-shaped) microstrip antenna printed on a ferrite substrate is shown in Figure 6.4 [9, 10]. Unlike the previous two designs, in this example the antenna was excited through electromagnetic coupling from a microstrip feed line printed on a dielectric substrate located below the ferrite layer. Both the lowest order mode (TM_{11}) with a broadside peak and higher order modes (TM_{21}, TM_{31}) with broadside nulls were investigated. Two different ferrite materials were also studied. The various parameters are listed for two cases in Figure 6.4. Biasing was carried out using a permanent magnet below the feed layer, with the magnetic fields normal to the substrate. As was the case with the

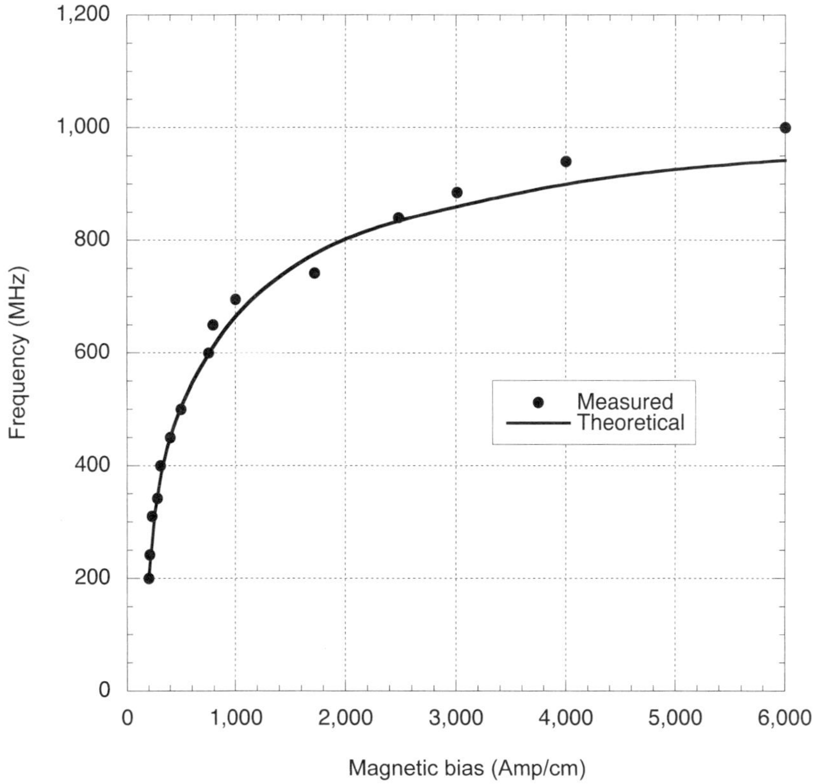

Figure 6.3 Resonant frequency versus applied magnetic bias for a rectangular microstrip patch antenna on a ferrite substrate. (From [8].)

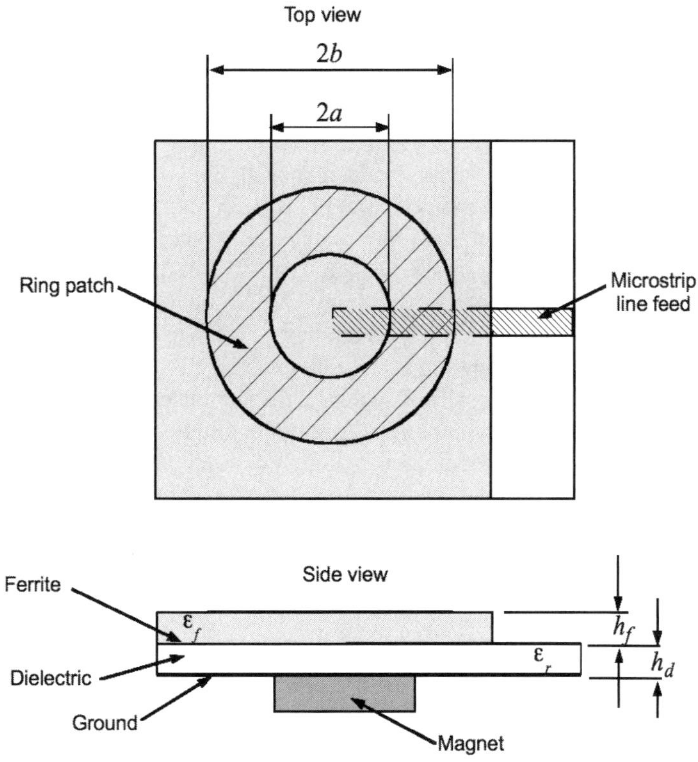

Figure 6.4 Annular microstrip patch antenna on a ferrite substrate: (case 1: a = 1.9 mm, b = 7.4 mm, h_f = 1.0 mm, $4\pi M_s$ = 348 G and ε_f = 13.69, h_d = 0.787 mm, and ε_r = 2.2; and case 2: a = 1.9 mm, b = 7.4 mm, h_f = 1.27 mm, $4\pi M_s$ = 1,730 G and ε_f = 14.78, h_d = 0.787 mm, and ε_r = 2.2). (After [9, 10].)

z-directed bias applied to the square patch, circular polarization was generated using a single feed point, with the sense of polarization depending on the polarity of the magnetic bias. As the magnetic bias was increased from zero, the resonant frequency of the RHCP monotonically increased while that of the LHCP initially decreased but then increased after a bias of about 1,300G. The results are plotted in Figure 6.5 for the TM_{11} and TM_{21} modes of the case 2 annular microstrip patch in Figure 6.4. The maximum tuning ranges achieved depend upon the case, the mode being generated, and the sense of polarization. Tuning ranges of up to TR = 18% were reported for bias values of up to 2,400 Oe [10]. There was, however, an accompanying deterioration in the impedance match and fluctuations in gain and axial ratio with applied magnetic bias strength.

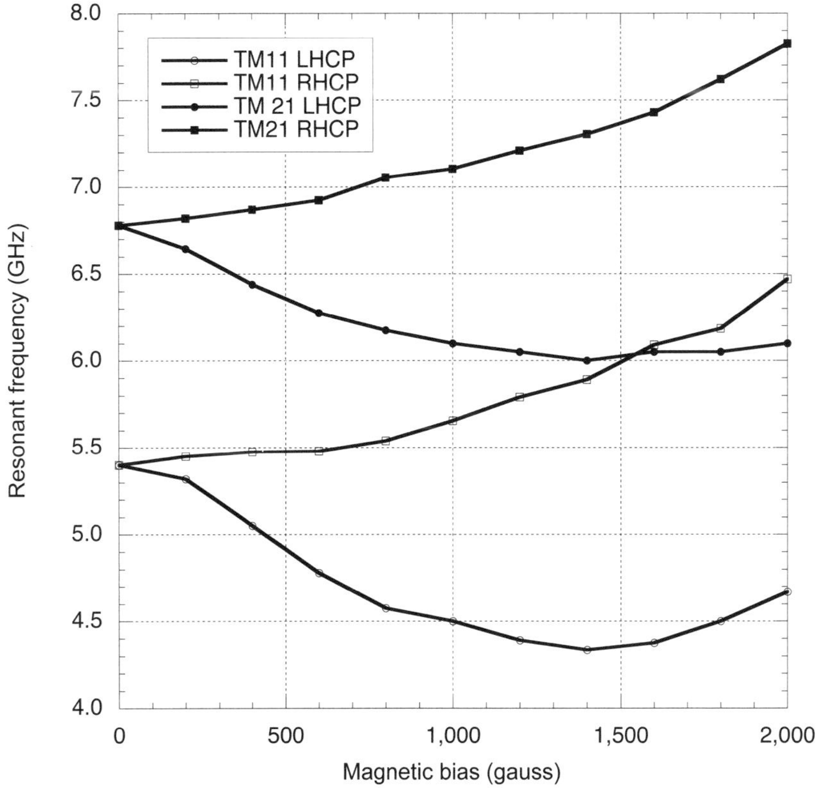

Figure 6.5 Resonant frequency versus magnetic bias for an annular microstrip patch antenna on a ferrite substrate. (From [9].)

A few prototype microstrip antennas have also been fabricated using thin-film ferrites. Rainville and Harackiewicz designed a microstrip antenna printed on a 0.5-mm gadolinium gallium garnet (GGG) sandwiched by 75-micron-thick gallium-doped YIG ferrite films with $4\pi M_s = 1{,}250$G [11]. These thin ferrite films are single-crystal in nature (as opposed to the polycrystalline structure of the previously used bulk ferrites) and are characterized by much narrower line widths on the order of a few oersteds. A square microstrip patch antenna with $L = W = 11$ mm was tuned from 5.75 to 6.03 GHz (TR = 4.8%) when a bias of up to 600 Oe was applied.

More recently, a microstrip patch antenna ($L = 22.2$ mm, $W = 30$ mm) printed on a 2-mm-thick alumina substrate coated with a 1-micron-thick layer of FeCoB ferrite film ($4\pi M_s = 16{,}000$G) has been reported [12]. A bias field of up to 50 Oe was applied both parallel and perpendicular to the resonant length L, resulting in small shifts in frequency, from 2,132 to 2,120 MHz (TR = 0.5%) for the parallel bias and up to 2,139 MHz for the perpendicular bias

(TR = 0.3%). These limited tuning ranges are a result of the small effect that such thin films can have on the resonance frequency compared to thicker bulk ferrites.

As was noted in [6], the relatively high dielectric constant of the ferrite substrates leads to increased power in the surface waves and thus a lowering of the radiation efficiency. To overcome some of these losses, an approach was proposed to use a multilayer structure consisting of sandwiching the ferrite substrate with low dielectric constant substrates, as shown in Figure 6.6 [13]. The prototype antenna consisted of a square patch printed on a combination of ferrite and dielectric multilayer substrates and fed by a microstrip line through a thin rectangular slot aperture in the ground plane. When a magnetic bias of up to H_o = 1,000 Oe was applied, the resonant frequency shifted from 5.5 to 6.6 GHz (TR = 18%). Although this is a more modest tuning range than what was achieved with single bulk ferrite substrates, the gain of the antenna was measured to be 8.6 dBi, which is considerably higher than for the bulk ferrite designs. A trade-off is therefore possible between the achievable tuning range and the radiation efficiency.

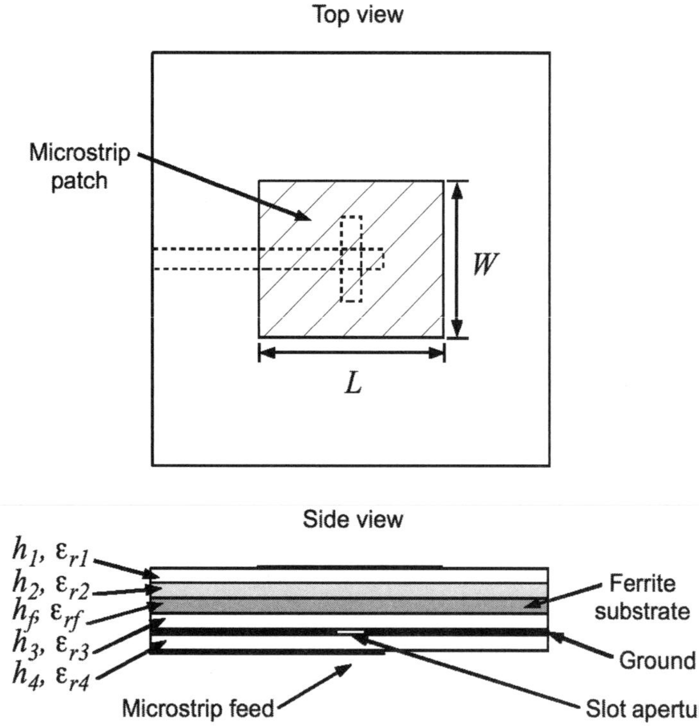

Figure 6.6 Rectangular microstrip patch antenna on a multilayered substrate (L = W = 15.2 mm, h_1 = 0.254 mm, ε_{r1} = 10.5, h_2 = 0.01 mm, ε_{r2} = 1.0, h_f = 0.635 mm, ε_{rf} = 15.3, h_3 = 1.15 mm, ε_{r3} = 1.07, h_4 = 0.635 mm, and ε_{r4} = 10). (After [13].)

The results of various published microstrip antenna designs on ferrite substrates are summarized in Table 6.1. The patches are predominately rectangular in shape, with the rest being annular. Tuning ranges vary from TR = 0.5% to 133% for maximum magnetic bias strengths varying from 100 to 6,250 Oe.

Table 6.1

Summary of Microstrip Antennas Printed on Ferrite Substrates

Patch Dimensions		Ferrite			Maximum Bias	Unbiased Frequency	Tuning Range	Reference
Rectangular		$4\pi M_s$	ε_r	h	H_o	f_o	TR (%)	
L (mm)	W (mm)	(G)		(mm)	(Oe)	(GHz)		
14	18	1,720	15	1.27	N/A	4.6	48.6	[6]
18	14	1,700	14	1.25	31	4.76	38	[14]
11	11	1,250	13	0.075 0.5 0.075	600	5.75	4.8	[11]
9.5	9.5	985	11.4	0.508	3,500	5.5	74	[15]
40	11.1	3,000	14.8	2.0	6,250	0.2	133	[8]
15.2	15.2	1,830	15.3	0.635	1,400	5.5	20	[13]
22.2	30	16,000	NA	0.001	100	2.13	0.5	[12]
Annular		$4\pi M_s$	ε_r	h	H_o	f_o	TR (%)	
r_o (mm)	r_i (mm)	(G)		(mm)	(Oe)	(GHz)		
6.5	3.0	1,730	14.8	1.27	1,800	5.4	15	[9]
10.1	0	650	N/A	N/A	1,000	3.0	28	[16]
7.4	1.9	348	13.7	1.0	1,800	4.5	18	[10]

6.1.2 Ferrite-Filled Cavity-Backed Slot Antenna

Work in the mid 1950s carried out on ferrite-filled waveguide cavities examined the use of magnetic bias for steering the beam peak of the radiation pattern [17, 18]. In 1967, an in-depth analysis of the radiation performance of ferrite- and dielectric-filled rectangular waveguide cavities was carried out [19]. As part of that study, the effects of an applied magnetic bias on a ferrite-filled cavity were measured. Although the details are somewhat sketchy, it appears that a probe-fed rectangular cavity of length a = 127 mm, height b = 63.5 mm, and depth c = 50.8 mm was filled with a ferrite material and that a magnetic bias of up to about 1,500 Oe was applied parallel to the b direction. The resonant frequency of the cavity shifted from 243 to 320 MHz (TR = 27%) with the input RL remaining above 10 dB at the resonance frequency over the tuning range. There was little mention of the effect on the radiation patterns.

More recently, investigations on cavity-backed slots, flush-mounted on an infinite ground plane and partially filled with magnitized ferrites, have been carried out [20–22]. Studies focused on probe-fed cavities, mainly rectangular, and partially filled with a ferrite slab, an example of which is shown in Figure 6.7. For this case, the cavity was excited in the TE_{10} mode and was magnetized in the direction parallel to the dominant electric fields. A prototype antenna was fabricated with dimensions given in Figure 6.7 and magnetized using one or two pairs of permanent magnets. The resonant frequency shifted from 0.735 GHz with 293 Oe bias (one magnet pair) to 0.84 GHz with 444 Oe bias (two magnet pairs). The estimated tuning range for the prototype was 0.7–1.1 GHz (TR = 21%). The radiation efficiency was determined to be between 60% and 90% for this antenna with gains of between 2.9 and 3.4 dBi.

6.1.3 Ferrite Resonator Antenna

Cylindrical and rectangular DRAs are examined in Section 3.8. They radiate patterns similar to short magnetic or electric dipoles, depending on the excited mode, and offer increased design flexibility over similar low-gain printed antennas

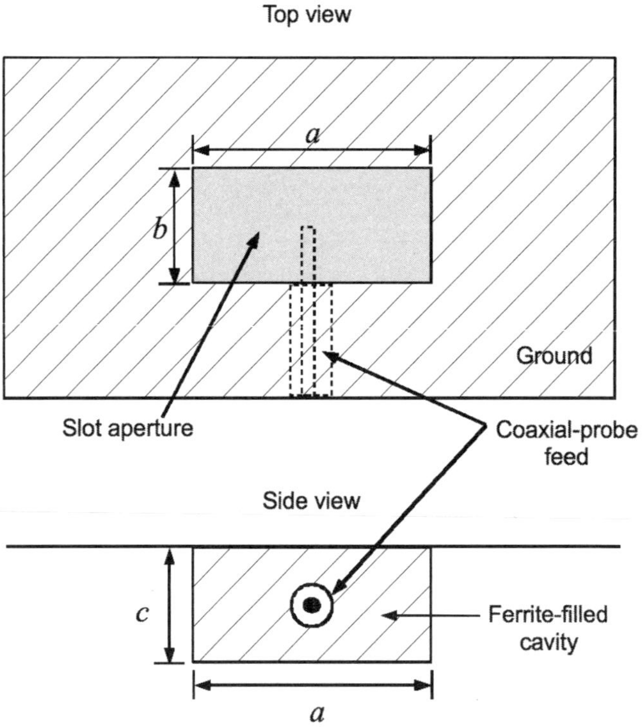

Figure 6.7 Cavity-backed rectangular slot antenna partially filled with ferrite (cavity: $a = b = c = 50.8$ mm; ferrite: 50.8 x 50.8 x 25.52 mm, $4\pi M_s = 800$G, and $\varepsilon_r = 13.9$). (After [20].)

such as microstrip patch antennas and printed dipoles. By replacing the dielectric material with ferrite, it was demonstrated that frequency agility could be achieved [23]. Figure 6.8 shows the geometry of the aperture-fed rectangular ferrite resonator antenna used to demonstrate frequency agility. By orienting a permanent magnet beneath the ground plane so that the poles were aligned parallel to the x-axis, the resonant frequency was tuned from 10.5 to 9.35 GHz (TR = 12%). Aligning the poles of the magnet parallel to the y-axis caused a shift upward in frequency from 10.5 to 11.45 GHz (TR = 9%). This behavior is shown in Figure 6.9. In both cases the maximum bias fields (measured in the location of the ferrite resonator antenna when the antenna was removed) were 1,250 Oe. As with the microstrip antennas on biased ferrites, there was an increase in the cross-polarization levels of the ferrite resonator antenna when the y-directed magnetic bias was applied. (This phenomenon was later used to design a polarization-agile ferrite resonator antenna [24].) However, since DRAs do not exhibit surface wave losses, their radiation efficiency remained high (> 97%).

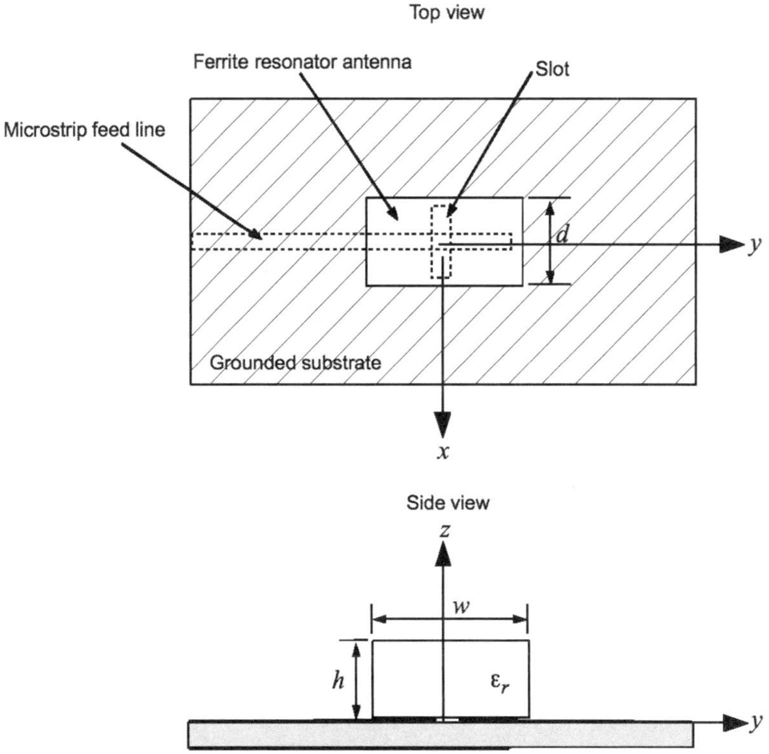

Figure 6.8 Aperture-fed ferrite resonator antenna (w = 10 mm, h = 2.5 mm, d = 5 mm, $4\pi M_s$ = 1,850G, and ε_r = 14.8). (After [23].)

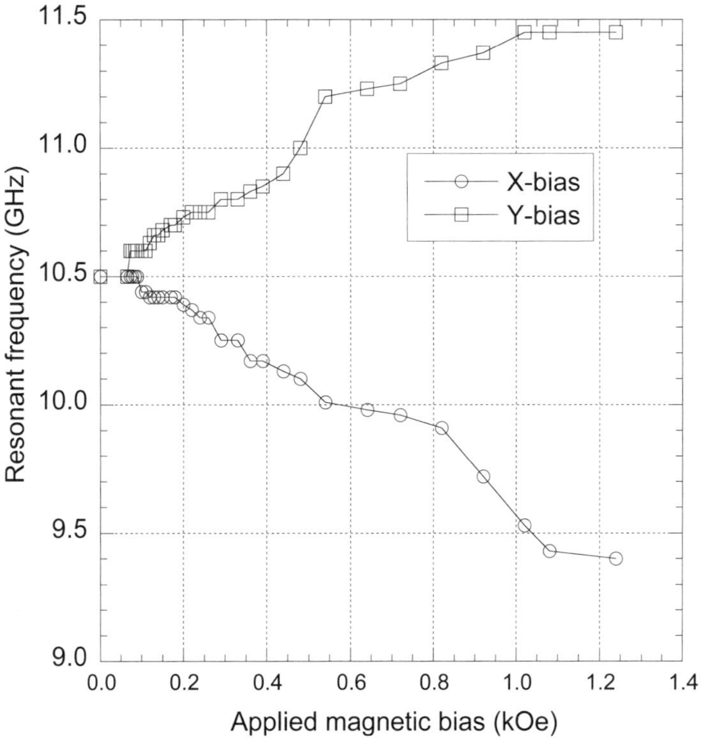

Figure 6.9 Resonant frequency versus applied magnetic bias of a ferrite resonator antenna (w = 10 mm, h = 2.5 mm, d = 5 mm, $4\pi M_s$ = 1,850G, and ε_r = 14.8). (After [23].)

It was shown in [25] that the maximum tuning range could be reasonably predicted by modifying the equations for the resonant frequency of DRAs [(3.90) in Section 3.8.2] by including the effective permeability:

$$k_x \tan(k_x d/2) = \sqrt{\mu_{eff}(\varepsilon_r - 1)k_o^2 - k_x^2} \tag{6.2}$$

where

$$k_x = \sqrt{\mu_{eff}\varepsilon_r k_o^2 - k_y^2 - k_z^2} \tag{6.3}$$

$$\mu_{eff} = \mu \pm \kappa \tag{6.4}$$

and the other variables remain the same as in (3.90). The "±" for μ_{eff} depends on the orientation of the applied bias magnetic field. The estimates for the maximum tuning range were determined by assuming that the ferrite was saturated and by substituting the values of μ and κ found from applying (4.16). The estimated maximum tuning ranges based on (6.2) to (6.4) for different frequencies and for different values of remnant magnetization were tabulated in [25] for rectangular ferrite resonator antennas of a given aspect ratio ($w = 2d = 4h$). Since, as seen from (4.16), the value of k is inversely proportional to frequency, the range of values of μ_{eff} (and hence the tuning range) also decreases with frequency. For the cases examined in [25], predicted values of tuning range varied from TR = 86% at 2 GHz down to TR = 45% at 20 GHz, for a ferrite with a remnant magnetization of $4\pi M_r$ = 4,000G. By considering ferrite resonators with other aspect ratios, wider tuning ranges could potentially be achieved.

6.2 FAAS DESIGNED USING FERROELECTRICS

As described in Section 4.2.2, ferroelectric materials (such as BST) exhibit a change in dielectric constant when a DC electric field is applied. Some research into the use of ferroelectrics for either beam-steering applications or frequency-tunable antennas was carried out in the late 1990s and early 2000s. Two configurations involving the integration of ferroelectrics in antennas for beam steering applications have been published. The first design consisted of an X-Band parallel-plate lens filled with layers of ferroelectric material [26]. Application of an increasing DC electric field bias between plates caused a progressive lowering of the dielectric constant of the ferroelectric layer within each pair of plates, leading to a linear phase shift progression that resulted in beam steering. Measurements carried out on a single pair of plates demonstrated that an adequate phase-shifting range could be achieved with the application of a DC voltage bias of up to 13.5 kV. The accompanying insertion loss was on the order of 2 to 2.5 dB. The second approach involved the use of ferroelectrics in CTS technology for achieving beam steering [27]. Simulations indicated that beam scanning of ±60° was possible with the proposed configuration. Again high-bias voltages would be required and the losses would be high.

The performance of a frequency-tunable microstrip antenna printed on a ferroelectric (BST) substrate (with an unbiased dielectric constant of $\varepsilon_r = 28$), as shown in Figure 6.10, was investigated in 1999 [28]. Although the authors estimated that a tuning range of up to 50% was theoretically possible with such a design, they were only able to achieve a shift from 850 to 825 MHz (TR = 4%) with a bias of 1.5 kV (6.38 kV/cm), as shown in Figure 6.11. The limitation of their design was in the DC blocking circuit, which had an operating limit of 2.0 kV. In order to achieve the full tuning range, a bias field of 20 kV/cm was needed, requiring a bias of 4.7 kV. (The radiation efficiency of the antenna

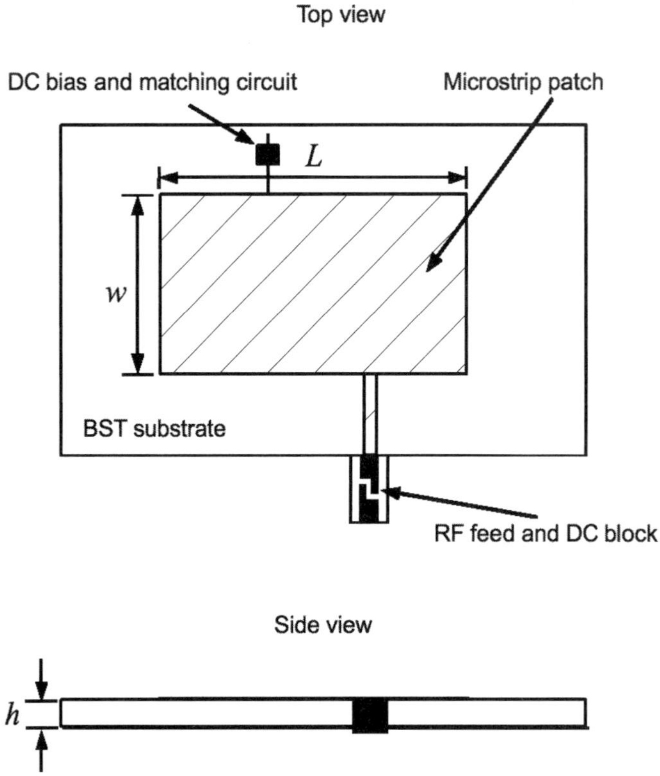

Figure 6.10 Microstrip antenna on ferroelectric substrate ($L = 46$ mm, $w = 33.5$ mm, and $h = 2.35$ mm). (After [28].)

was not reported.) This example illustrates the main limitation to the use of ferroelectrics: the high DC electric field strength required to alter the dielectric constant, which, for thick substrates, translates to the need for high values of bias voltage.

One potential solution to this problem is to use thin-film ferroelectrics, where the field strength can be achieved using much lower values of DC voltage. This approach was used in the development of thin-film ferroelectric tunable microwave components such as varactors, tunable ring resonators, phase shifters, and tunable filters, where the required bias voltages were on the order of a few hundred volts [29, 30].

The performance of a Ka-band coplanar folded slot antenna incorporating a thin film of ferroelectric material, as shown in Figure 6.12, was investigated in [31]. The BST ferroelectric had a dielectric constant varying from $400 < \varepsilon_r < 1,200$. Simulations indicated a shift in the resonant frequency from

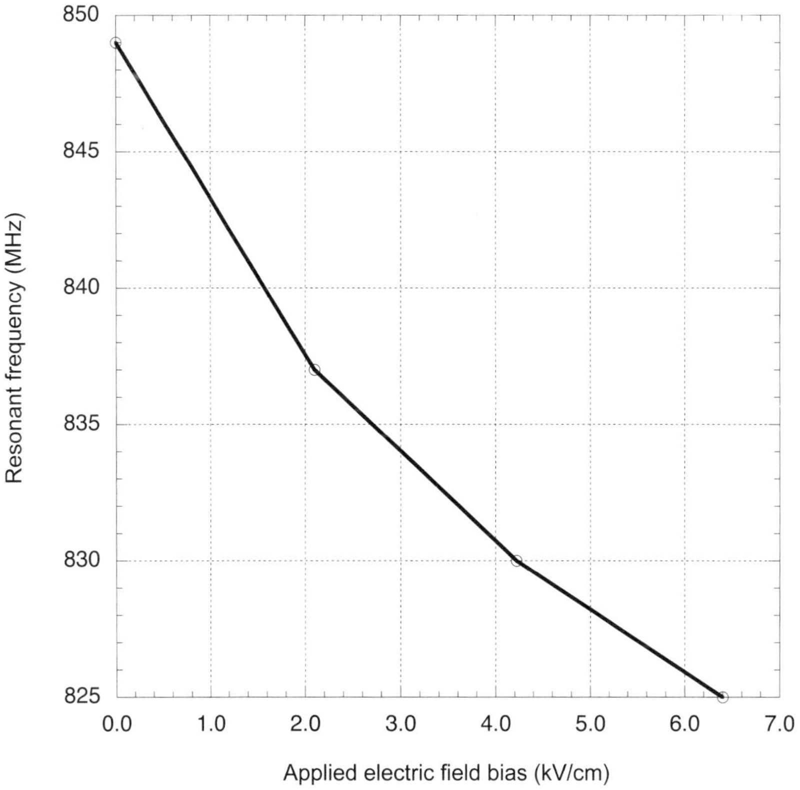

Figure 6.11 Microstrip antenna on ferroelectric substrate (L = 46 mm, w = 33.5 mm, and h = 2.35 mm). (From [28].)

27.46 to 25.65 GHz (TR = 6.8%) with an average gain of 3.77 dBi. The required bias voltage was not mentioned but should be similar to the values used for the thin-film ferroelectric microwave components.

A final example of a ferroelectric-based antenna is shown in Figure 6.13, which depicts a stacked microstrip patch antenna [32]. The lower microstrip patch is supported on a regular dielectric substrate while the upper patch is placed on a ferroelectric substrate. The RF signal was fed directly to the lower patch while the upper patch was electromagnetically coupled. By biasing the upper patch with a DC voltage, the resonant frequency of the antenna could be tuned. This design offers the advantage that no DC blocking capacitors are required, since there is no DC contact between upper and lower patches. Although the antenna dimensions, the amount of bias voltage strength, and the radiation efficiency were not disclosed in [32], it was stated that this dual-band antenna could be tuned between European (890–960 MHz) and North

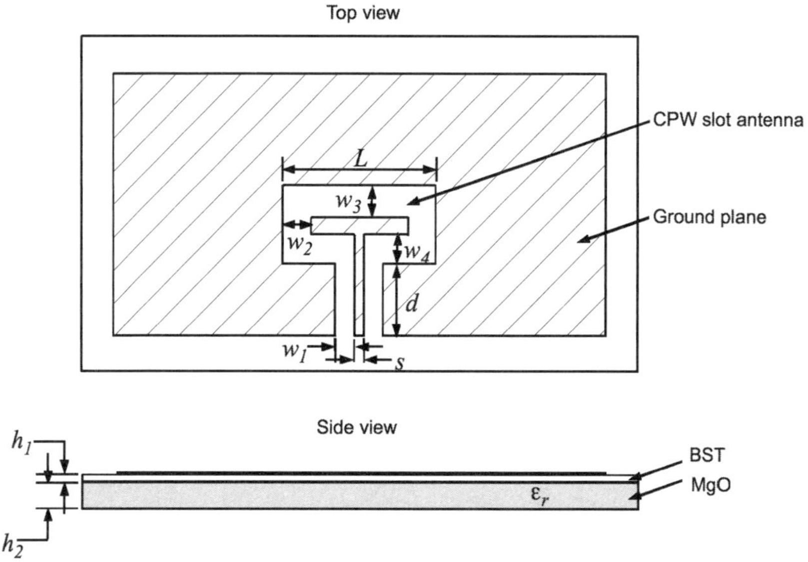

Figure 6.12 CPW folded slot antenna on thin ferroelectric substrate ($L = 2$ mm, $s = d = 0.15$ mm, $w_1 = w_2 = w_3 = 0.25$ mm, $w_4 = 0.16$ mm, $h_1 = 0.4$ μm, $h_2 = 0.305$ mm, and $\varepsilon_r = 9.8$). (After [31].)

American (824–894 MHz) Global System for Mobile Communications (GSM) bands and digital communications systems (DCS1800) bands (1,710–1,785 MHz and 1,805–1,880 MHz).

Since the early 2000s, research on the use of ferroelectrics for FAAs appears to have halted. The high-bias voltages required, the high value of dielectric constant, and the relatively high losses are current barriers preventing the use of ferroelectrics for practical designs of FAAs. Advances in the material science of ferroelectrics will be required before they can be considered viable candidates for FAAs. In the meantime, some of these barriers can be overcome by using liquid crystals, which will be considered next.

6.3 LIQUID CRYSTAL-BASED FAAS

As with ferroelectrics, liquid crystals share the property that their dielectric constant at microwave frequencies can be controlled by the application of a DC (or low-frequency AC) electric field, as described in Section 4.2.3. The advantages of liquid crystals over ferroelectrics include the much lower bias voltages required and the inherently low values of the dielectric constant. Starting in the early 1990s,

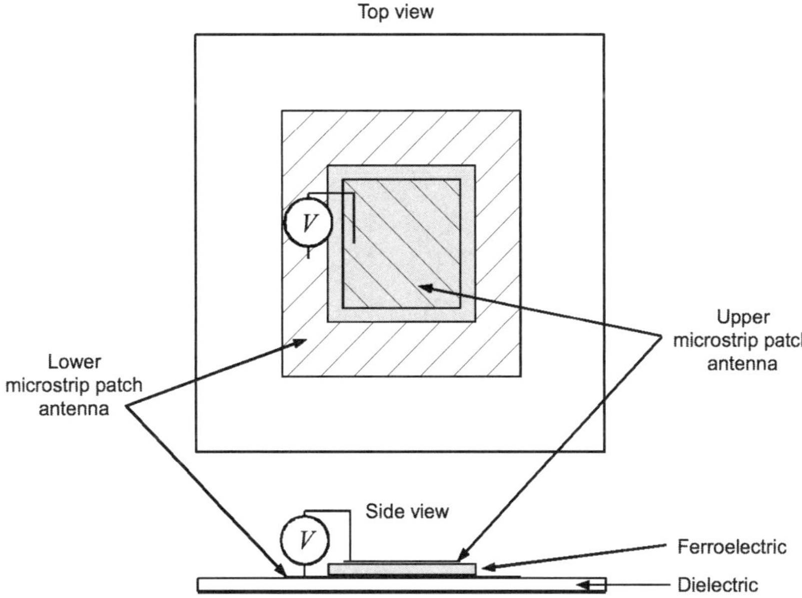

Figure 6.13 Stacked microstrip patch antenna with thin ferroelectric substrate. (After [32].)

the use of liquid crystals has been investigated at various frequencies in the microwave and millimeter-wave bands to implement tunable phase shifters, tunable capacitors, and beam steering lenses [33–43]. Research on tunable antennas using liquid crystals began in the early 2000s and has been solely focused on developing a tunable microstrip antenna. Figure 6.14 shows an example of a rectangular microstrip antenna where the substrate has partially been replaced with liquid crystal (Merck K15, with $2.7 < \varepsilon_r < 2.9$) [44]. By applying a bias voltage of up to 20V, the frequency of the antenna was shifted from 4.74 down to 4.6 GHz (TR = 3%). Details of the biasing network or radiation efficiency were not provided.

A second example is shown in Figure 6.15, where the liquid crystal (Merck E7), with a dielectric constant ranging from $2.72 < \varepsilon_r < 3.17$) is present underneath the entire microstrip antenna [45]. A bias of 0–10V produced a frequency shift from 5.44 to 5.66 GHz (TR = 4%), similar in magnitude to the tuning range in the first example. The radiation efficiency of the antenna varied from 20% to 40%, depending on the bias voltage.

Recent research has focused on the use of tunable microstrip antennas using liquid crystals as elements in reflectarrays to achieve beam steering [46–51].

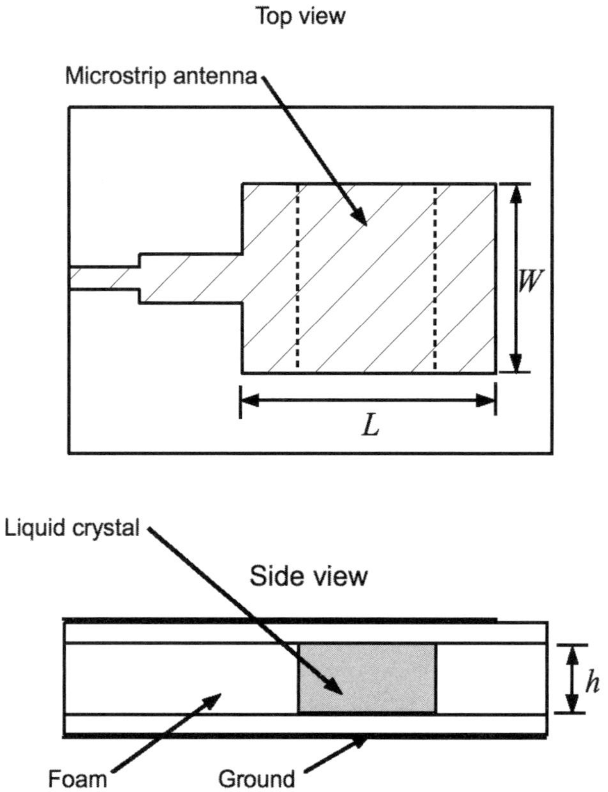

Figure 6.14 Microstrip antenna with liquid crystal (L = 20.7 mm, W = 20 mm, and h = 0.5 mm). (After [44].)

Designs have been carried out at 9 GHz [46, 49], 35 GHz [47], and 77 GHz [48, 51]. Although the results show that beam steering is possible with low-bias voltages, the high dielectric losses in the liquid crystal resulted in radiation efficiencies of only between 10% and 20%. Another disadvantage of liquid crystals is the narrow range of temperatures over which their tunable dielectric properties can be maintained (about 20–35°C), which precludes their use in most outdoor applications unless temperature control systems are included.

6.4 BULK SEMICONDUCTOR-BASED FAAs

The properties of bulk semiconductors are reviewed in Section 4.2.4. The change in conductivity as a function of light intensity is the property that will be considered for use in FAAs. Such a change in value of conductivity of semiconductors can be used to potentially design antennas whose frequencies are

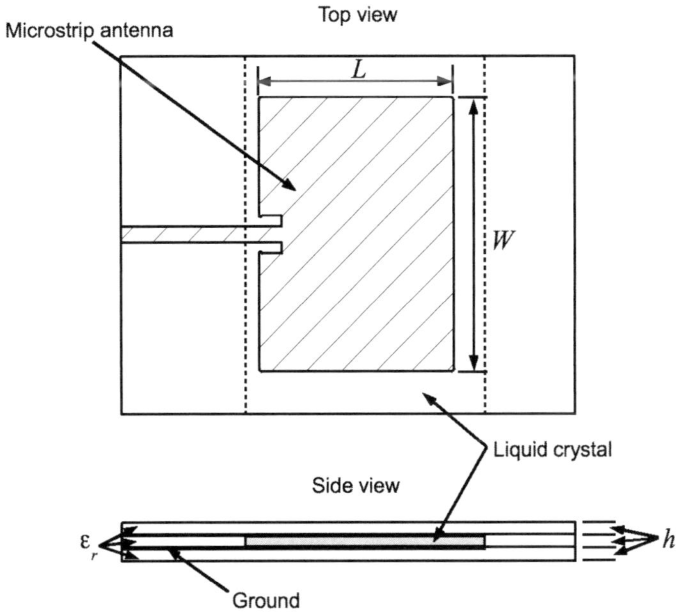

Figure 6.15 Microstrip antenna with liquid crystal (L = 15 mm, W = 20 mm, h = 0.51 mm, and ε_r = 2.2). (After [45].)

optically tunable. An example is shown in Figure 6.16, where a bowtie antenna element was fabricated using a silicon semiconductor substrate with a bulk resistivity of 7,000Ω cm and a thickness of 0.635 mm [52, 53]. Light was shone onto the dipole from a laser output fiber through a thin quartz slab waveguide (labeled light tank delivery system in Figure 6.16), located beneath the bowtie. The laser light propagating in the light tank waveguide leaked out of the waveguide into the bowtie, due to the higher refractive index of the silicon. By shining laser light through optical fibers into the cavity beneath the silicon, the semiconductor becomes conductive. The photoconductance of the semiconductor could thus be increased as a function of the optical power. For the prototype in Figure 6.16, by increasing the applied optical power intensity from 0 to 1 W/cm^2, the relative gain of the silicon bowtie increased from -25 to -5 dB compared to a conventional copper bowtie. This approach could be used to selectively illuminate increasing areas of the semiconductor, altering the size of the antenna to tune its frequency. Due to the relatively low conductivity achieved, the radiation efficiency of the silicon bowtie was quite poor, and the use of bulk semiconductors for FAAs appears quite limited. A more promising approach is to use optically controlled silicon-based devices (PINs or varactors) for designing FAAs. Examples of such antennas will be shown in Chapter 8.

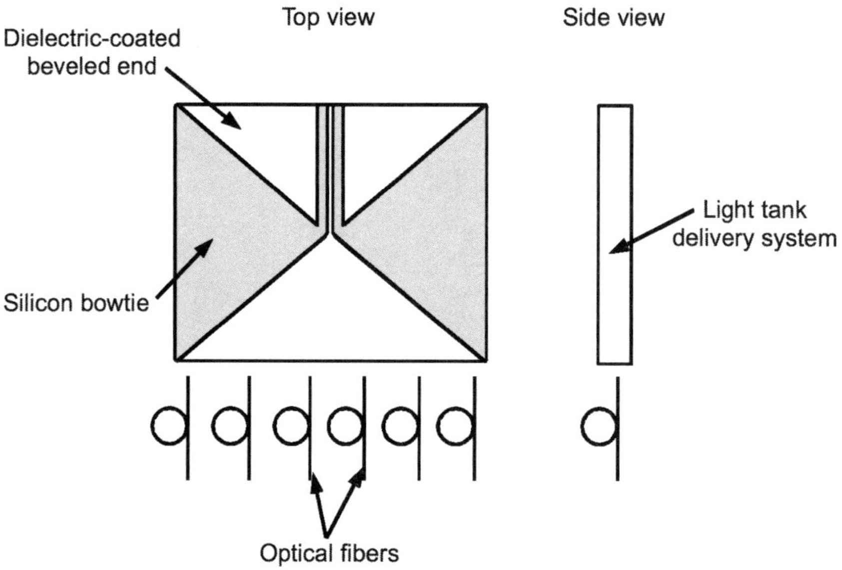

Figure 6.16 Silicon bowtie antenna. (After [52, 53].)

6.5 THIN-FILM SEMICONDUCTOR-BASED FAAs

The final example of an FAA based on a tunable substrate consists of microstrip patch antennas fabricated using an organic semiconductor polymer [54]. As the name implies, organic semiconductors are organic-based materials that exhibit properties similar to those of semiconductors. The material used in this example, P3HT (3-hexylthiophene), undergoes a change in dielectric constant with a change in the optical illumination power density. Figure 6.17 shows two microstrip patch antennas fabricated from a thin layer of P3HT instead of the conventional copper. An ordinary white light projector with a tunable light intensity from 0.5 to10 W/m^2 was used to illuminate the antennas. The shifts in resonant frequency of the square microstrip patch antenna and the square ring microstrip antenna are plotted in Figure 6.18. The resonant frequency of the square microstrip patch antenna shifted from 1.85 to 2.8 GHz (TR = 40.9%) while for the square ring microstrip antenna, the shift in resonant frequency was from 1.1 to 2.6 GHz (TR = 81.1%). The maximum shift in resonant frequency was achieved at an illumination intensity of approximately 5.5 W/m^2 for both antennas. Further increases in illumination intensity did not result in any further changes in resonant frequency.

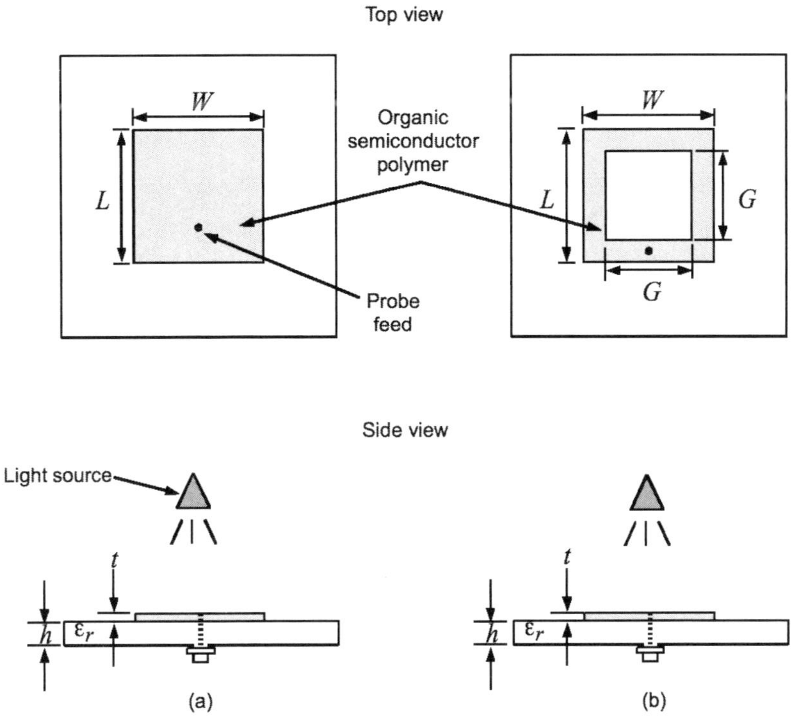

Figure 6.17 Organic semiconductor polymer-based microstrip antennas: (a) square-patch microstrip antenna and (b) square ring microstrip antenna ($L = W = 50$ mm, $G = 30$ mm, $h = 1.8$ mm, $\varepsilon_r = 2.4$, and $t = 0.1$ mm). (After [54].)

The average measured gains for the square patch and square ring antennas were 3.6 and 3.1 dBi, respectively, corresponding to radiation efficiencies of 51% and 42%. These results were compared to a conventional copper-based microstrip patch antenna designed at 2.1 GHz, which demonstrated a gain of 7.2 dBi and corresponding radiation efficiency of 97%. The thickness of the P3HT was determined to be only half of the skin depth, thus increasing the thickness could help improve radiation efficiency in future designs.

6.6 SUMMARY

This chapter considers the use of tunable substrates for the design of various FAAs. To date, FAAs based on ferrite substrates appear to offer the best performance among the tunable substrates. Microwave ferrite materials are a

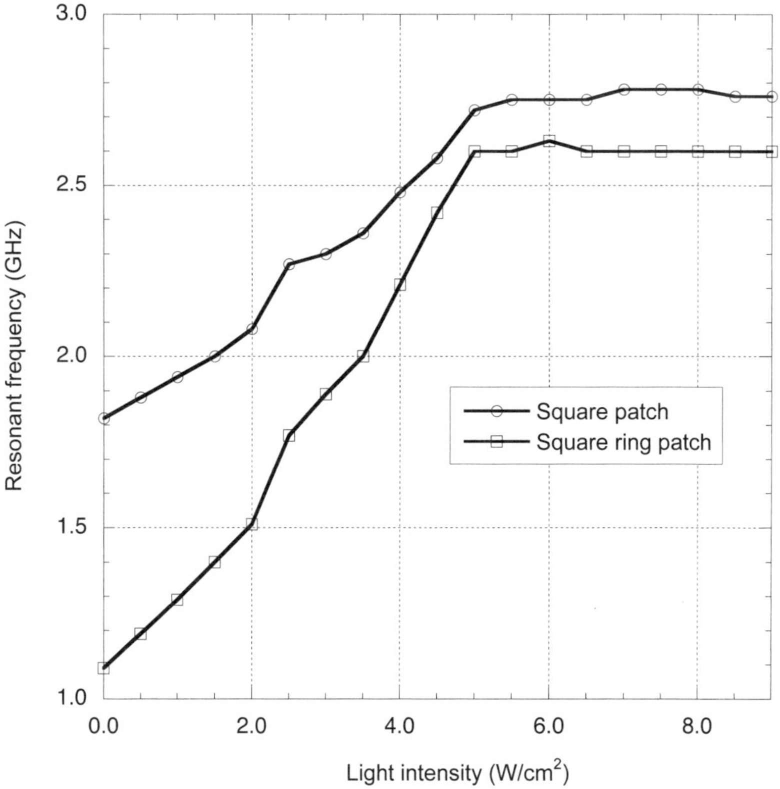

Figure 6.18 Resonant frequency of the organic semiconductor polymer-based microstrip antennas versus light intensity. (From [54].)

mature technology and low-loss substrates are readily available commercially. The ferrite-based FAAs offer the widest tuning range (up to over 100%) and highest radiation efficiencies. The main drawback of the ferrites is the need for a DC H-field bias. This requires either an electromagnetic, which draws constant current, or a permanent magnetic, which must be physically moved to alter the bias field strength.

The FAAs based on ferroelectrics are biased with an E-field, which is much simpler than for ferrites and draws no current. Unfortunately, ferroelectrics have fairly high losses and require high-bias fields in order to have a sufficient change in dielectric constant. Ferroelectric-based FAAs have achieved maximum tuning ranges of less than 10%. The high dielectric constant of ferroelectrics is also not an attractive property for many antenna designs, leading to narrow bandwidths or increased surface waves for certain microstrip antenna designs.

An electric field is also used to alter the dielectric constant of liquid crystals at microwave frequencies. These materials are more attractive than ferroelectrics in that their dielectric constant is quite low, and they require significantly lower levels of bias field intensity to alter the dielectric constant. The drawbacks of the liquid crystal designs are their limited tuning range (up to 4%), the relatively high losses at lower microwave frequencies, and their high temperature sensitivity.

To date, bulk semiconductors seem to have been the least used in the design of FAAs, due to the relatively poor radiation efficiency and the relatively high DC power consumption required. However, recent developments in organic semiconductors appear promising and may prove to be viable candidates as tuning mechanisms for FAAs.

References

[1] Das, N., and J. S. Chatterjee, "Microstrip Antenna on a Ferrimagnetic Substrate," *Electronics Letters,* Vol. 17, No. 13, June 1981, pp. 441-442.

[2] Das, N., and S. K. Chowdhury, "Microstrip Rectangular Resonators on Ferrimagnetic Substrates," *Electronics Letters,* Vol. 16, No. 21, October 1980, pp. 817-818.

[3] Das, S. N., and S. K. Chowdhury, "Rectangular Microstrip Antenna on Ferrite Substrate," *IEEE Transactions on Antennas and Propagation,* Vol. 30, No. 1, May 1982, pp. 499-502.

[4] Das, N., S. K. Chowdhury, and J. S. Chatterjee, "Circular Microstrip Antenna on Ferrimagnetic Substrate," *IEEE Transactions on Antennas and Propagation,* Vol. 31, No. 1, January 1983, pp. 188-190.

[5] Araki, K., D. I. Kim, and Y. Naito, "A Study on Circular Disk Resonators on a Ferrite Substrate," *IEEE Transactions on Microwave Theory and Techniques,* Vo. 30, No. 2, February, 1982, pp. 147-154.

[6] Pozar, D. M., and V. Sanchez, "Magnetic Tuning of a Microstrip Antenna on a Ferrite Substrate," *Electronics Letters,* Vol. 24, June, 1988, pp. 729-731.

[7] Pozar, D. M., "Radiation and Scattering Characteristics of Microstrip Antennas on Normally Biased Ferrite Substrate, *IEEE Transaction on Antennas and Propagation,* Vol. 40, No. 9, September 1992, pp. 1084-1092.

[8] Mishra, R. K., S. S. Pattnaik, and N. Das, "Tuning of Microstrip Antenna on Ferrite Substrate," *IEEE Transaction on Antennas and Propagation,* Vol. 41 No. 2, February 1993, 230-233.

[9] Tsang, K. K., and R. J. Langley, "Annular Ring Microstrip Antennas on Biased Ferrite Substrates," *Electronics Letters,* Vol. 30, No. 16, August 1994, pp. 1257-1258.

[10] Tsang, K. K., and R. J. Langley, "Design of Circular Patch Antennas on Ferrite Substrates," *IEE Proceeding on Microwaves, Antennas and Propagation,* Vol. 145, No. 1, February 1998, pp. 49-55.

[11] Rainville, P. J., and F. J. Harackewiez, "Magnetic Tuning of a Microstrip Patch Antenna Fabricated on a Ferrite Film," *IEEE Microwave and Guided Wave Letters,* Vol. 2, No. 12, December 1992, pp. 483-485.

[12] Sun, N. X, et al., "Electronically Tunable Magnetic Patch Antennas with Metal Magnetic Films," *Electronics Letters,* Vol. 43, No. 8, April 2007, pp. 434-435.

[13] Leon, G., et al., "Characteristics of Aperture Coupled Microstrip Antennas on Magnetized Ferrite Substrates," *IEEE Transactions on Antennas and Propagation,* Vol. 53, No. 6, June 2005, pp. 1957-1966.

[14] How, H., et al., "Radiation Frequencies of Ferrite Patch Antennas," *Electronics Letters,* Vol. 28, No. 15, July 1992, pp. 1405-1406.

[15] How, H., T. M. Fang, and C. Vittoria, "Intrinsic Modes of Radiation in Ferrite Patch Antennas," *IEEE Transactions on Microwave Theory and Techniques,* Vol. 42, No. 6, June, 1994, pp. 988-994.

[16] Batchelor, J. C., G. Classen, and R. Langley, "Microstrip Antennas on Ferrites," *Proc. International Conference on Antennas and Propagation ICAP 97*, Edinburgh, Scotland, April 14-17, 1997, Vol. 1, pp. 30-33.

[17] Angelakos, D. J., and M. M. Korman, "Radiation from Ferrite-Filled Apertures," *Proceedings of the IEEE,* Vol. 44, No. 10, October, 1956, pp. 1463-1468.

[18] Tyras, G., and G. Held, "Radiation from a Rectangular Waveguide Filled with Ferrite," *IRE Transactions on Microwave Theory and Techniques,* Vol. 6, No. 3, July 1958, pp. 268-277.

[19] Adams, A. T, "Flush Mounted Rectangular Cavity Slot Antennas – Theory and Design," *IEEE Transactions on Antennas and Propagation,* Vol. 15, No. 3, March, 1967, pp. 342-351.

[20] Polycarpou, A. C., et al., "Radiation and Scattering from Ferrite-Tuned Cavity-Backed Slot Antennas: Theory and Experiment," *IEEE Transactions on Antennas and Propagation,* Vol. 46, No. 9, September 1998, pp. 1297-1306.

[21] Vouvakis, M. N., et al., "Ferrite-Loaded Cavity-Backed Antennas Including Nonuniform and Nonlinear Magnetization Effects," *IEEE Transactions on Antennas and Propagation,* Vol. 51, No. 5, May 2003, pp. 1000-1010.

[22] Kononov, V. G., et al., "Non-Uniform Field Modeling of Ferrite-Loaded Cavity-Backed Slot Antennas," *IEEE Tranasctions on Antennas and Propagation,* Vol. 57, No. 10, October 2010, pp. 3402-3405.

[23] Petosa, A., et al., "Magnetically Tunable Ferrite Resonator Antenna," *Electronics Letters,* Vol. 30, No. 13, June 1994, pp. 1021-1022.

[24] Petosa, A., et al., "Switchable LP/CP Ferrite Disk Resonator Antenna," *Electronics Letters,* Vol. 31, No. 3, February, 1995, pp. 163-164.

[25] Petosa, A., *Dielectric Resonator Antenna Handbook*, Norwood, MA: Artech House, 2007.

[26] Rao, J. B. K., D. P. Patel, and V. Krichevsky, "Voltage-Controlled Ferroelectric Lens Phased Arrays," *IEEE Tranasactions on Antennas and Propagation,* Vol. 47, No. 3, March, 1999, pp. 458-468.

[27] Iskander, M. F., et al., "Design of a Low-Cost 2-D Beam-Steering Antenna Using Ferroelectric Material and the CTS Technologoy, *IEEE Transactions on Microwave Theory and Techniques,* Vol. 49, No. 5, May 2001, pp. 1000-1003.

[28] Jose, K. A., V. K. Varadan, and V. V. Varadan, "Experimental Investigations on Electronically Tunable Microstrip Antennas," *Microwave and Optical Technology Letters,* Vol. 20, No. 3, February 1999, pp. 166-169.

[29] Jiang, H., et al., "Frequency Agile Microstrip Patch Antenna Using Ferroelectric Thin Film Varactor Technology," *Proc. IEEE International Symposium on Antennas and Propagation,* Charleston, SC, June 1-5, 2009.

[30] Van Keuls, F. W., et al., "Current Status of Thin Film (Ba,Sr)TiO3 Tunable Microwave Components for RF Communications," NASA/TM, 2000-210233, 2000.

[31] Castro-Vilaro, A. M., and R. A. Rodriquez Solis, "Tunable Folded-Slot Antenna with Thin Film Ferroelectric Material," *Proc. IEEE International Symposium on Antennas and Propagation,* Columbus, OH, June 22-27, 2003, pp. 549-553.

[32] Varadan, V. K., "Tunable Dual-Band Ferroelectric Antenna," US Patent 6,329,959 B1, December 11, 2001.

[33] Dolfi, D., et al., "Liquid Crystal Microwave Phase Shifter," *Electronics Letters,* Vol. 29, No. 10, May 1993, pp. 926-928.

[34] Lim, K. C., J. D. Margerum, and A. M. Lackner, "Liquid Crystal Millimeter Wave Electronic Phase Shifter," *Applied Physics Letters,* Vol. 62, No. 10, March 1993, 1065-1067.

[35] Yang, F., and J. R. Sambles, "Microwave Liquid Crystal Wavelength Selector," *Applied Physics Letters,* Vol. 79, No. 2, November 2001, pp. 3717-3719.

[36] Fujikake, H., et al., "Thick Polymer-Stabilized Liquid Crystal Films for Microwave Phase Control," *Journal of Applied Physics,* Vol. 89, No. 10, May 2001, pp. 5295-5298.

[37] Kuki, T., H. Fujikake, and T. Nomoto, "Microwave Variable Delay Line Using Dual-Frequency Switching-Mode Liquid Crystal," *IEEE Transactions on Microwave Theory and Techniques,* Vol. 50, No. 11, November 2002, pp. 2604-2609.

[38] Weil, C., G. Luessem, and R. Jakoby, "Tunable Inverted-Microstrip Phase Shifter Device Using Neumatic Liquid Crystals," *Proc. IEEE Microwave Theory and Techniques,* Seattle, Washington, June 2-7, 2002, p. 367-370.

[39] Tanaka, M., and S. Sato, "Focusing Properties of Liquid Crystal Lens Cells With Stack-Layered Structure in the Millimeter-Wave Region," *IEEE Microwave and Wireless Components Letters,* Vol. 12, No. 5, May 2002, pp. 163-165.

[40] Martin, N., et al., "Improvement of an Inverted Microstrip Line-Based Microwave Tunable Phase-Shifter Using Liquid Crystal," *Proc. 33rd European Microwave Conference,* Munich, Germany, October 7-9, 2003, pp. 1417-1420.

[41] Muller, S., et al., "Tunable Passive Phase Shifter for Microwave Applications Using Highly Anisotropic Liquid Crystals," *Proc. IEEE International Microwave Symposium,* Fort Worth, Texas, June 6-11, 2004, pp. 1153-1156.

[42] Yang, F., and J. R. Sambles, "Microwave Liquid-Crystal Variable Phase Grating," *Applied Physics Letters,* Vol. 85, No. 11, Sept. 2004, pp. 2041-2043.

[43] Yeh, J. A., et al., "Microwave Characteristics of Liquid-Crystal Tunable Capacitors," *IEEE Electron Device Letters,* Vol. 26, No. 7, July 2005, pp. 451-453.

[44] Martin, N., et al., "Patch Antenna Adjustable in Frequency Using Liquid Crystal," *Proc. 33rd European Microwave Conference,* Munich, Germany, October 7-9, 2003, pp. 699-702.

[45] Liu, L., and R. J. Langley, "Liquid Crystal Tunable Microstrip Patch Antenna," *Electronics Letters,* Vol. 44, No. 20, September, 2008, pp. 1179-1180.

[46] Hu, W., et al., "Tunable Liquid Crystal Reflectarray Patch Element," *Electronics Letters,* Vol. 42, No. 9, April 2006, pp. 509-511.

[47] Moessinger, A., et al., "Electronically Reconfigurable Reflectarrays with Nematic Liquid Crystals," *Electronics Letters,* Vol. 42, No. 16, August 2006, pp. 899-890.

[48] A. Gaebler, A., et al., "Liquid Crystal – Reconfigurable Antenna Concepts for Space Applications at Microwave and Millimetewaves," *Proc. European Conference on Antennas and Propagation EuCAP 2006,* Nice, France, November 6-10, 2006.

[49] Ismail, M. Y., et al., "Phase Agile Reflectarray Cells Based on Liquid Crystals," *IET*

Microwaves, Antennas and Propagation, Vol. 1, No. 4, 2007, pp. 809-814.

[50] Liu, L., and R. Langley, "Study on Tunable Electrically Small Antennas," *Proc. Loughborough Antennas and Propagation Conference LAPC,* Loughborough, UK, November 16-17, 2009, pp. 805-808.

[51] Marin, R., et al., "77 GHz Reconfigurable Reflectarray with Nematic Liquid Crystal," *Proc. European Conference on Antennas and Propagation, EuCAP 2007,* Edinburgh, Scotland, November 11-16, 2007.

[52] Liu, D., et al., "Structurally-Embedded Photoconductive Silicon Bowtie Antenna," *IEEE Photonics Technology Letters,* Vol. 10, No. 5, May 1998, pp. 716-718.

[53] VanBlaricum, M. L., "A Brief History of Photonic Antenna Reconfiguration," *Proc. International Topical Meeting on Microwave Photonics MWP,* Oxford, UK, September 11-13, 2000, pp. 9-12.

[54] Sathi, V., N. Ehteshami, and J. Nourinia, "Optically Tuned Frequency-Reconfigurable Microstrip Antenna," *IEEE Antennas and Wireless Letters,* Vol. 11, 2012, pp. 1018-1020.

Chapter 7

FAAs With Integrated Devices and Continuous Tuning

This chapter explores the use of integrated electronic devices to realize antennas with continuous frequency-tuning capabilities. The electronic devices are divided into four categories: diodes, transistors, MEMS, and varactors, the last of which forms the bulk of the examples. These electronic devices are integrated into various antenna elements including slot antennas, PIFAs, CPW antennas, DRAs, and a variety of printed antennas including loops, dipoles, patches, rings, monopoles, and helices. Representative examples of the various configurations are illustrated in this chapter to provide an indication of achievable performance. The TR and, where available, the TS[1] of these designs are provided, along with bias conditions and antenna performance.

7.1 DIODE-LOADED FAAs

This section examines two antenna elements, the rectangular microstrip patch antenna and the rectangular slot antenna, which are loaded with various types of diodes (Schottky, tunnel, or photodiodes) in order to achieve continuous frequency tuning.

---

1 A 10-dB return loss is used for determining TS, and for certain designs TS could be less than TR.

7.1.1 Microstrip Patch Antenna Loaded With a Schottky Diode

Figure 7.1 shows a prototype antenna consisting of a stacked rectangular microstrip patch antenna into which a Schottky diode was embedded [1]. The RF source to the antenna was provided by a transistor, connected in the common-base configuration, with the signal applied to the emitter. (The bias circuitry is not shown in Figure 7.1.) In this configuration, the transistor oscillates, and the antenna acts as the path for the feedback loop. The Schottky diode was embedded into the stacked patch at a position that was determined empirically to produce the widest tuning range. The DC bias voltage for the diode was applied via a pin located at the impedance null at the center of the patch, with the return path provided by the ground plane. By varying the bias voltage between 0 and 30V, the resonant frequency of the microstrip antenna was tuned from 2.19 to 2.28 GHz (TR = 4%), as shown in Figure 7.2. By comparing this FAA to a passive prototype, it was determined that the presence of the diode resulted in minimal pattern degradation in terms of pattern shape and cross-polarization level. However, the introduction of the diode resulted in increased losses, between 1 and 1.5 dB. A detailed analysis of the performance and tuning range of a microstrip patch antenna loaded with tunnel diodes was carried out in [2, 3]. For a patch designed at 50 GHz (with length L = 1.5 mm, width W = 1.8 mm,

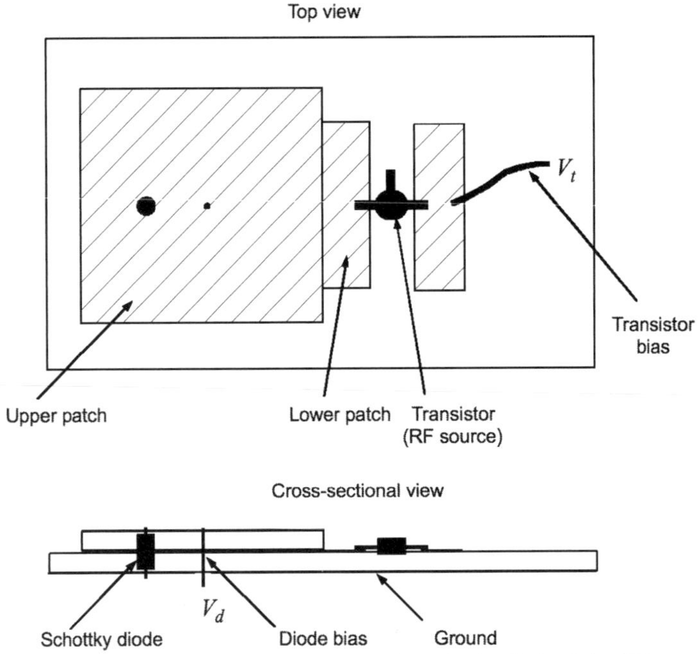

Figure 7.1 Stacked microstrip patch antenna loaded with a Schottky diode. (After [1].)

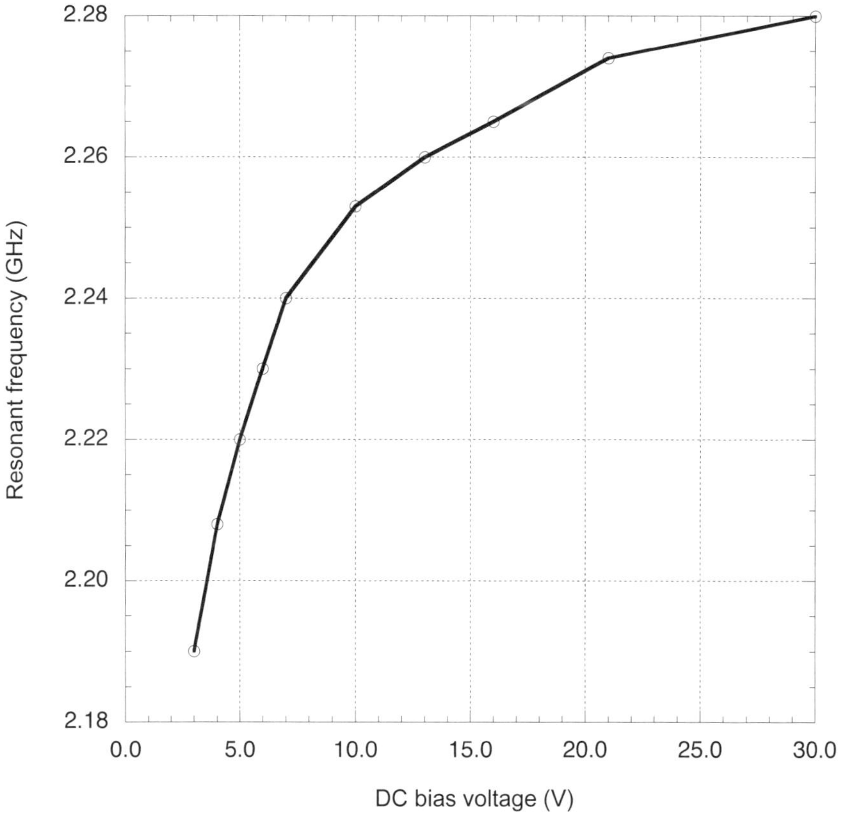

Figure 7.2 Tuning performance of the Schottky diode-loaded microstrip stacked. (From [1].)

height $h = 0.6$ mm, and dielectric constant $\varepsilon_r = 2.55$), and loaded with a Ge tunnel diode, the theoretical tuning range was determined to decrease from 54 to 51 GHz (TR = 5.7%), for a bias voltage of 0–0.35V [2]. Similar results were reported in [3], where a coaxial-probe fed microstrip patch designed at 0.85 GHz with two tunnel diodes was determined to have a theoretical tuning range of TR = 6.7%, for a bias voltage of up to 0.55V.

7.1.2 Slot Antenna Loaded With a Photodetector Diode

In Section 3.5.1, slot antennas are shown to radiate patterns similar to dipole antennas. An example of a frequency-tunable slot antenna based on a photodetector diode is shown in Figure 7.3 [4]. The antenna consists of a rectangular slot fed by a microstrip transmission line with a photodetector connected to the open end of the line. The biasing light source was generated by a laser photodiode and was carried to the photodetector via a 125-micron

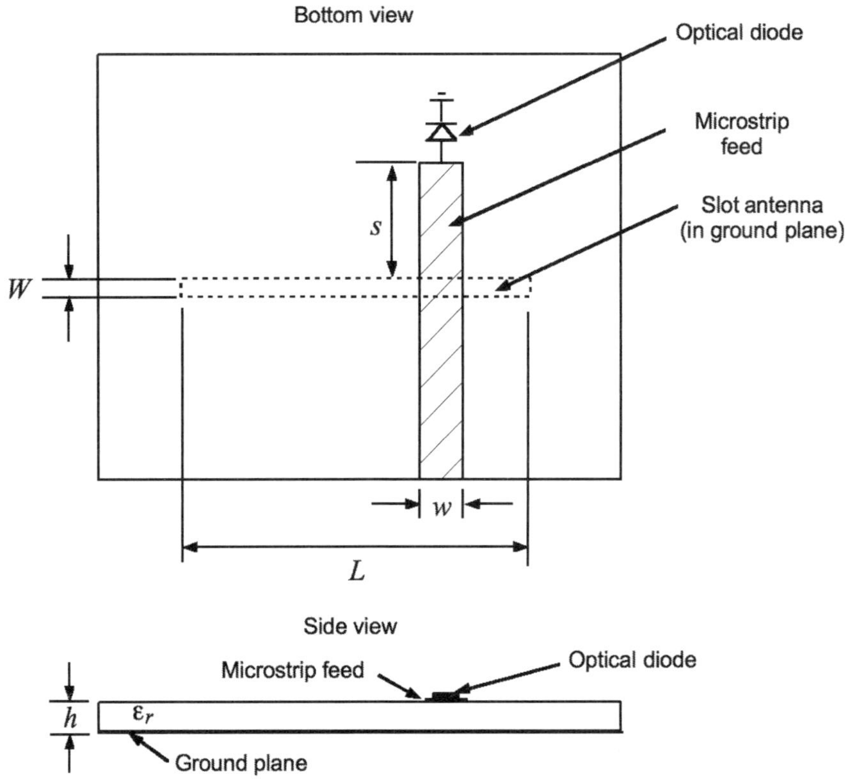

Figure 7.3 Microstrip-fed slot antenna loaded with a photodiode (L = 79.06 mm, W = 0.4 mm, h = 0.508 mm, ε_r = 2.2, s = 15 mm, and w = 1.56 mm). (After [4].)

multimode fiber. The light bias caused a change in the load impedance that the photodetector presented to the slot antenna, thereby altering its resonant frequency. By increasing the laser diode current from 0 to 88.1 mA, the resonant frequency of the slot antenna decreased from 2.85 to 2.08 GHz (TR = 31.2%), as shown in Figure 7.4. The antenna achieved a total spectrum of TS > 33%. The effect on the radiation patterns and the overall radiation efficiency were not reported.

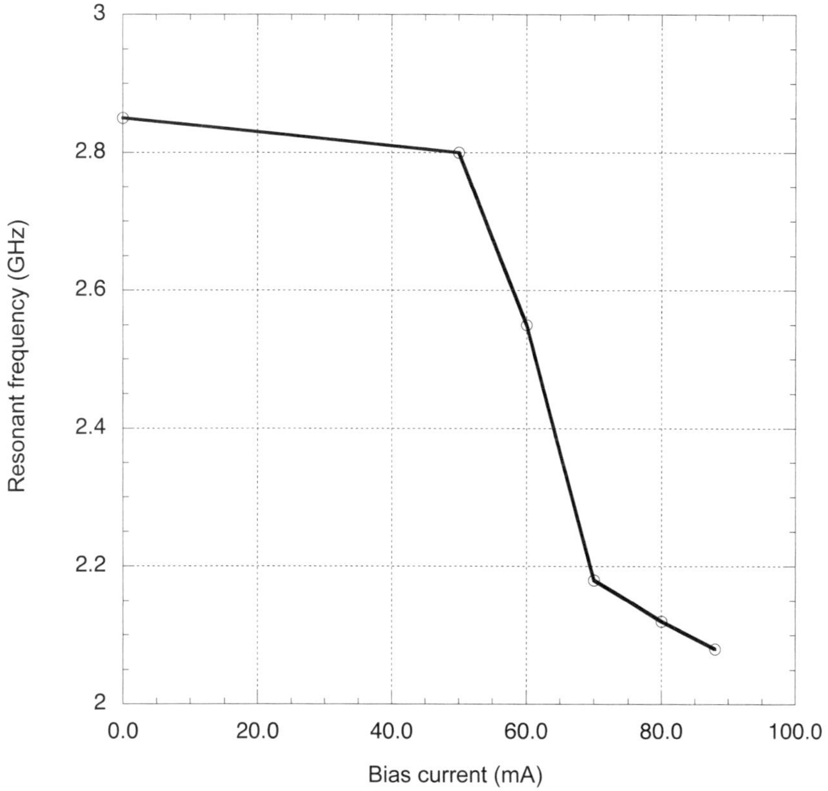

Figure 7.4 Tuning performance of the photodetector-loaded slot antenna. (From [4].)

7.2 TRANSISTOR-LOADED FAAs

The operation of FETs as microwave switches is reviewed in Section 4.3.3. Switching between the On and Off states was achieved by lowering the gate bias voltage from 0V to the pinchoff (V_p) value. However, if the gate voltage is varied between 0 and V_p, the resultant loading impedance of the transistor can be used to tune the resonant frequency of an antenna, as illustrated in the following two cases.

An example of a frequency-agile slot antenna is shown in Figure 7.5 [5]. The rectangular slot was fed at its center by a microstrip line and was loaded at each end by a microstrip stub connected to a packaged MESFET. By adjusting the gate-source voltage V_{gs} over the range -0.6V $\leq V_{gs} \leq$ 0V and drain-source voltage V_{ds} between 0V $\leq V_{ds} \leq$ 0.4V, the resonant frequency of the slot was tuned between 9.22 and 10.22 GHz (TR = 10.3%), with a total spectrum of TS = 20%. There was little observed change in the radiation patterns over the tuning range.

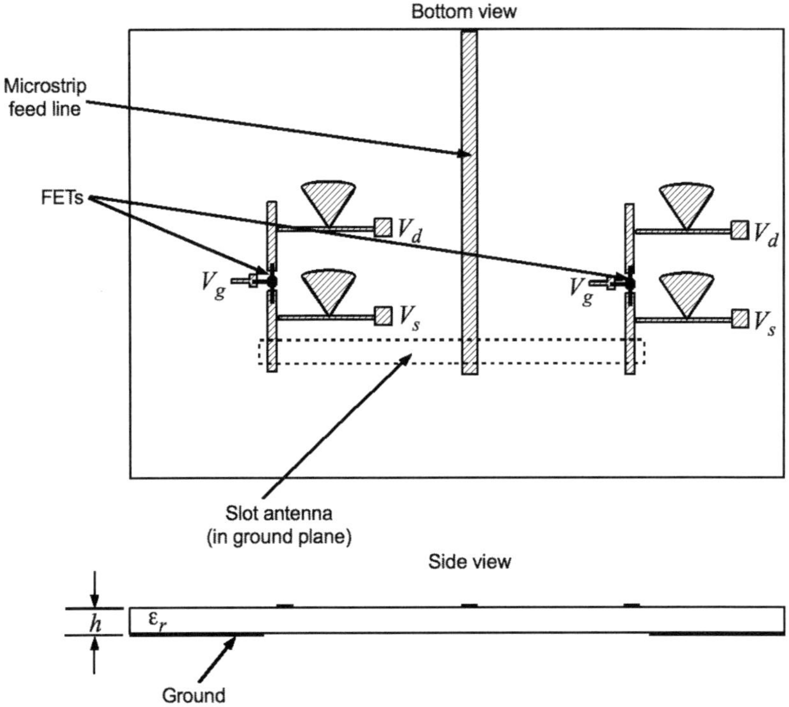

Figure 7.5 Microstrip-fed slot antenna loaded with MESFETs ($L = \lambda_g$ at 10 GHz, h = 0.76 mm, and ε_r = 2.55). (After [5].)

As discussed in Section 4.4.3, FETs draw negligible current and thus consume nearly no DC power, making them an attractive tuning method for applications requiring low power consumption.

A second example of a similar design is shown in Figure 7.6 [6]. In this case, the antenna consists of two slots in a Yagi-Uda array configuration, with the first slot being directly fed by the microstrip line and the second slot being electromagnetically coupled and acting as the director element. The director slot was loaded at its two edges by microstrip lines, each connected to a GaAs MESFET. The gate-source voltage was altered over a range of $-1.54\text{V} \leq V_{gs} \leq 0\text{V}$ while the drain-source voltage was adjusted between $0\text{V} \leq V_{ds} \leq 0.36\text{V}$, resulting in a frequency shift from 10.1 to 11 GHz (TR = 8.5%) and a total spectrum of TS = 12%. For this design, there was also a significant change in the radiation pattern. Without DC bias, the measured front-to-back ratio was 3 dB, while with a DC bias of V_{gs} = -1.535V and V_{ds}, = 0.35V, a front-to-back ratio of 10 dB was observed.

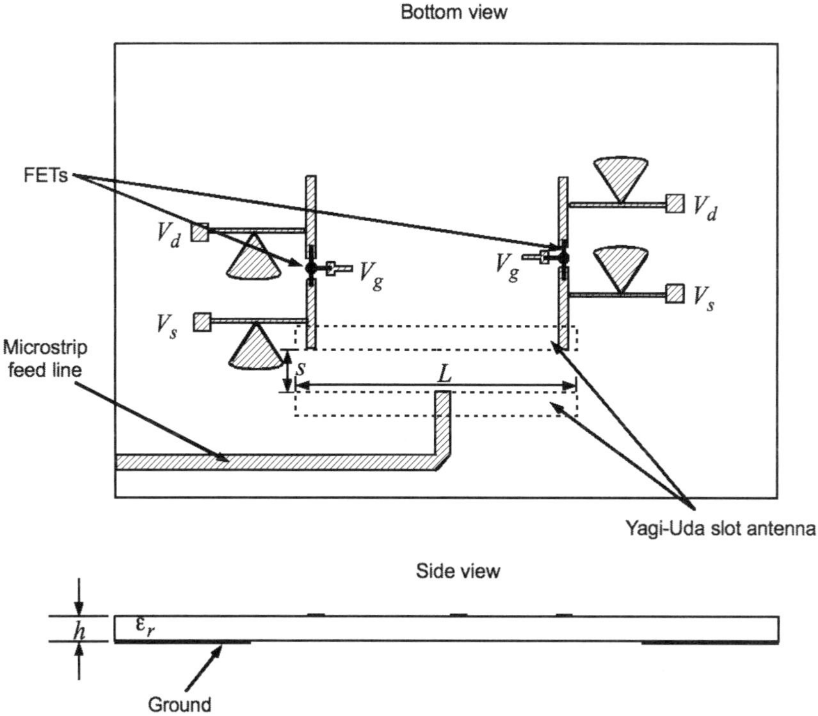

Figure 7.6 Microstrip-fed slot antenna loaded with MESFETs ($L = \lambda_g$ @ 10 GHz, $s = 7.5$ mm, $h = 0.76$ mm, and $\varepsilon_r = 2.33$). (After [6].)

7.3 MEMS-LOADED FAAs

An overview of MEMS devices is provided in Section 4.3.2. This section examines two examples of the use of MEMS capacitors for achieving frequency agility. The first example features a rectangular ring slot antenna while the second is based on a microstrip antenna.

7.3.1 CPW-Fed Rectangular Ring Slot Antenna

Annular slot antennas are examined in Section 3.5.2. A variation of the annular slot is a rectangular ring slot, as shown in Figure 7.7, which produces similar radiation patterns [7, 8]. In this example, the rectangular slot was fed by a

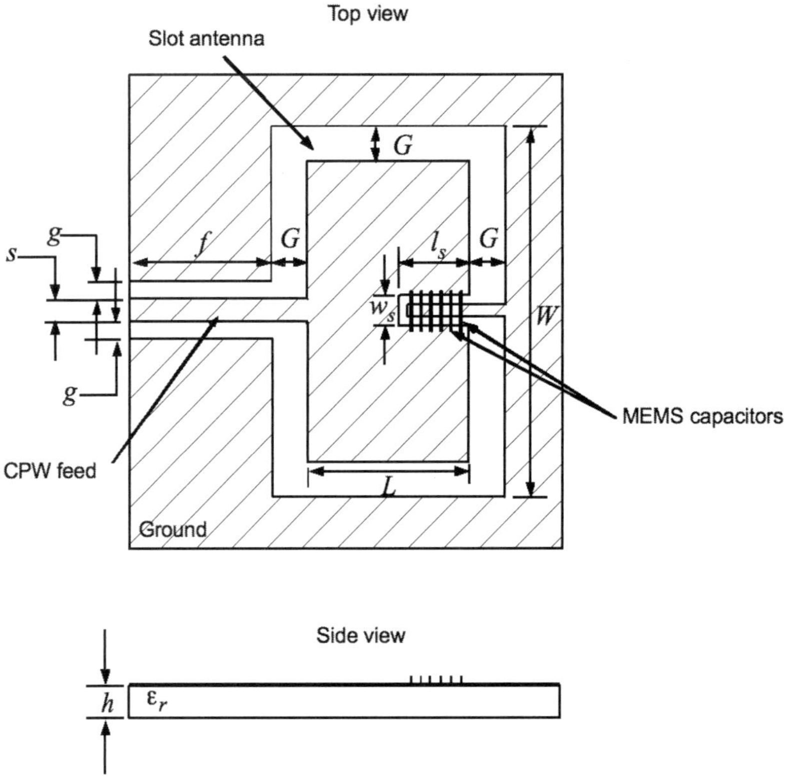

Figure 7.7 CPW-fed slot antenna loaded with MEMS capacitors (L = 5.6 mm, W = 13 mm, G = 0.5 mm, f = 5.6 mm, g = 25 μm, s = 210 μm, ε_r = 4.6, l_s = 2.975 mm, w_s = 0.1 mm, and h = 0.5 mm). (After [7, 8].)

CPW line and loaded with a stub, across which MEMS capacitors were connected. By varying the applied DC voltage across the set of MEMS capacitors, the stub impedance was altered, thereby shifting the resonant frequency of the slot antenna. Both cantilever [7] and fixed-fixed beam [8] MEMS capacitors were investigated. Since the lengths of the cantilever arms had to be fairly long to provide sufficient capacitance to tune the resonant frequency, there was a significant stress gradient developed in the fabrication process, and the resultant cantilevers were curled, leading to poorer performance. This difficulty was resolved with the fixed-fixed beam design in [8]. With the application of a DC bias voltage of up to 25V, the MEMS capacitors deflected from an original height of 2 microns down to 1.4 microns. The increased loading capacitance resulted in a change in the characteristic impedance of the stub. This, in turn, provided a shift in the resonant frequency of the rectangular slot antenna.

By using six MEMS capacitors, the frequency shifted from 8.7 to 7.7 GHz (TR_1 = 12%) for the lowest order mode and from 10.57 down to 10.22 GHz (TR_2 = 3.4%) for the next resonant mode, as plotted in Figure 7.8. The corresponding values for the total spectrum for each of the two modes were TS_1 = 10% and TS_2 = 18%. No adverse effects on pattern shape were reported; however, there was no mention of the resultant radiation efficiency of the antenna.

7.3.2 CPW-Fed Rectangular Microstrip Patch Antenna

A second example of a MEMS-loaded antenna is shown in Figure 7.9, consisting of a CPW-fed stub-loaded microstrip patch antenna [9]. Five MEMS capacitors, spaced 2 mm apart, were bridged across the CPW feed line. By varying the DC bias voltage across the MEMS capacitors from 0 to 11.9V, a variable load was presented to the resonant edge of the microstrip patch, and the resultant shift in resonant frequency of the antenna decreased from 16.05 to 15.75 GHz (TR = 1.9%).

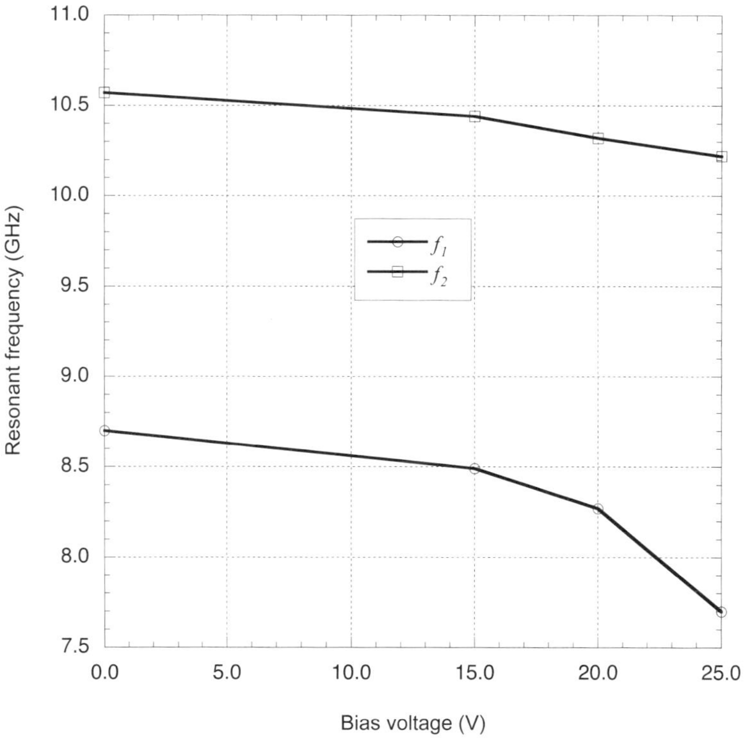

Figure 7.8 Tuning performance of the CPW-fed slot antenna loaded with MEMS capacitors. (From [8].)

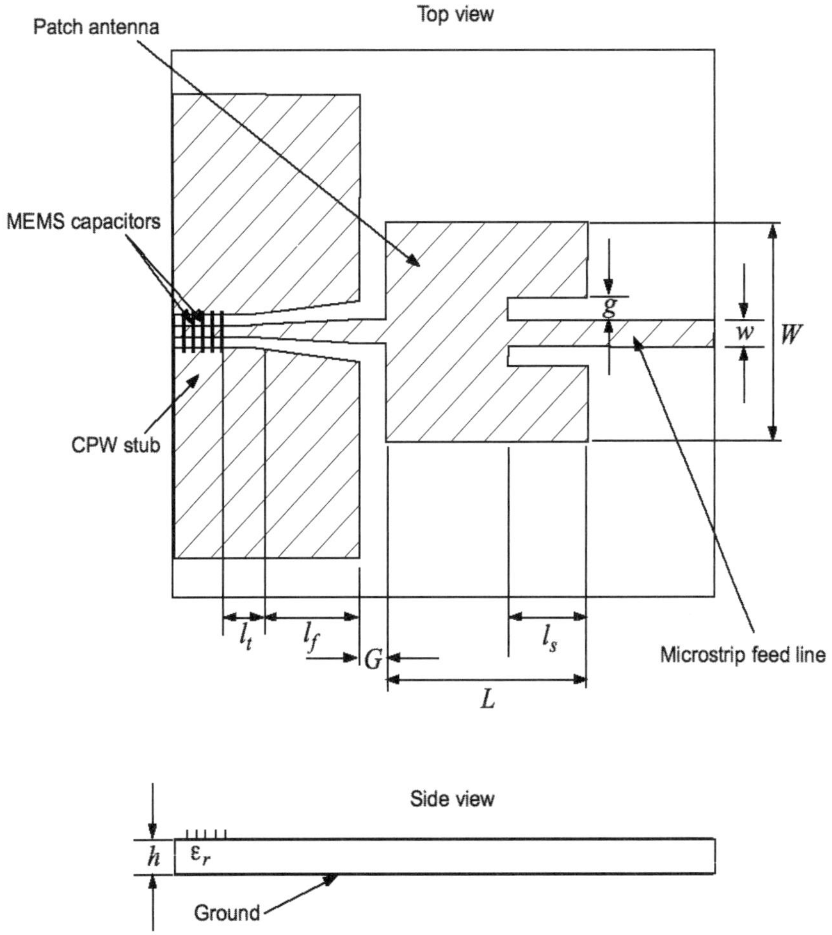

Figure 7.9 CPW-stub loaded microstrip patch antenna loaded with MEMS capacitors (L = 4.3 mm, W = 4.3 mm, l_s = 3 mm, g = 0.45 mm, w = 0.9 mm, G = 1 mm, l_f = 3 mm, l_t = 0.55 mm, h = 0.5 mm, and ε_r = 4.6). (After [9].)

The DC bias was applied to the center conductor of the CPW line via a bias-T attached to the antenna connector. The DC ground was supplied to the ground planes of the CPW by shorting plates at the edge of the CPW stub connecting the upper and lower grounds. The authors report no negative effects on the radiation pattern when the DC bias was applied, but neither the gain nor the radiation efficiency was mentioned.

7.4 VARACTOR-LOADED FAAs

Varactor diodes, described in Section 4.3.5, have been by far the most commonly used electronic device for achieving FAAs with continuous tuning. This section illustrates numerous examples where varactors are integrated with various antenna elements including loop antennas, dipoles, microstrip patch antennas, slot antennas, helical antennas, IFAs, PIFAs, and DRAs.

7.4.1 Loop Antennas

This section examines three loop antennas loaded with a varactor diode. The first example, shown in Figure 7.10, consists of a semicircular loop etched on a thin substrate (0.786-mm), designed for UHF band mobile applications [10]. The substrate was placed perpendicularly on a square ground plane (300 mm x 300 mm x 4 mm). One end of the loop was excited using electromagnetic coupling from a short section of microstrip line, while the other end of the loop was shorted to ground. The varactor bridged across a small gap in the printed loop, and the bias voltage was applied to one side of the varactor through an RF choke. A DC block prevented the bias voltage from mixing with the RF source. By varying the bias voltage from 0 to 5V, the resonant frequency of the printed loop was tuned from 470 to 675 MHz (TR = 35.8%), as shown in Figure 7.11, with a total spectrum of TS = 39%. The measured gain of the antenna varied from -1.5 to 2.8 dBi over the tuning range.

The second example consists of two crossed-wire loops, as depicted in Figure 7.12(a) and as shown schematically in Figure 7.12(b) [11]. The wires were galvanically connected at the intersection. As indicated in the schematic

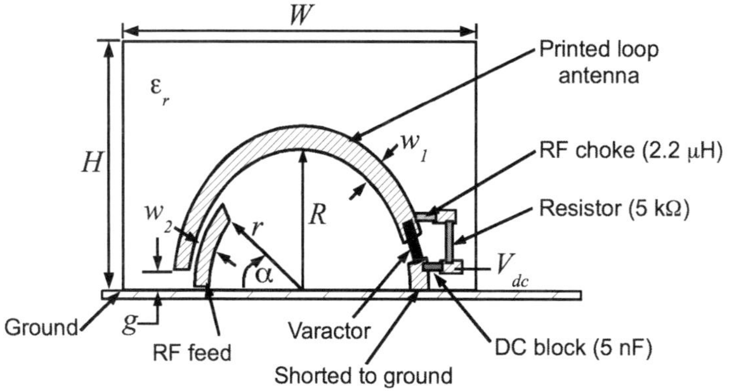

Figure 7.10 Varactor-loaded printed loop antenna (R = 37.5 mm, r = 34.5 mm, w_1 = 1 mm, w_2 = 2 mm, g = 2 mm, a =43.5°, W = 80 mm, H = 40 mm, and ε_r = 3). (After [10].)

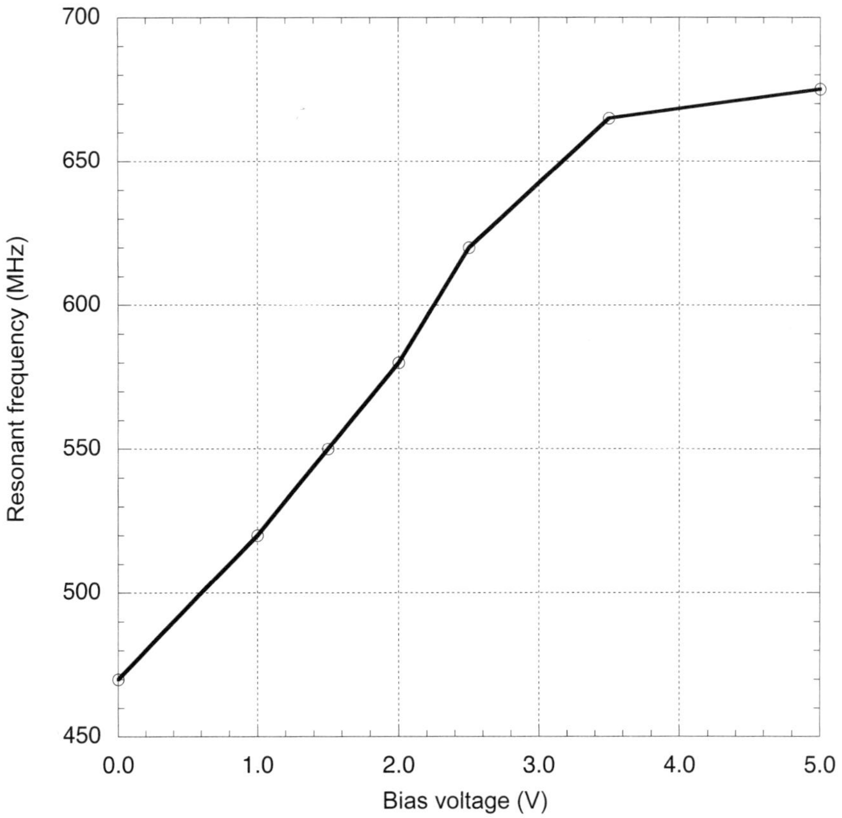

Figure 7.11 Tuning performance of the varactor-loaded printed loop antenna. (From [10].)

figure, the antenna was excited at Port 1, short-circuited at Port 2, match-terminated in a 50Ω load at Port 3, and capacitively loaded at Port 4. Varying the capacitance value from 0 to 6.8 pF by connecting various chip capacitors to Port 3 between the wire and ground resulted in the resonant frequency of this compact antenna shifting from 1,585 to 310 MHz (TR = 134.6%), as shown in Figure 7.13. The 10-dB return loss fractional bandwidth ranged from 1.9% to 3.2% over the tuning range. The relative accepted power (i.e., the power not reflected at Port 1 or absorbed at Port 2) ranged from 72% at 1,585 MHz to 12% at 310 MHz. The radiation efficiency at 1,585 MHz was also measured and found to be about 60%. This compact antenna could become frequency-agile by replacing the fixed-chip capacitors with a varactor. An approximate equation for determining the resonant frequency of the crossed-loop antenna was [11]

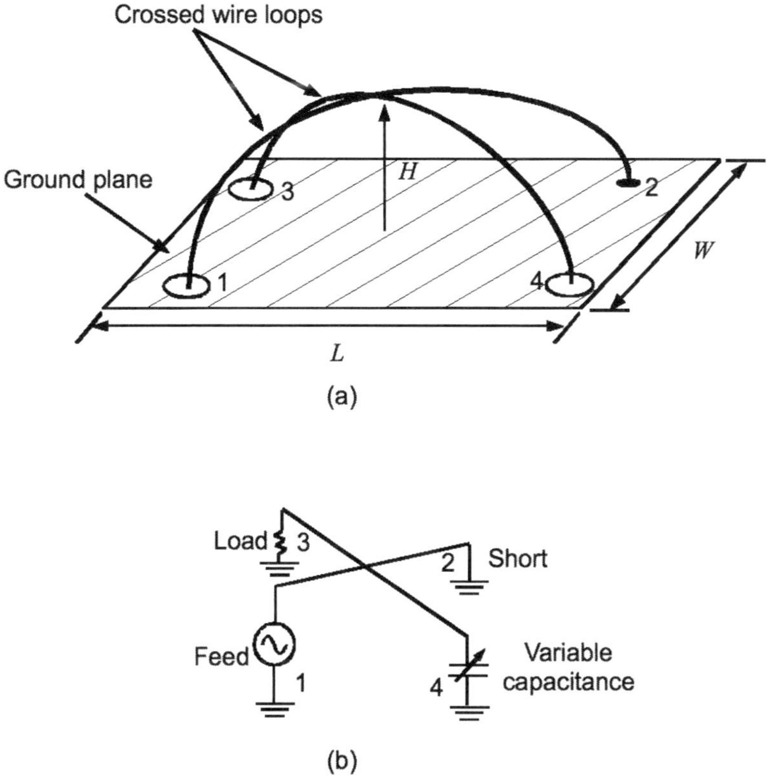

Figure 7.12 Compact crossed-loop antenna with variable capacitive load: (a) antenna geometry and (b) antenna schematic ($L = W = 30$ cm and $H = 19$ cm). (After [11].)

$$f_o \approx \frac{0.85}{2\pi\sqrt{LC}} \tag{7.1}$$

where the value of L was determined by curve fitting to experimental data to be 28 nH for the parameters in Figure 7.12, where 0.8-mm-diameter copper wires were used, and C is the value of the load capacitor at Port 4.

The final example of a varactor-loaded loop antenna is the printed half-loop shown in Figure 7.14 [12]. The half-loop has a rectangular shape and was etched on a 0.787-mm substrate ($\varepsilon_r = 2.2$) that was mounted on top of a large ground plane. Three pairs of varactors were connected across small gaps on the vertical and horizontal segments of the loop. The DC bias signals were provided to the varactors through thin (high-impedance) lines loaded with inductors (RF chokes) to prevent the RF entering the DC circuit. DC blocks were also added near the bottom ends of the loop to prevent the bias signals from mixing with the RF signal. The loop could operate at various resonant modes, where modes with $n < 0$ result in conical shaped patterns and where the $n = 0$ mode produced a

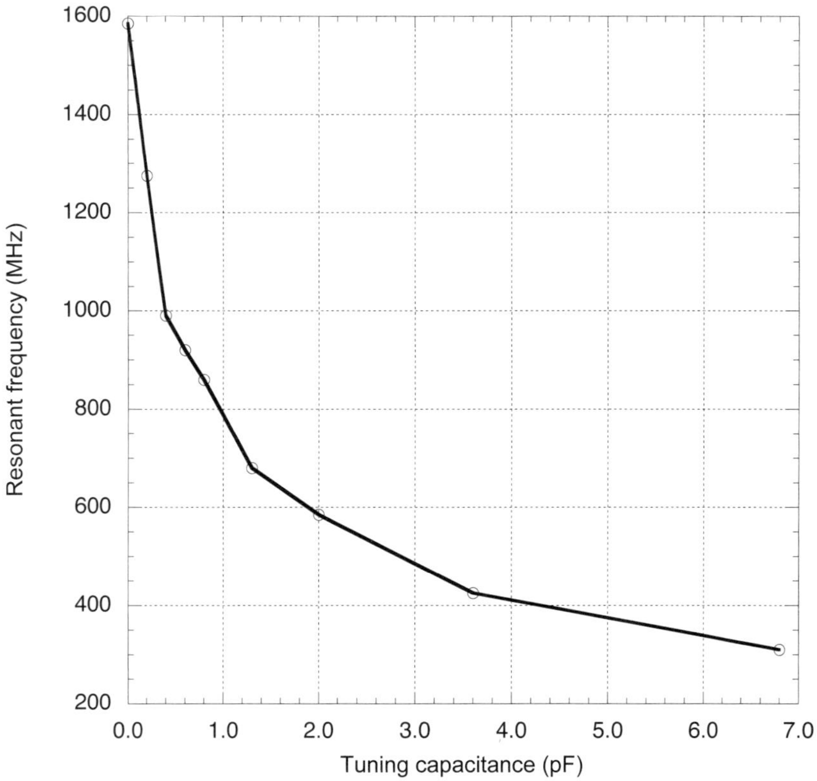

Figure 7.13 Tuning performance of the compact crossed-loop antenna. (From [11].)

broadside pattern. By varying the bias voltage in the range 2–22V (corresponding to a varactor capacitance range of 1.73–0.39 pF), the resonant frequency of the $n = 0$ mode shifted from 584.8 to 1,376 MHz (TR = 80.7%) with TS = 48%. The radiation efficiency of the antenna was measured to be greater than 54% over the frequency range of 840–1,375 MHz.

7.4.2 Printed Monopole Antennas

Printed coplanar monopole antennas are briefly introduced in Section 3.2. They can be designed to operate over a wide frequency range and are easily amenable to integration of electronic devices. This section examines a coplanar printed monopole antenna loaded with a varactor diode for continuous frequency tunability. The antenna, shown in Figure 7.15, consists of a rectangular shaped flat monopole fed by a CPW line [13]. Introducing the slots in the monopole results in dual-band operation, where the slot radiation accounted for operation

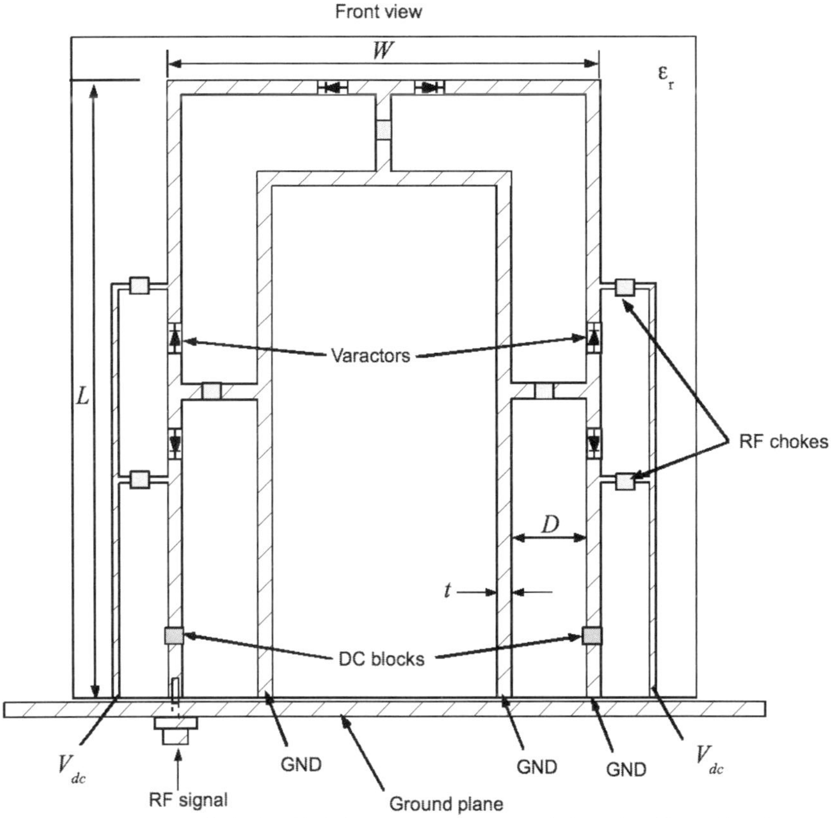

Figure 7.14 Varactor-loaded printed half-loop antenna on a ground plane (L = 50 mm, W = 40 mm, D = 10 mm, t = 0.9 mm, and ε_r = 2.2). (After [12].)

at the lower band and the monopole radiation operated at the upper band [14]. Connecting a varactor diode across one of the slots resulted in tuning of the lower resonant frequency, while the upper resonant frequency remained relatively unchanged. The selected diode had a capacitance that varied from 0.6 to 0.075 pF when a DC voltage bias between 2 to 20V was applied. The DC bias was provided through the CPW feed line using a bias-T. An RF choke provided the DC return path, and an AC coupling capacitor was used to bridge the slot for the RF input signal. Varying the voltage from 1.2 to 23.7V resulted in a shift in the resonant frequency from 1.4 to 2.23 GHz (TR = 45.7%), as plotted in Figure 7.16. The antenna achieved a total spectrum of TS = 34%. The radiation efficiency of the antenna ranged from 20% to 60% over the band of 1.6 to 2.23 GHz. The low efficiency is mainly attributed to the small electrical size of the monopole antenna.

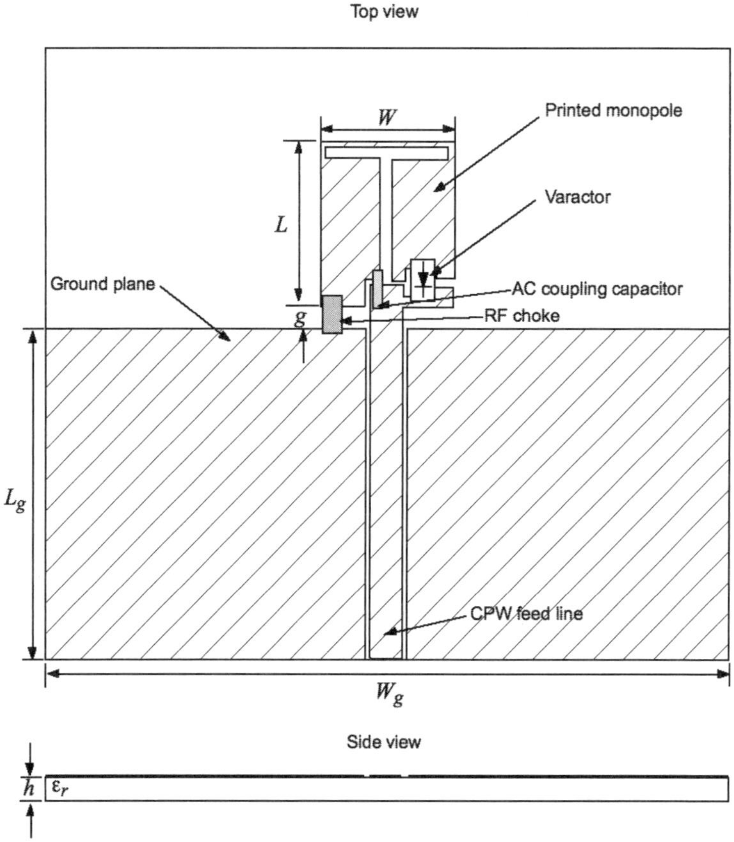

Figure 7.15 Varactor-loaded printed monopole antenna (L = 7.5 mm, W = 6.3 mm, g = 1 mm, L_g = 15 mm, W_g = 32 mm, h = 1.59 mm, and ε_r = 4.35). (After [13].)

7.4.3 Dipole Antennas

The traditional wire dipole antenna is reviewed in Section 3.1. This section examines two variations on the wire dipole, a folded dipole and a printed dipole, with integrated varactors. Figure 7.17 shows the configuration of a coaxial probe-fed asymmetric folded dipole placed atop a ground plane with the varactor diode connecting the folded dipole to the ground [15]. By applying between 10- and 29-V DC bias, the resonant frequency was altered from 2.78 to 3.45 GHz (TR = 21.5%), as shown in Figure 7.18. (DC bias voltages below 10V resulted in a poor match for this particular design.) The antenna achieved a total spectrum of TS = 23%. It was also mentioned that the antenna suffered from poor gain, but the actual value was not specified in [15].

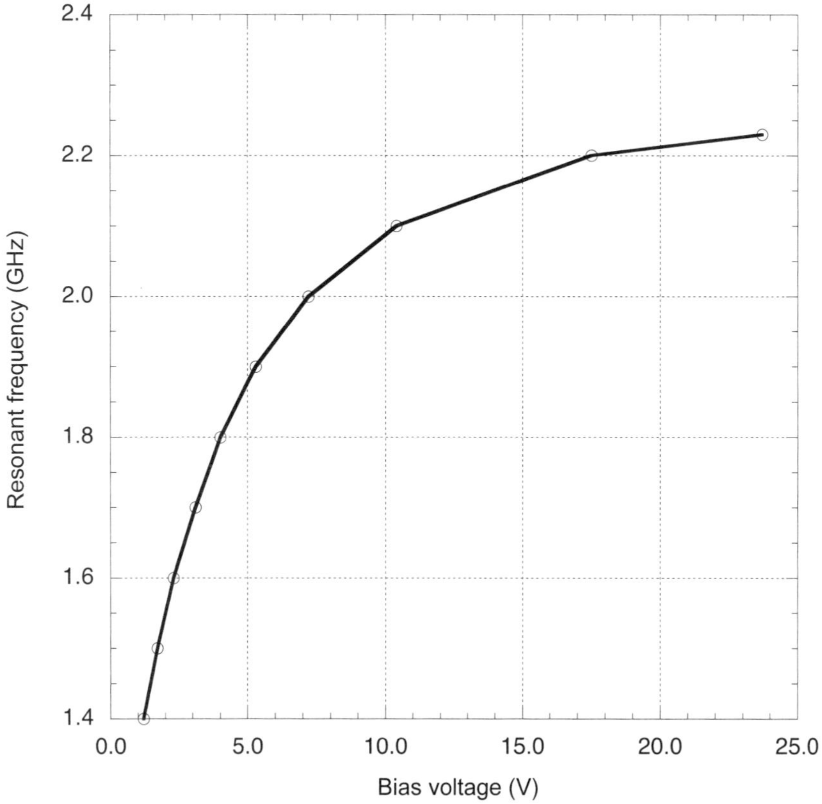

Figure 7.16 Tuning performance of the varactor-loaded printed monopole antenna. (From [13].)

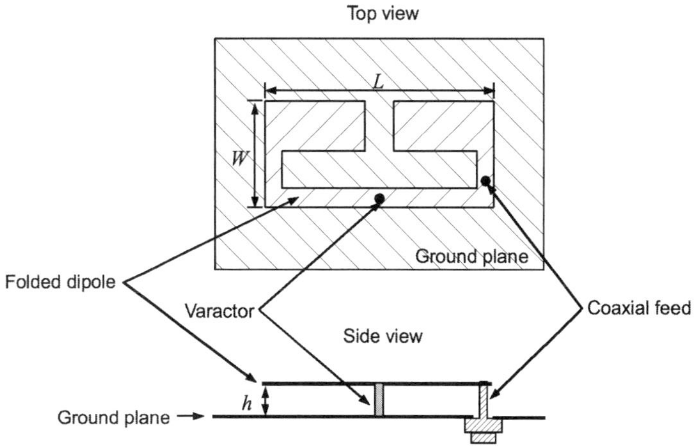

Figure 7.17 Varactor-loaded folded dipole (L = 32 mm, W = 14 mm, and h = 1.2 mm). (After [15].)

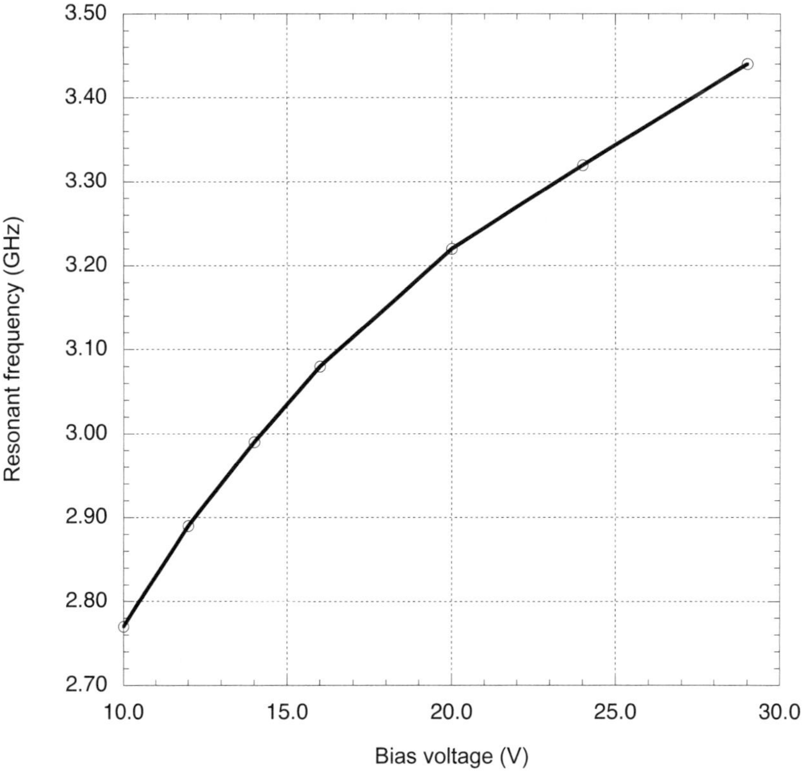

Figure 7.18 Tuning performance of the varactor-loaded folded dipole. (From [15].)

The second example is a printed dipole excited by CPW feed, as shown in Figure 7.19 [16]. Two small gaps were symmetrically located on each dipole arm across which a varactor diode was connected. The polarity of the varactors is indicated by the positive and negative bias voltages in Figure 7.19. The introduction of the small gaps resulted in dual-frequency behavior. As the bias voltage was varied between 0 and 30V, the two resonant frequencies varied from 2.065 to 2.365 GHz (TR_1 = 13.5%) and 3.955 to 3.505 GHz (TR_2 = 12.1%), respectively. The corresponding values of total spectrum were TS_1 = 25% and TS_2 = 21%. Neither gain nor radiation efficiency were reported for this antenna.

The final example is a printed quasi-Yagi antenna, as shown in Figure 7.20, consisting of the printed dipole as the driver element, a single dipole director, and the ground plane acting as the reflector [17]. A varactor diode was connected across a small gap in either arm of the driven dipole, and a third varactor was connected across a small central gap in the director dipole. The DC bias voltage was provided through thin printed wires, loaded with high-value chip resistors (100-kΩ) to prevent the RF signals from flowing into the DC lines. Different

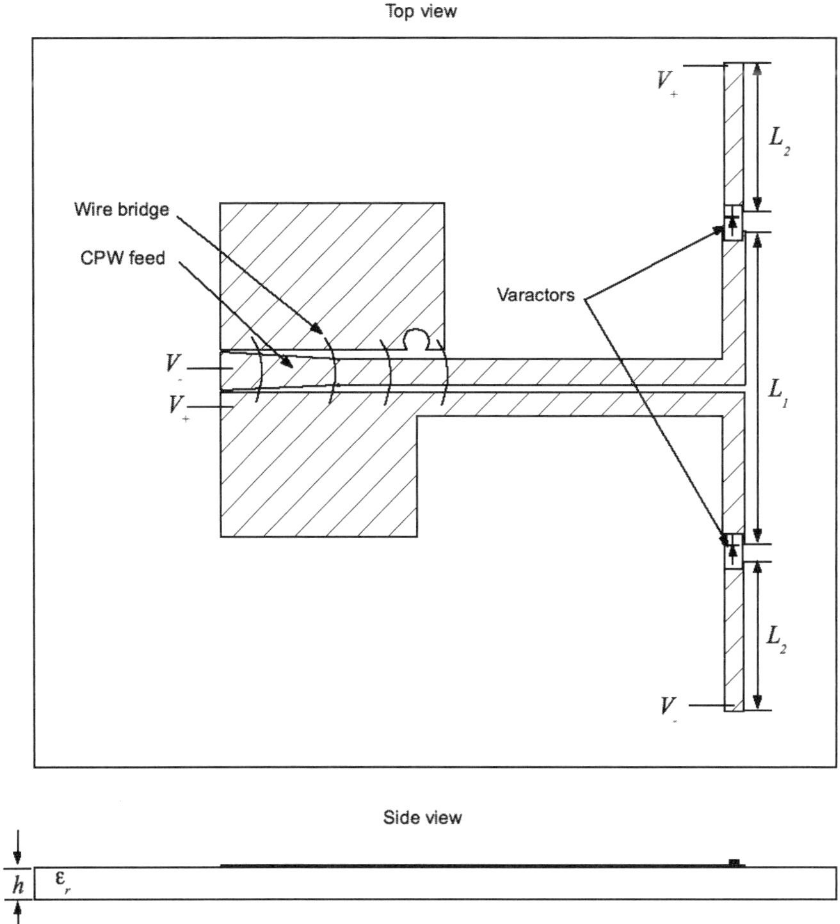

Figure 7.19 Varactor-loaded dipole antenna (L_1 = 18 mm, L_2 = 10 mm, h = 1.575 mm, and ε_r = 2.2). (After [16].)

combinations of the bias voltages V_1 and V_2 (up to 16.6V) on the drive and director dipoles, respectively, resulted in shifts in the resonant frequency, from 1.8 to 2.45 GHz (TR = 30.6%), with a corresponding total spectrum of TS = 40%. Measured realized gains varied from 5.5 to 7.5 dBi, with the antenna efficiency ranging from 75% to 95%. The bias lines were reported to have only a marginal effect on the radiation patterns.

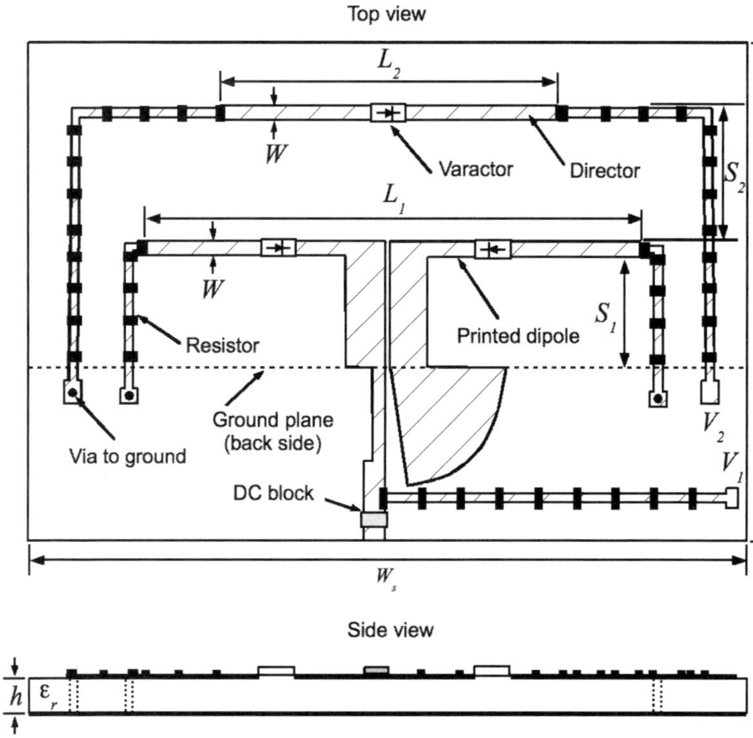

Figure 7.20 Varactor-loaded quasi-Yagi dipole antenna (L_1 = 83.5 mm, L_2 = 56 mm, S_1 = 19 mm, S_2 = 21.5 mm, L_s = 87 mm, W_s = 102 mm, W = 1.5 mm, h = 0.813 mm, and ε_r = 3.55). (After [17].)

7.4.4 Compact Helical Antennas

The monofilar helix antenna is covered in Section 3.9, and an example of a mechanically tuned quadrifilar helix is examined in Section 5.2. In Figure 7.21, two small (five-turn) monofilar helical antennas (made from 0.5-mm diameter copper wire), each with an inductance of 1.2 nH, were loaded with varactor diodes in the design of a compact UHF antenna for digital video broadcasting-handheld (DVB-H) applications [18]. The ground plane dimensions ($L \times W$) used are typical for a handset. The monofilar helix examined in Section 3.9 radiated an endfire pattern that was circularly polarized. In this case, however, the helix is much smaller than a wavelength and behaves more like an inductively loaded dipole.

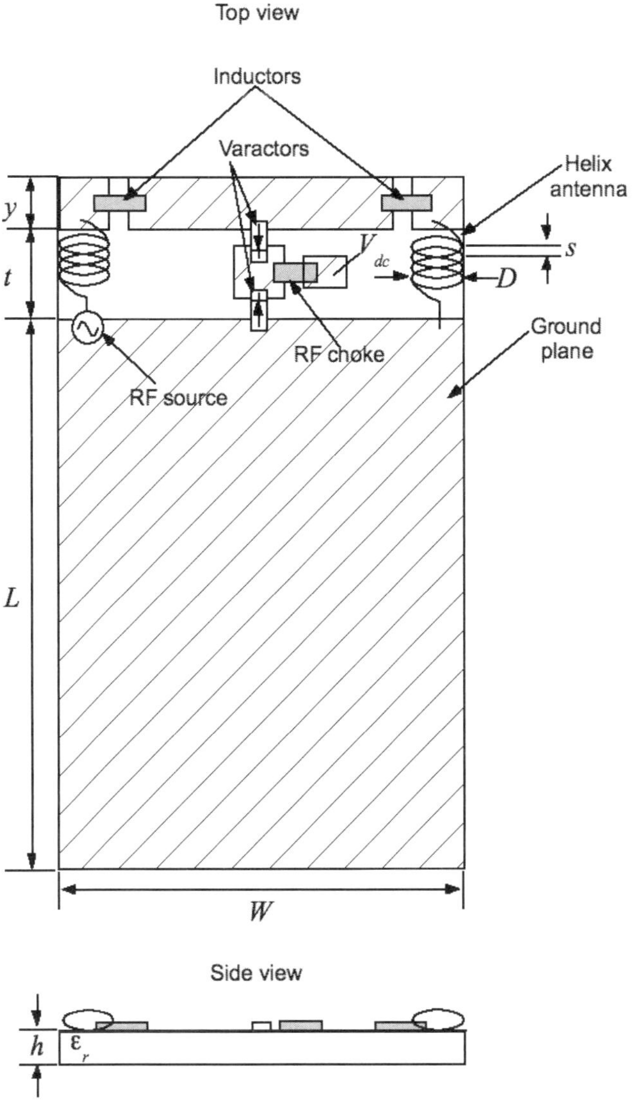

Figure 7.21 Varactor-loaded helical radiators (L = 95 mm, W = 50 mm, y = 1 mm, t = 4 mm, D = 5 mm, s = 1 mm, h = 1 mm, and ε_r = 4.4). (After [18].)

The circuit model for the antenna is shown in Figure 7.22, where the variable capacitance is used to tune the resonant frequency. By treating the circuit in Figure 7.22 as a unit cell of a periodic structure and applying the Bloch-Floquet theory, the resonant frequency of the antenna can be estimated by the following expression [18]

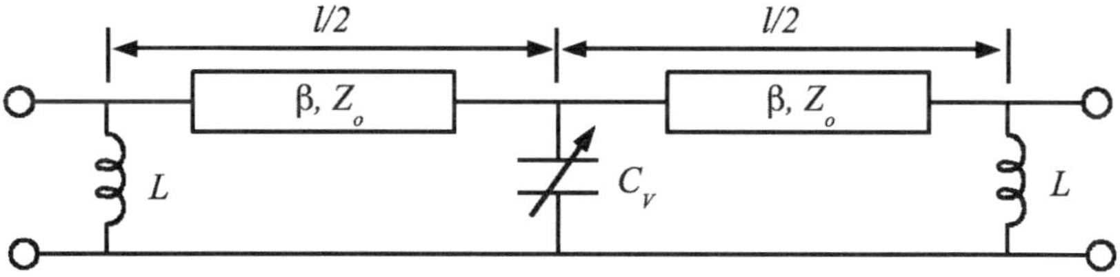

Figure 7.22 Circuit model of the DVB-H antenna of Figure 7.17. (After [18].)

$$\cos(\beta_{eff} l) = \cos(\beta l)\left[1 + \frac{Z_o^2 C_v}{2L}\right] + \sin(\beta l)\left[\frac{Z_o}{2\pi f L} - Z_o \pi f C_v\right] - \frac{Z_o^2 C_v}{2L} \quad (7.2)$$

where β_{eff} is the propagation constant of the periodically loaded unit cell, and β, Z_o, and l are the propagation constant, the characteristic impedance, and the length per unit cell, respectively. From (7.2), the resonant frequency can be determined by solving for f when $\beta_{eff} = 0$. This resonant frequency can be tuned by varying the value of C_v. Using two varactors connected in series, whose capacitance ranged from 4.75 to 1.2 pF as the DC bias voltage was increased from 0.3 to 2.7V, the resonant frequency of the antenna was tuned from 474 to 702 MHz (TR = 38.8%), as shown in Figure 7.23. The antenna achieved a total spectrum of TS = 7%. A significant drop in gain (over 10 dB) was reported over the tuning range, due to the increased losses in the varactors at low values of bias voltage. However, the achieved gain was still deemed acceptable for the intended DVB-H application.

A second design is shown in Figure 7.24, where the turns of the compact helix antenna are wound around a rectangular dielectric bar and then placed on a thin dielectric sheet [19]. The varactor was connected to the base of a long stub, whose other end was connected to the top of the helix to improve the impedance match. The DC control signal for the varactor was provided through a bias network, and the RF signal was injected at the base of the helix. By varying the bias voltage from 0 to 3.5V, the resonant frequency of the helix shifted from 178 to 210 MHz (TR = 16.5%) with a corresponding total spectrum of TS = 18%. Although the gain and radiation efficiency were not reported, the performance of the helix was comparable to that of a 120-mm-long monopole antenna.

7.4.5 IFAs

Wire and printed versions of the IFAs are examined in Section 3.6. This section looks at two examples of printed IFAs with varactor loading for achieving frequency agility. Printed IFAs are well-suited for integration in printed circuit boards (PCBs) and are commonly used antennas in cellular phones or smart phones, due to their compact size and low cost.

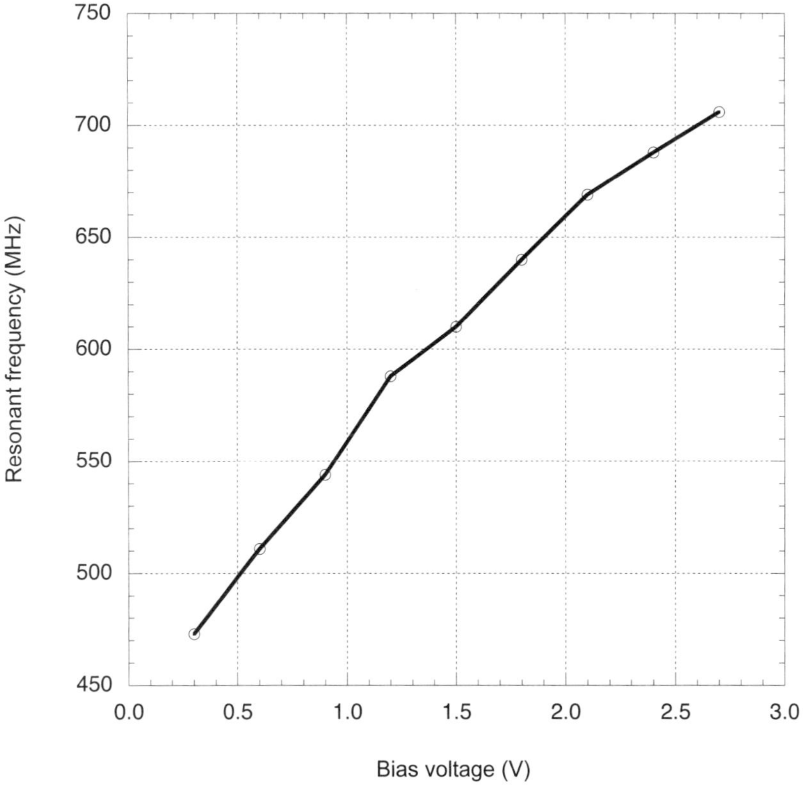

Figure 7.23 Tuning performance of the varactor-loaded helical radiators. (From [18].)

The first example, shown in Figure 7.25, represents a design for DCS/PCS/UMTS bands [20]. The IFA was printed on a small FR4 PCB whose dimensions are typical of those of a cellular phone. The IFA was located at one end of the PCB where the ground plane was removed, and it was fed by a 50-Ω microstrip line. The varactor was connected across a small gap in the horizontal arm of the IFA. The varactor was biased through a small printed trace, which included a low-pass filter T-type DC bias circuit (not shown) with two series 47-nH inductors and a shunt 15-pF capacitor inserted between the DC supply and the bias trace to act as the RF choke and DC block, respectively. As the DC bias voltage was varied from 0 to 9.5V, the resonant frequency of the antenna increased from 1.69 to 2.12 GHz (TR = 22.6%), as shown in Figure 7.26. The antenna achieved a total spectrum of TS = 36% with an instantaneous 10-dB fractional bandwidth maintained at greater than 10%. The measured radiation efficiency of the antenna varied from about 67% to 74% over the tuning range. Variations on the basic design for achieving dual- and tri-band FAAs were also proposed in [20] and are sketched in Figure 7.27. The dual-band IFA was

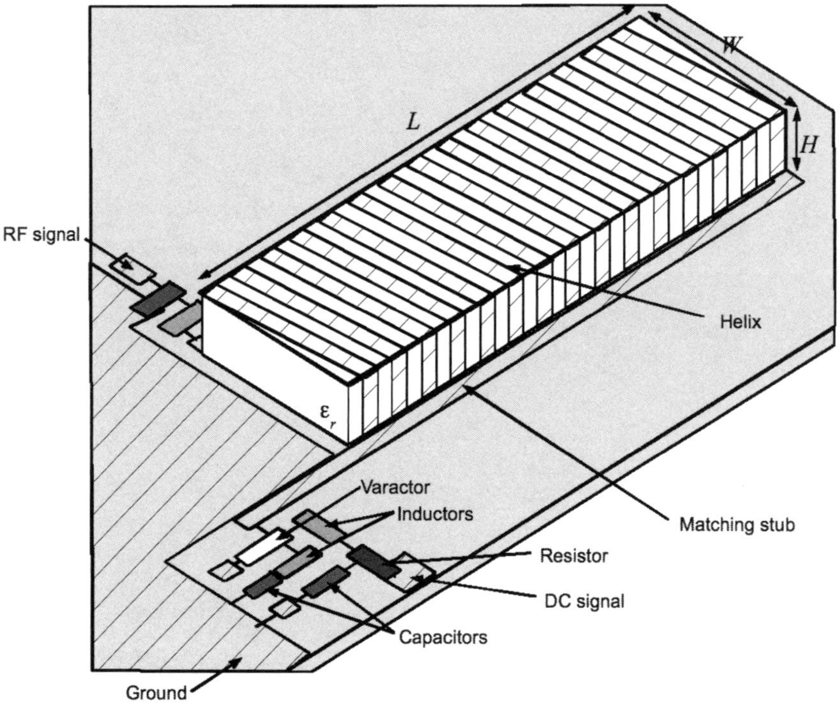

Figure 7.24 Varactor-loaded compact helix (L = 30 mm, W = 10 mm, H = 3 mm, and ε_r = 4.5). (After [19].)

designed for the 900/1,800-MHz bands, while the tri-band design covered the 900/1,800/1,900-MHz bands.

A second example of a varactor-loaded IFA is shown in Figure 7.28 [21]. This antenna was designed to cover the DVB-H band (470–862 MHz) and was also printed on a PCB board whose dimensions are typical of those of a mobile phone. The varactor selected had a capacitance whose values varied from 0.2 to 2 pF for a bias voltage ranging from 0 to 22V. In this case, the DC bias voltage was supplied to the varactor through the use of a bias-T connected to the RF coaxial probe used to excite the IFA. By varying the bias voltage from 2 to 22V, the resonant frequency was tuned between 480 and 850 MHz (TR = 55.6%), as shown in Figure 7.29. The total spectrum achieved was TS = 58%. The measured peak realized gain varied between -6 and 1 dBi for the range of bias voltages, which is above the typical gain requirements for DVB-H applications, which vary linearly from -10 to -4 dBi over the frequency band.

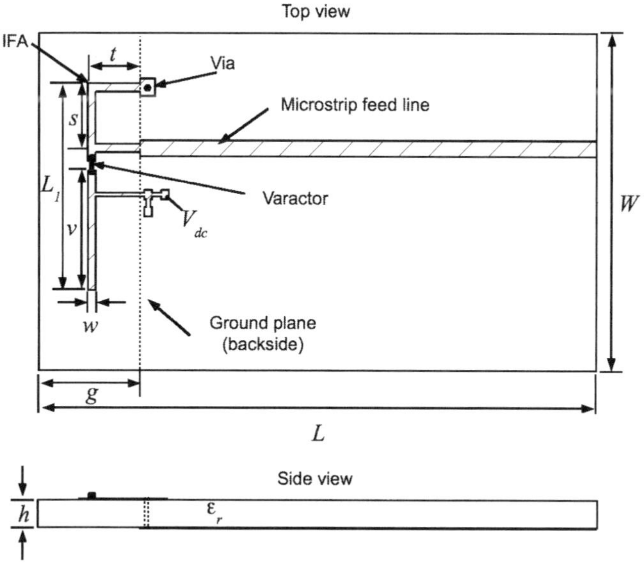

Figure 7.25 Varactor-loaded IFA antenna (L = 80 mm, W = 50 mm, g = 15 mm, L_l = 32 mm, $t = s$ = 7 mm, v = 20 mm, $\varepsilon_r \sim 4.4$, w = 1 mm, and h = 1.575 mm). (After [20].)

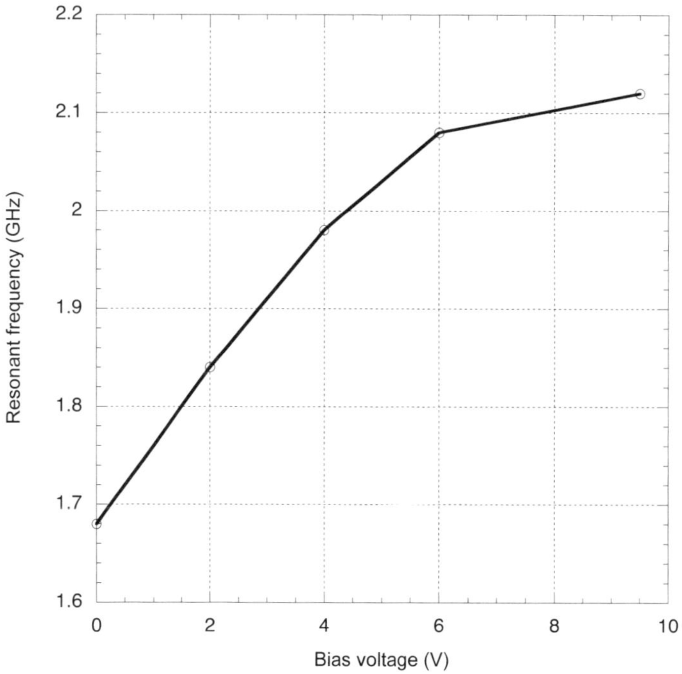

Figure 7.26 Tuning performance of the varactor-loaded IFA. (From [20].)

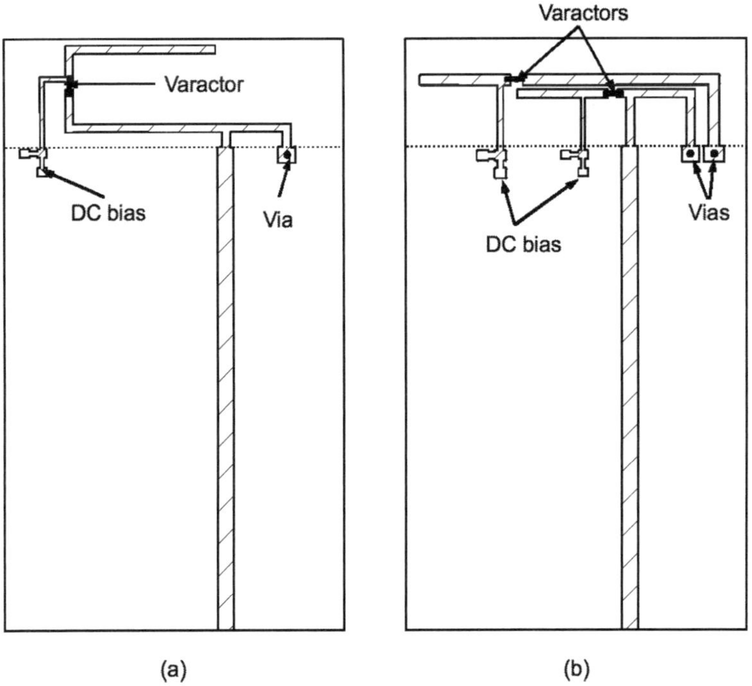

Figure 7.27 (a) Dual- and (b) tri-band varactor-loaded IFAs. (After [20].)

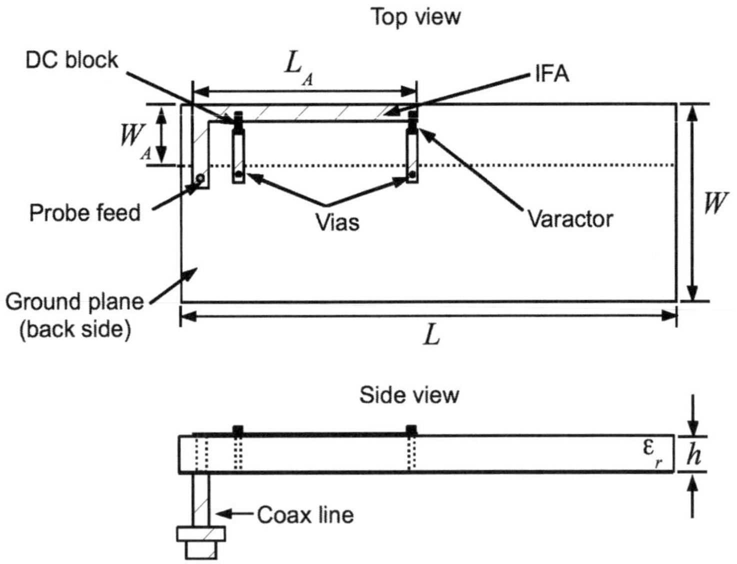

Figure 7.28 Varactor-loaded IFA. (L = 90 mm, W = 35 mm, L_A = 45.6 mm, W_A= 20 mm, h = N/A, and ε_r = 4.9). (After [21].)

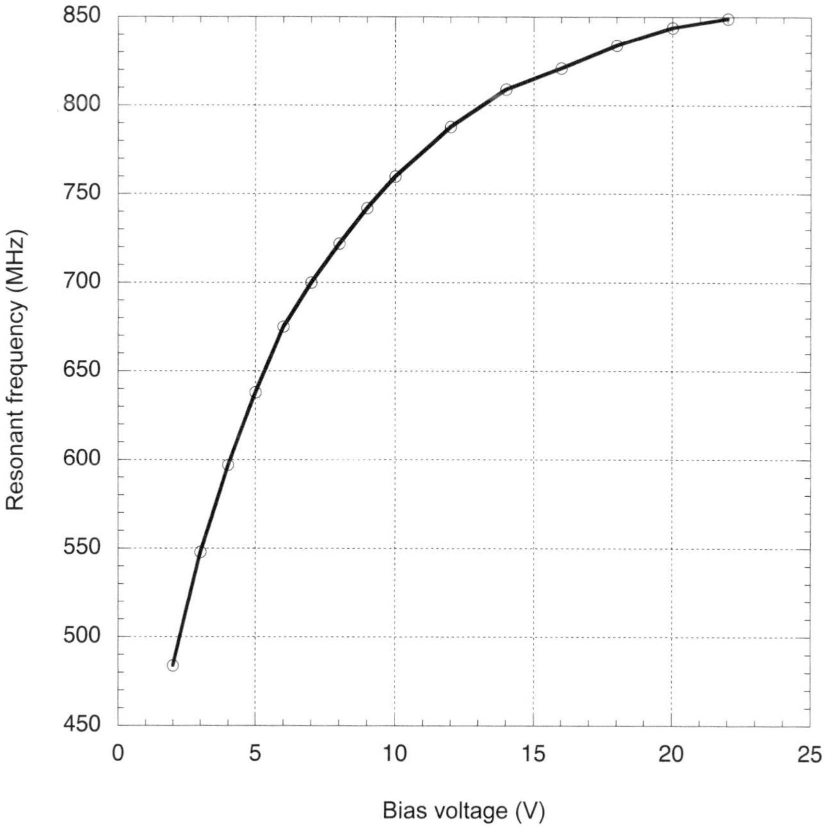

Figure 7.29 Tuning performance of the varactor-loaded IFA. (From [21].)

7.4.6 PIFAs

PIFAs, closely related to IFAs, are reviewed in Section 3.7. Two examples of varactor-loaded PIFAs are examined in this section. The first example, designed for a personnal communications handset, is shown in Figure 7.30 [22]. The probe-fed PIFA was located in one corner of a small ground plane, whose dimensions are similar to those of a handset. Slots were cut out of the PIFA in order to excite three resonances at 2, 5.35, and 5.8 GHz (the latter two frequencies used for WLAN applications). The varactor was connected across one of the slots (controlling the 2-GHz resonance) with a DC block (a 15-pF chip capacitor) and an RF choke (a 68-nH inductor) preventing the DC bias and RF excitation signals from mixing. By varying the DC bias voltage from 0 to 18.4V, the resonant frequency of the PIFA was tuned from 1.65 to 2.4 GHz (TR = 37%), as shown in Figure 7.31, with a corresponding total spectrum of TS = 41%.

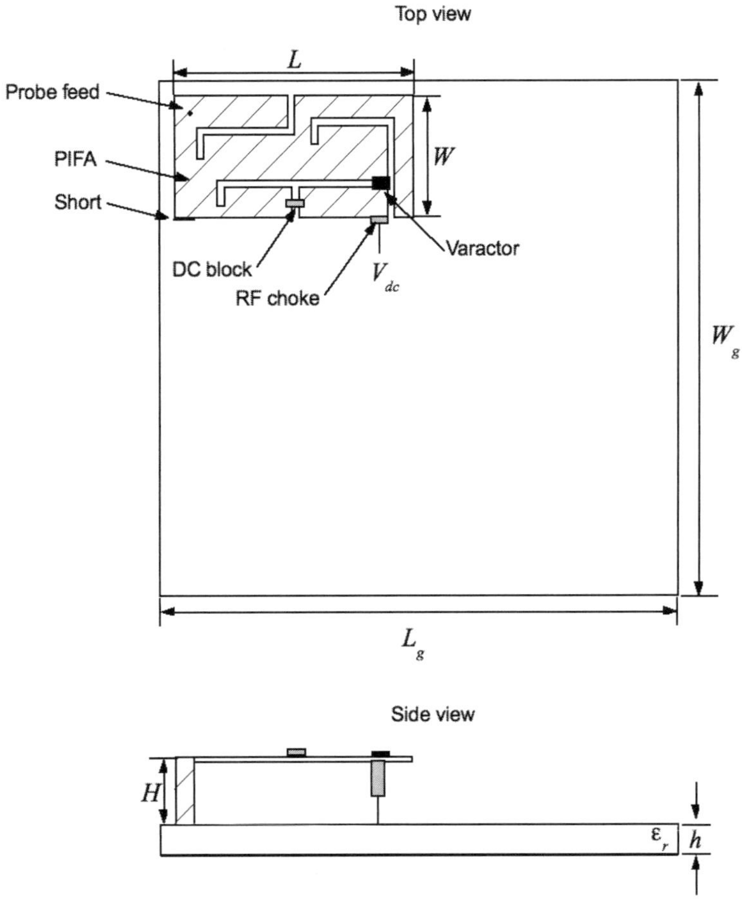

Figure 7.30 Varactor-loaded PIFA antenna (L = 19.5 mm, W = 9.5 mm, H = 4 mm, L_g= 80 mm, W_g = 45 mm, h = 1.0 mm, and ε_r = 4.6). (After [22].)

This tuning range covers several communication bands including DCS (1.71–1.88 GHz), PCS (1.88–1.99 GHz), UMTS (1.9–2.17 GHz), WiBro (2.3–2.39 GHz) and Bluetooth (2.4–2.48 GHz). There was only a small change in the 5.3-GHz resonance frequency and little change in the 5.8-GHz resonant frequency over the range of applied DC voltage bias. Measured peak gain values varied from about 1.7–2.4 dBi across the tunable frequency range.

The second example of a varactor-loaded PIFA is shown in Figure 7.32 [23]. This PIFA was designed to cover the Integrated Services Digital Broadcasting-Terrestrial (ISDB-T) band (470–770 MHz). The compact PIFA (L_1 = 40 mm, W_1 = 24 mm) was made of a thin (0.2-mm) copper sheet and was mounted at the

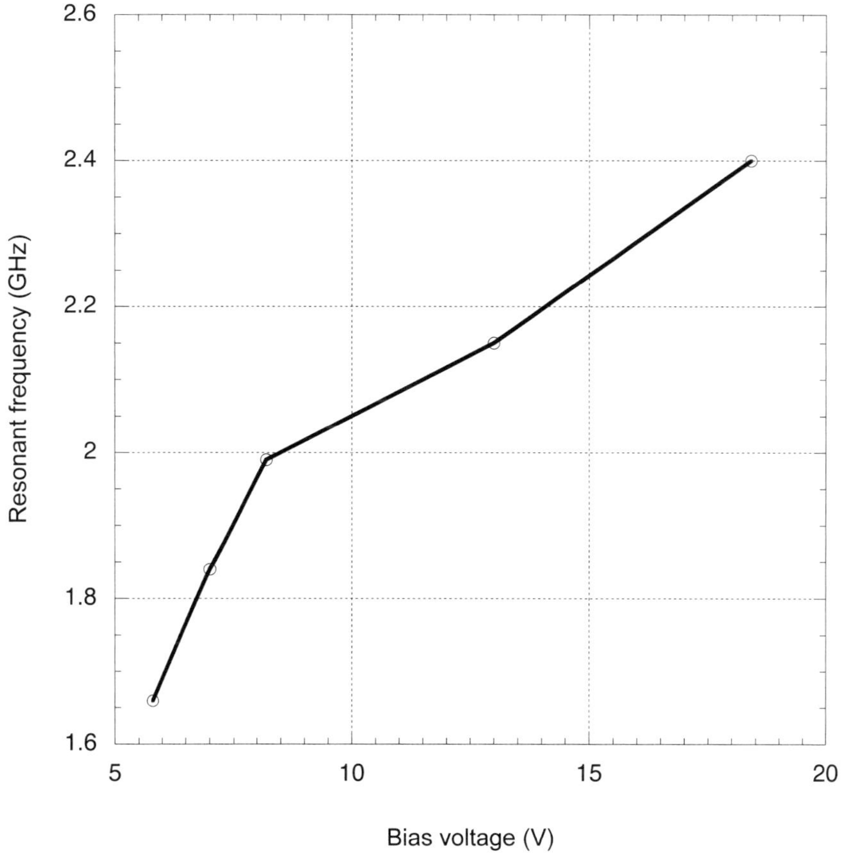

Figure 7.31 Tuning performance of the varactor-loaded PIFA. (From [22].)

top of a small PCB substrate (L = 100 mm, W = 60 mm). In this case, the slots in the PIFA were used to lower the resonant frequency, resulting in a more compact design. The varactor used had a capacitance range from 8 to 0.8 pF for a voltage bias range of 0 to 5V, and was connected to one end of the PIFA. A 100-pF DC blocking capacitor and two 100-nH inductors were used to isolate the RF and DC signals. By varying the DC bias voltage from 0 to 5V, the resonant frequency of the PIFA was tuned from 410 to 770 MHz (TR = 61%), as shown in Figure 7.33; however a 10-dB return loss was only maintained over a small tuning range resulting in a total spectrum of TS ~ 15%. The measured peak gain of the PIFA ranged from -9 to -1 dBi over the tuning range, which was still deemed suitable for the ISDB-T application. Other examples of varactor-loaded PIFAs can be found in [24–31].

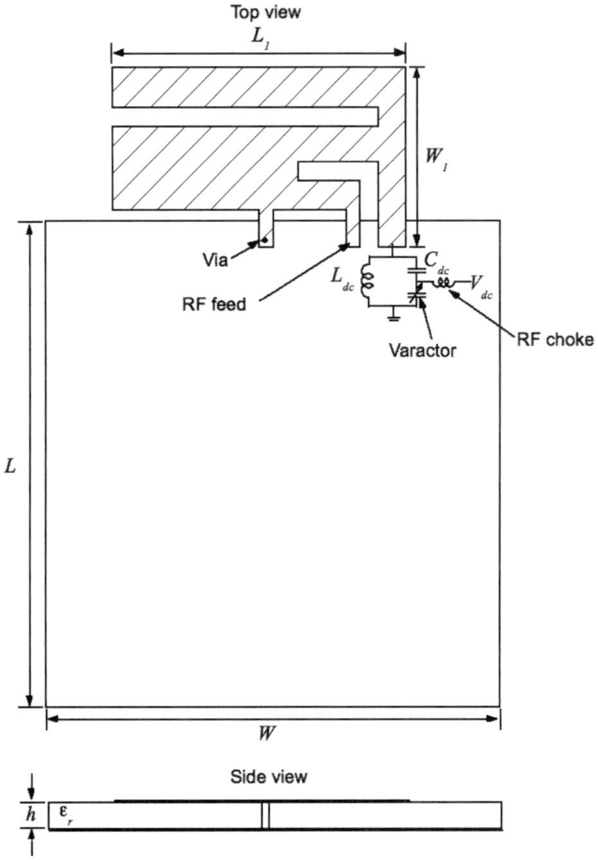

Figure 7.32 Varactor-loaded PIFA (L = 100 mm, W = 60 mm, L_1= 40 mm, W_1= 24 mm, h = 0.8 mm, and ε_r= 2.65). (After [23].)

7.4.7 Slot Antennas

Rectangular and annular slot antennas are examined in Section 3.5. This section looks at several examples of varactor-loaded slot antennas designed for frequency agility, including circular and rectangular annular slots, rectangular and cross-shaped slots, and a cavity-backed rectangular slot.

7.4.7.1 Stripline-Fed Circular Annular Slot

The first example of a varactor-loaded slot antenna is the electrically small circular annular slot, shown in Figure 7.34 [32]. The passive stripline-fed slot was

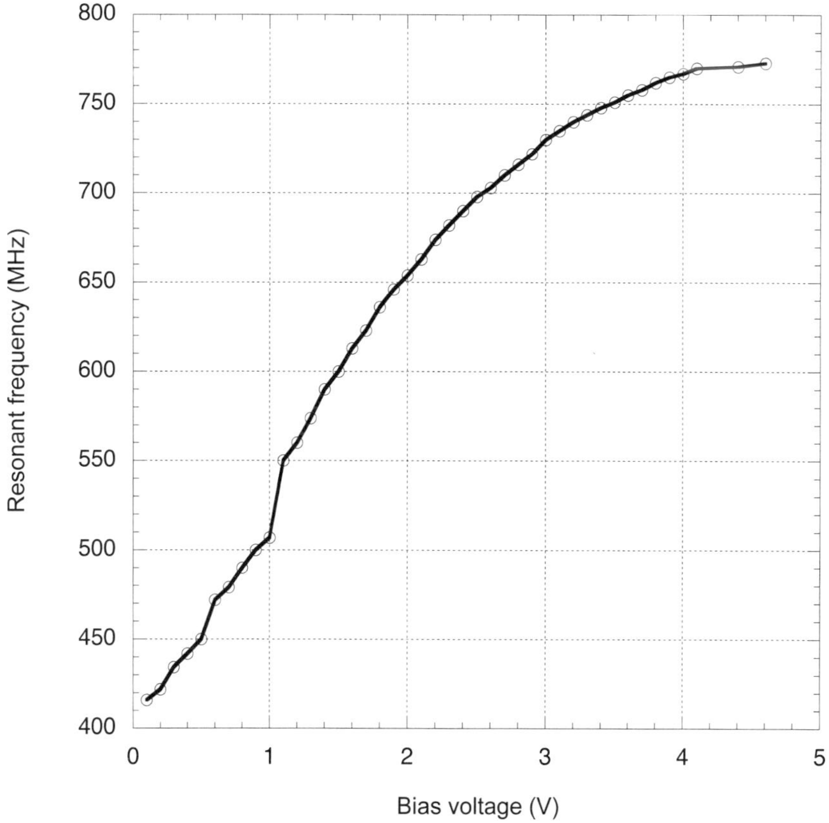

Figure 7.33 Tuning performance of the varactor-loaded IFA. (From [23].)

designed to operate at about 2 GHz and radiate like a short vertically mounted monopole antenna. The 10-dB return loss bandwidth of the passive slot was, however, much narrower than that of the corresponding monopole (10 MHz as opposed to >150 MHz). The hyperabrupt junction varactor diode selected had a capacitance-to-voltage ratio of 1.44 pF/V and was connected between the top and bottom ground planes at the center of the circular annular slot. The resonant frequency of the antenna varied from 1.92 to 2.55 GHz (TR = 28.2%) as the DC bias voltage was increased from 0 to 15V, as shown in Figure 7.35. The total spectrum achieved was TS = 30%. The measured radiation efficiency of the slot varied from 27% to 45% as the DC bias voltage was increased from 0 to 5V, with a corresponding gain of between -1 and 2.5 dBi. The radiation efficiency (e_{rad}) for the varactor-loaded annular slot can be determined using [30]

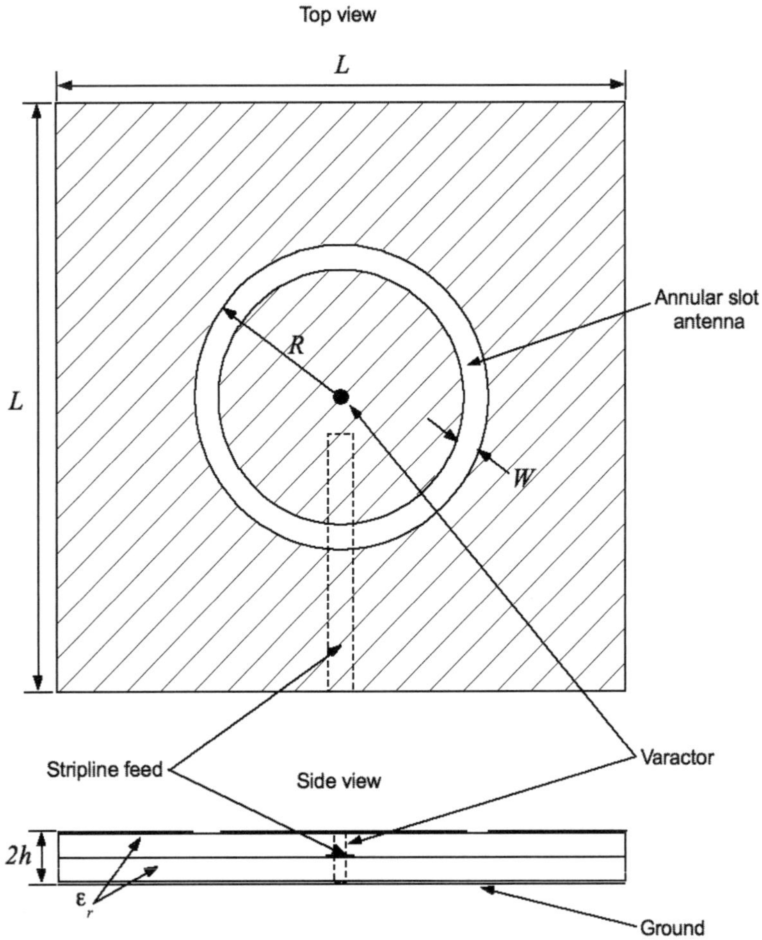

Figure 7.34 Varactor-loaded annular slot antenna (R = 9 mm, w = 0.5 mm, L = 60 mm, h = 1.59 mm, and ε_r = 2.2). (After [32].)

$$e_{rad} = \frac{P_r}{P_r + P_f + P_v + P_c + P_d} \tag{7.3}$$

where P_r is the radiated power, P_f is the power dissipated in the feed network, P_v is the power loss in the varactor, P_c is the copper loss, and P_d is the dielectric loss. For this antenna, the various losses were estimated to be $P_c \sim 0.03$ dB, $P_d \sim 1$ dB, $P_f \sim 1$ dB, and 2.8 dB > P_v > 1.4 dB as the bias voltage varied between 0 to 5V. Use of less lossy devices and materials would result in improved efficiency for the antenna.

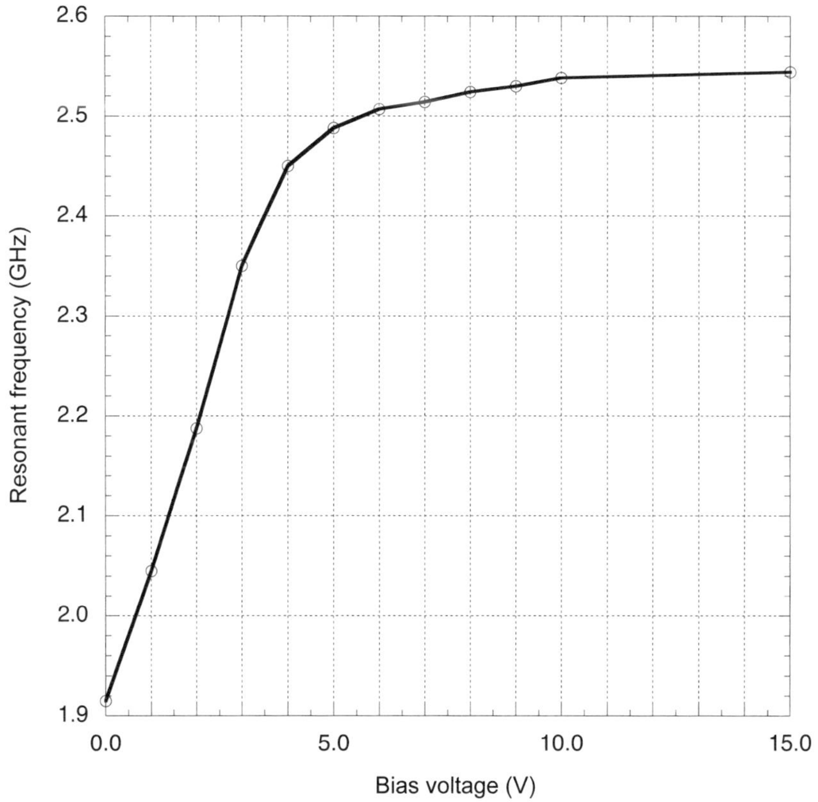

Figure 7.35 Tuning performance of the varactor-loaded annular slot antenna. (From [30].)

7.4.7.2 CPW-Fed Rectangular Annular Slot

A varctor-loaded rectangular annular slot antenna fed by a CPW line and designed to operate around 5 GHz is shown in Figure 7.36 [33]. The varactor was connected across the top slot. A sketch of the E-fields across the slot is shown in Figure 7.37. The E-fields over the top and bottom slots are in-phase and therefore contribute to the radiated field, while the fields across the two side slots are out-of-phase, canceling each other and contributing little to the radiated field. The top and bottom slots are thus considered the radiating edges while the two side slots are considered nonradiating. Placing the varactor across the top radiating edge and increasing its capacitance value had the effect of increasing the effective length of the antenna, thereby decreasing the resonant frequency. The varactor capacitance ranged from 1.4 to 0.15 pF for a DC bias voltage ranging from 0.5 to 20V. The resonant frequency of the antenna varied from 4.92 to 5.4 GHz (TR = 9.7%), as shown in Figure 7.38, when the bias voltage was varied from

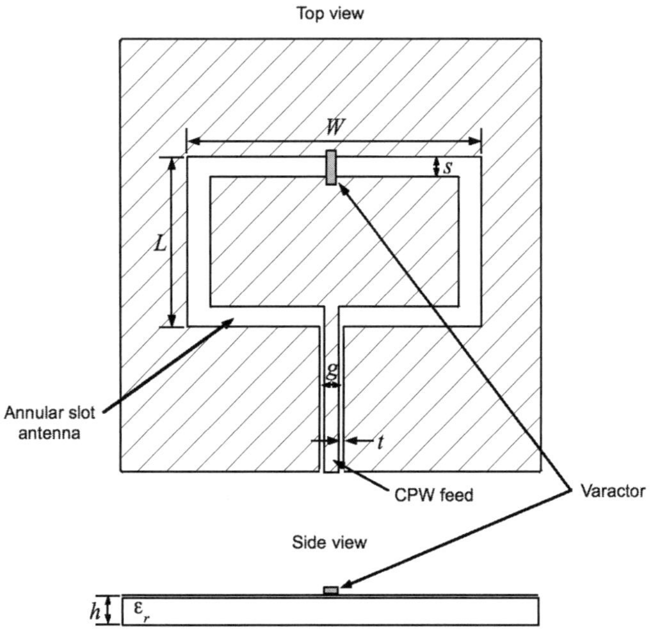

Figure 7.36 Varactor-loaded CPW-fed annular slot antenna (L = 25.4 mm, W = 34.4 mm, s = 2.2 mm, g = 0.1 mm, t = 1.59 mm, h = 0.762 mm, and ε_r = 2.94). (After [33].)

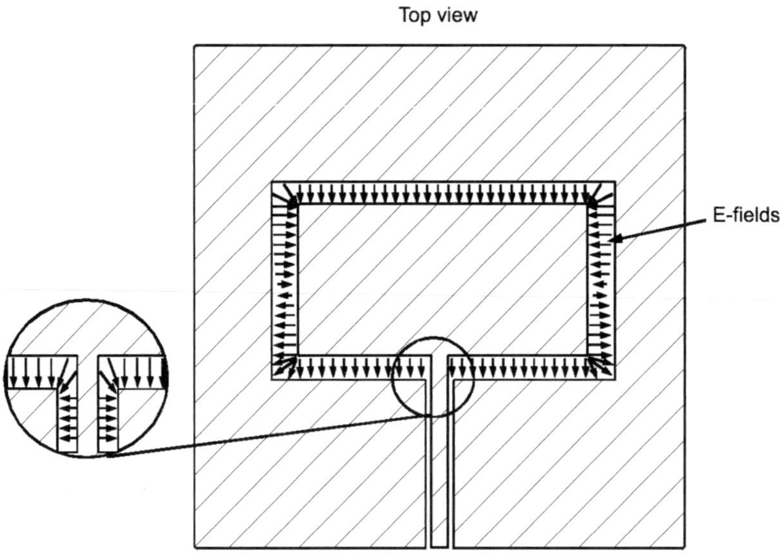

Figure 7.37 Varactor-loaded CPW-fed annular slot antenna. (After [33].)

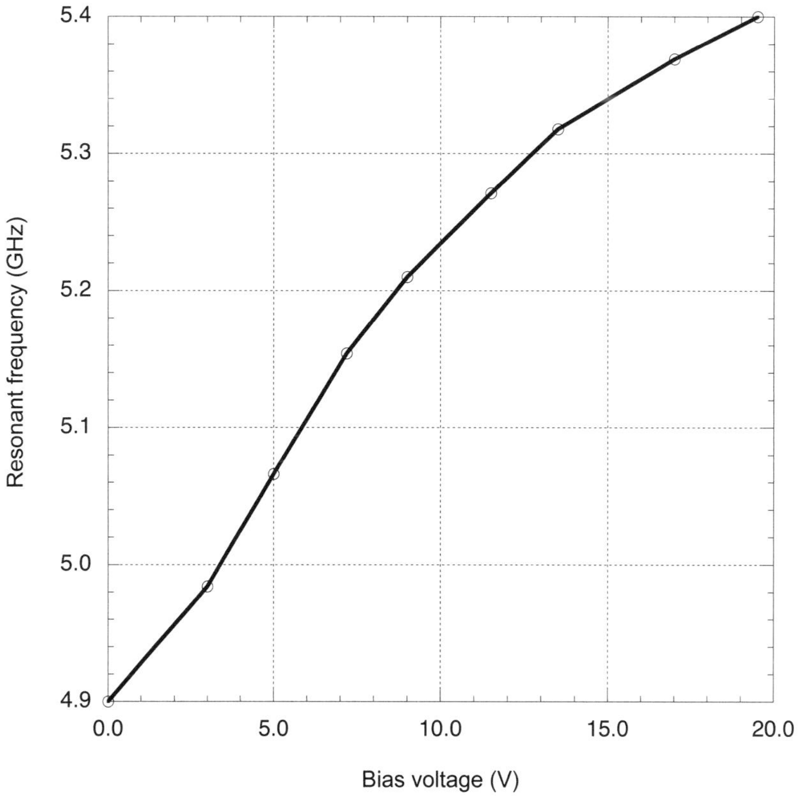

Figure 7.38 Tuning performance of the varactor-loaded annular slot antenna. (From [33].)

0 to 19.5V. The total spectrum for the antenna was TS > 14%. Due to the capacitive loading effect of the varactor on the top radiating edge, the phase of the radiating field from the top edge somewhat lagged that of the bottom radiating edge, resulting in a beam tilt of the radiation pattern in the E-plane. No information was provided on the gain or radiation efficiency of this design.

7.4.7.3 Microstrip-Fed Rectangular Annular Slot

A rectangular annular slot antenna fed by a microstrip line and etched on a square substrate (70 mm x 70 mm) is shown in Figure 7.39 [34]. Two varactor diodes with capacitance values ranging from 2.5 to 0.45 pF for an applied bias voltage of 0–20V were connected across the two sides of the slot antenna perpendicular to the microstrip feed line. Biasing of the varactor diodes was achieved by applying a DC voltage to the microstrip line using a bias-T that was connected

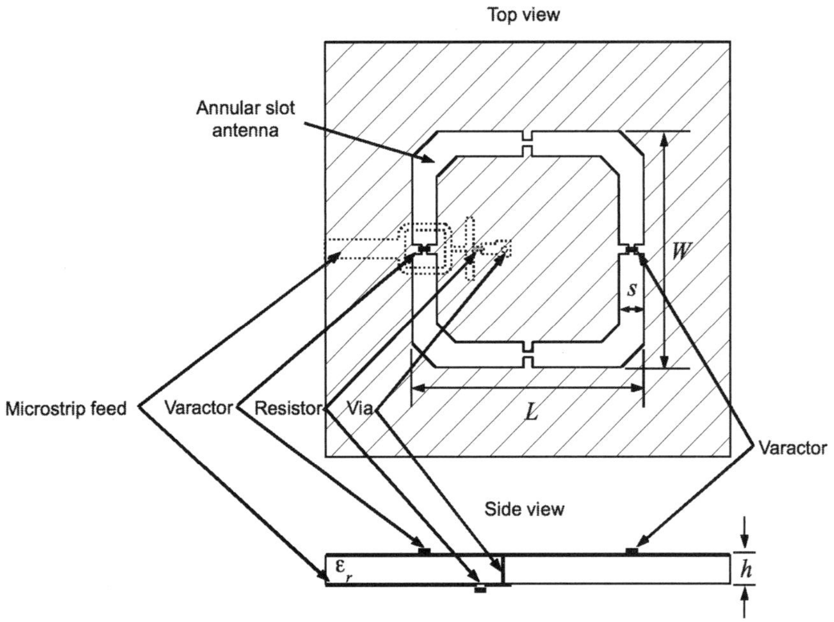

Figure 7.39 Varactor-loaded annular slot antenna ($L = W = 29$ mm, $s = 3$ mm, $h = 0.787$ mm, and $\varepsilon_r = 2.2$). (After [34].)

to the center of the annular slot through a 100-kΩ resistor and a via hole. The resonant frequency of the antenna ranged from 0.93 to 1.72 GHz (TR = 59.6%), as shown in Figure 7.40, with a total spectrum of TS = 65%. The measured gain of the antenna ranged from -5 to 2 dBi corresponding to a radiation efficiency between about 12% and 67%.

By adding a second microstrip feed in the vertical direction and two more varactor diodes, a dual-polarized design was achieved, where the resonant frequency for each polarization could be controlled nearly independently, with a tuning range of 0.93–1.6 GHz (TR = 53%). The isolation between the two ports remained better than 14 dB and measured radiation effiency ranged from 20% to 85%.

7.4.7.4 Microstrip-Fed Rectangular Slot

This section examines two rectangular shaped slot antennas fed by microstrip lines and loaded with varactor diodes for continuous frequency tuning. The first

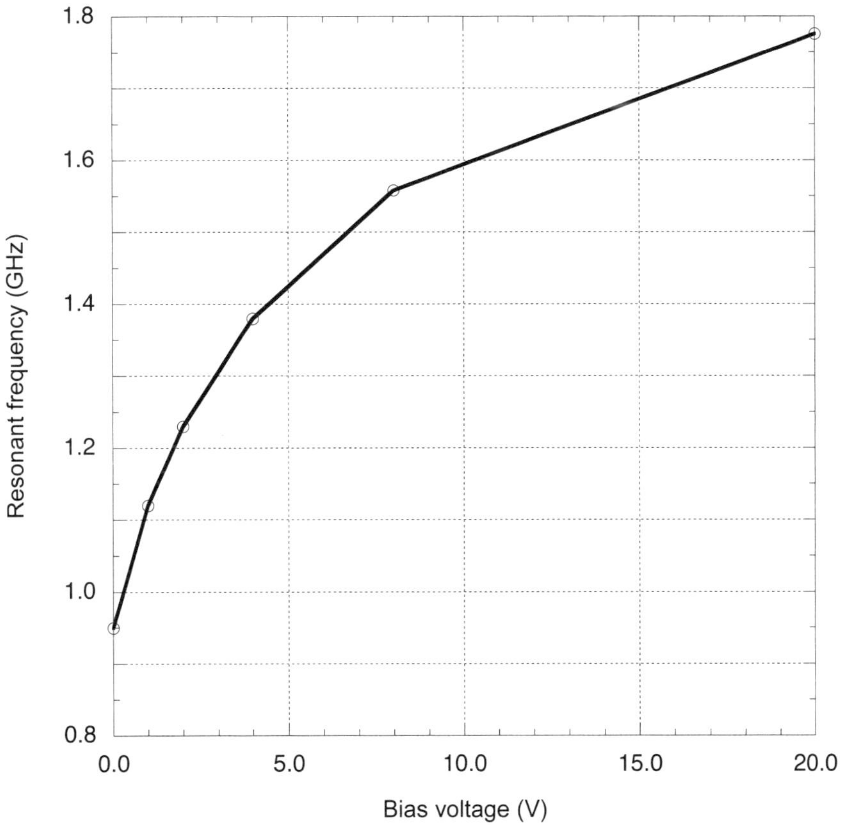

Figure 7.40 Tuning performance of the varactor-loaded annular slot antenna. (From [34].)

is shown in Figure 7.41 and consists of a thin rectangular slot across which a varactor diode was connected at one end [35]. The DC bias voltage was provided through an RF choke, and two DC blocking capacitors were added to prevent the bias voltage from interacting with the RF signal, which was fed onto the microstrip line. The GaAs hyperabrupt varactor diode used had a capacitance that varied from 1.2 to 0.2 pF when the DC bias increased from 0 to 20V. The resonant frequency of the slot shifted from 0.78 to 2.09 GHz (TR = 91.3%) as the bias to the varactor increased from 0 to 20V, as shown in Figure 7.42. The antenna achieved a total spectrum of TS = 91%. The gain or radiation efficiency of the varactor-loaded antenna was not reported.

The second slot antenna, shown in Figure 7.43, was designed for dual-band operation [36]. The antenna consists of a rectangular slot, folded at one end and fed by a microstrip line. Two varactor diodes were connected across the slot, whose proper placement allowed for independent control of the two resonant

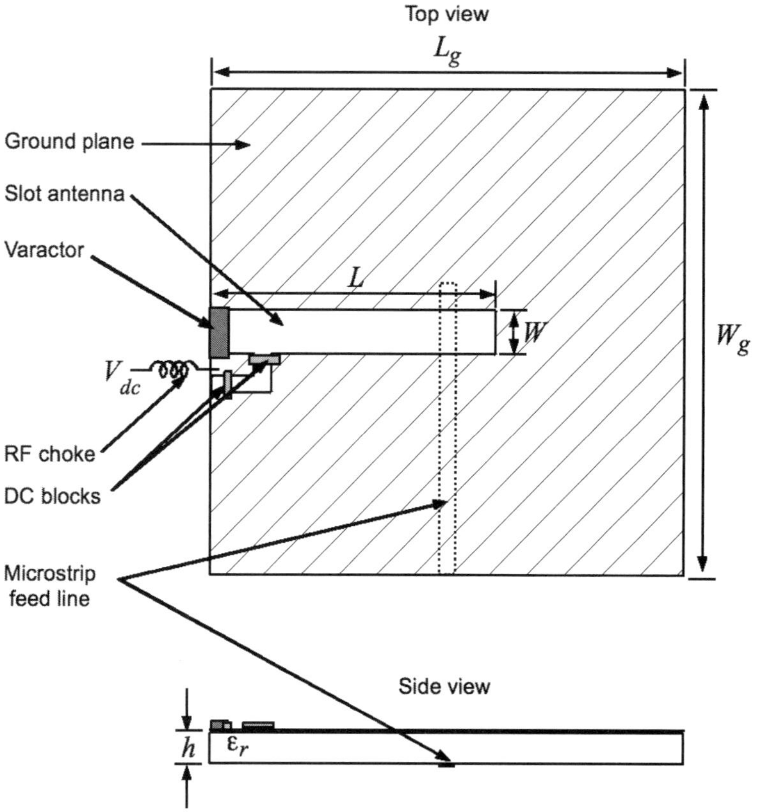

Figure 7.41 Varactor-loaded rectangular slot antenna (L_g = 30 mm, W_g = 80 mm, L = 20 mm, W = 2 mm, h = 0.8 mm, and $\varepsilon_r \sim 4.2$). (After [35].)

frequencies. The varactors had a capacitance ranging fom 2.25 to 0.5 pF when the applied bias increase from 0 to 30V. Bias was provided to the varactors via thin printed lines. RF chokes and DC blocks were added, as illustrated in Figure 7.43, to isolate the DC and RF signals.

To understand the behavior of this antenna, the slot can be modeled as a transmission line, as shown in Figure 7.44. By using the transverse resonance technique, the impedance looking to the left (Z_L) and to the right (Z_R) of the circuit at any arbitrary location along the transmission line must satisfy the following condition

$$Z_R + Z_L = 0 \tag{7.4}$$

where

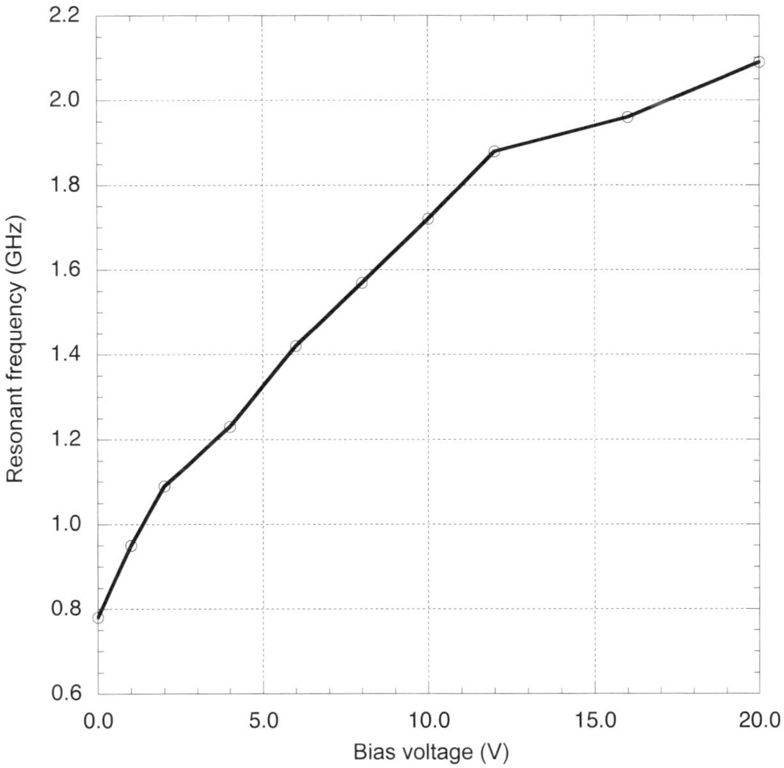

Figure 7.42 Tuning performance of the varactor-loaded rectangular slot antenna. (From [35].)

$$Z_R = \frac{jZ_o \tan(\theta - \theta_2)}{1 - \omega C_2 \tan(\theta - \theta_2)} \tag{7.5}$$

$$Z_L = \frac{\tan(\theta_1) + \tan(\theta_2 - \theta_1)[1 - \omega C_1 Z_o \tan(\theta_1)]}{1 - \omega C_1 Z_o \tan(\theta_1) - \tan(\theta_1)\tan(\theta_2 - \theta_1)} \tag{7.6}$$

and $\theta = \beta(\omega)L$, $\theta_1 = \beta(\omega)L_1$, $\theta_2 = \beta(\omega)L_2$, $\omega = 2\pi f$. The values for β and Z_o are found in [37] and are repeated in Appendix A. Substituting (7.5) and (7.6) into (7.4) results in

$$\begin{aligned} &\{\tan(\theta_1) + \tan(\theta_2)[1 - \omega C_1 Z_o \tan(\theta_1)]\}[1 - \omega C_2 Z_o \tan(\theta - \theta_2)] \\ &+ [1 - \omega C_1 Z_o \tan(\theta_1) - \tan(\theta_2)\tan(\theta_1)]\tan(\theta - \theta_2) = 0 \end{aligned} \tag{7.7}$$

The resonant frequencies of the slot are determined by solving for the roots of (7.7). For given slot dimensions, substrate thicknesses, and dielectric constants,

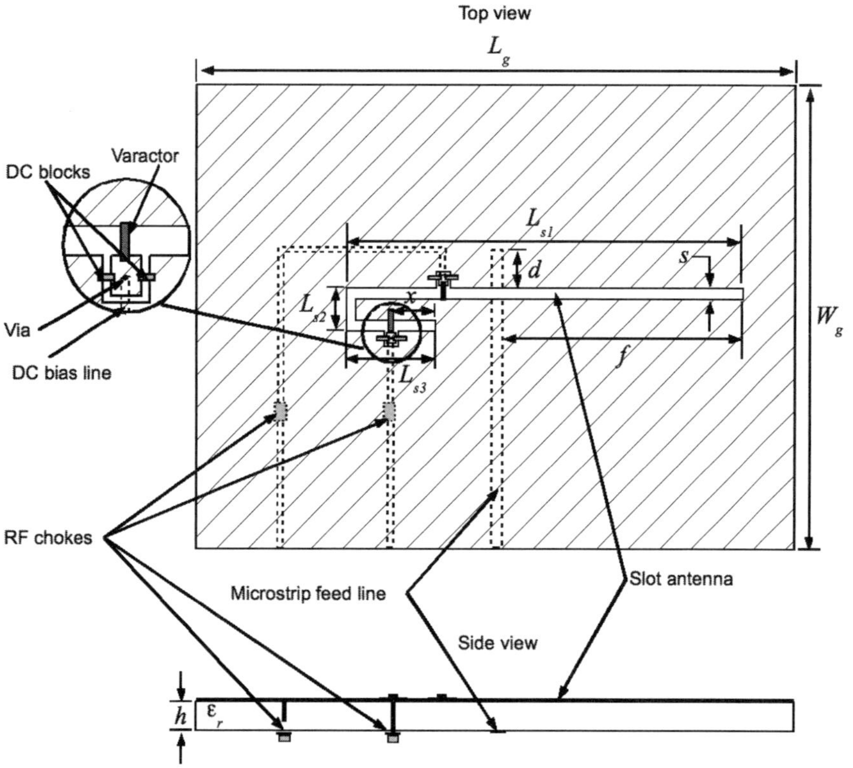

Figure 7.43 Varactor-loaded slot antenna (L_g = 150 mm, W_g = 110 mm, L_{s1} = 45 mm, Ls_2 = 4 mm, L_{s3} = 13 mm, s = 1 mm, f = 27.5 mm, d = 3.2 mm, x = 4.5 mm, h = 0.5 mm, and ε_r = 3.4). (After [36].)

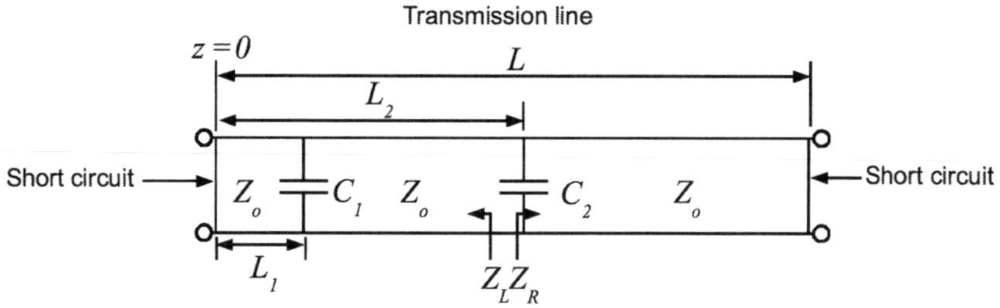

Figure 7.44 Transmission line model of the varactor-loaded slot antenna. (After [36].)

the resonant frequencies will be dependent on the location of the varactors (L_1, L_2) and on their capacitance values (C_1, C_2). The varactor locations can thus be optimized to maximize the tunability range of the two resonant frequencies.

The reason for adding a bend in the slot is explained with the aid of Figure 7.45. Figure 7.45(a) shows a sketch of the magnetic currents and voltage magnitude of the first (f_1) and second (f_2) resonances in a straight slot. The first resonance has a voltage peak in the center of the slot, resulting in a radiation pattern with a broadside beam, while the second resonance has a voltage null, resulting in a broadside null in the radiation pattern. Since in most applications it is desirable to have a constant pattern shape over the tuning range, the straight slot is not a suitable candidate. By introducing a bend in the slot, whose length is $\lambda_g/4$ at f_2, the resultant magnetic currents and voltage magnitude sketched in Figure 7.45(b) arise. The magnetic current in the bent portion of the slot has equal magnitude but opposite direction than the portion directly above it, so that their respective contributions to the radition pattern will be canceled. The resultant voltage is that of a shortened straight slot, as sketched in the right-hand portion of Figure 7.45(b), which will radiate a pattern with a broadside peak, similar to that at f_1. Since, at f_1, the bent portion of the slot will be significantly less than $\lambda_g/4$, it will only have a small impact on the radition pattern at this lower frequency.

For the antenna designed in Figure 7.43, the resonant frequency f_1 could be independently controlled by maintaining the DC bias voltage V_1 constant at 30V and varying V_2 from 0 to 30V. The resultant shift in f_1 was from 1.1 to 1.45 GHz (TR_1 = 27.5%), as shown in Figure 7.46. The total spectrum for this mode was TS_1 = 30%. To independently vary f_2 required a combination of values for V_1 and

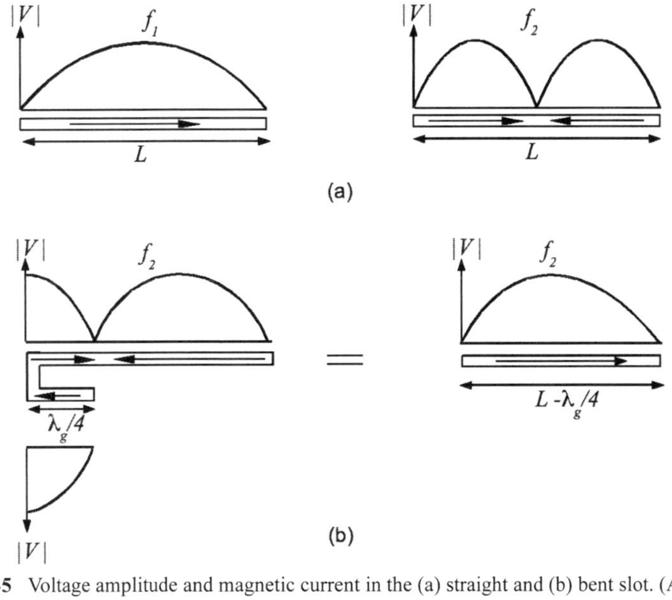

Figure 7.45 Voltage amplitude and magnetic current in the (a) straight and (b) bent slot. (After [36].)

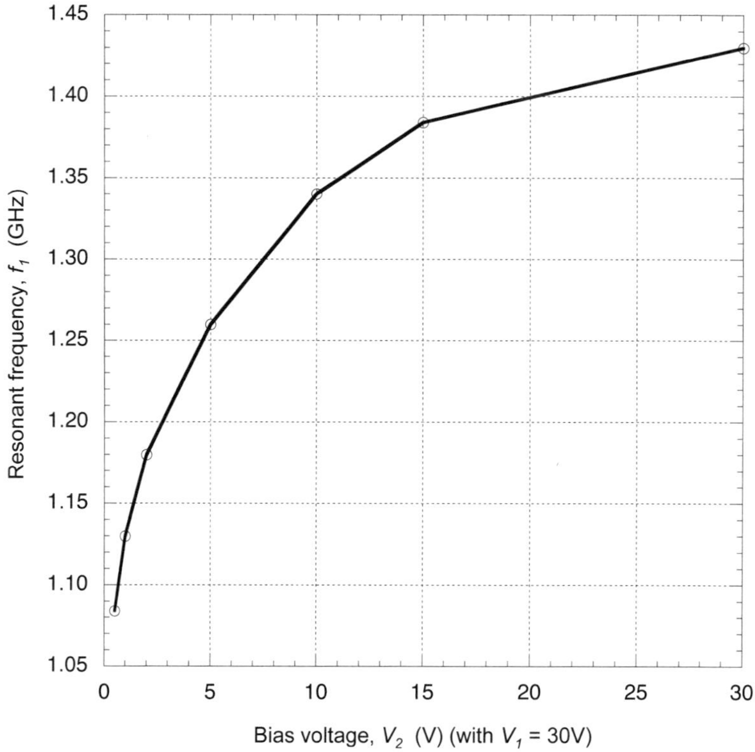

Figure 7.46 Tuning performance of f_1 for the varactor-loaded slot antenna. (From [36].)

V_2, as shown in Figure 7.47, with a frequency shift from 1.75 to 2.92 GHz (TR_2 = 50%) with a total spectrum of TS_2 = 54%. The measured gains for this antenna varied from -0.6 to 1.2 dBi for f_1 and -0.4 to 1.8 dBi for f_2. The corresponding radiation efficiency from 58% to 87% for f_1 and 59% to 92% for f_2.

7.4.7.5 Microstrip-Fed Cross-Slot Antenna

An example of a varactor-loaded cross-slot antenna designed for operation around 7.5 GHz is illustrated in Figure 7.48 [38]. The cross-slot was fed with a microstrip line that also incorporated the varactor diode. The application of this design was as a unit cell for an active grid array where the horizontal slot received the RF signal while the vertical slot received the local oscillator (LO) signal, which would then be combined by a mixer. (By appropriate selection of the lengths of the two slots, the cross-slot antenna could also be designed for circular polarization.) A bias-T was used to provide the DC voltage to the anode terminal of the varactor via the microstrip line, while the cathode terminal was

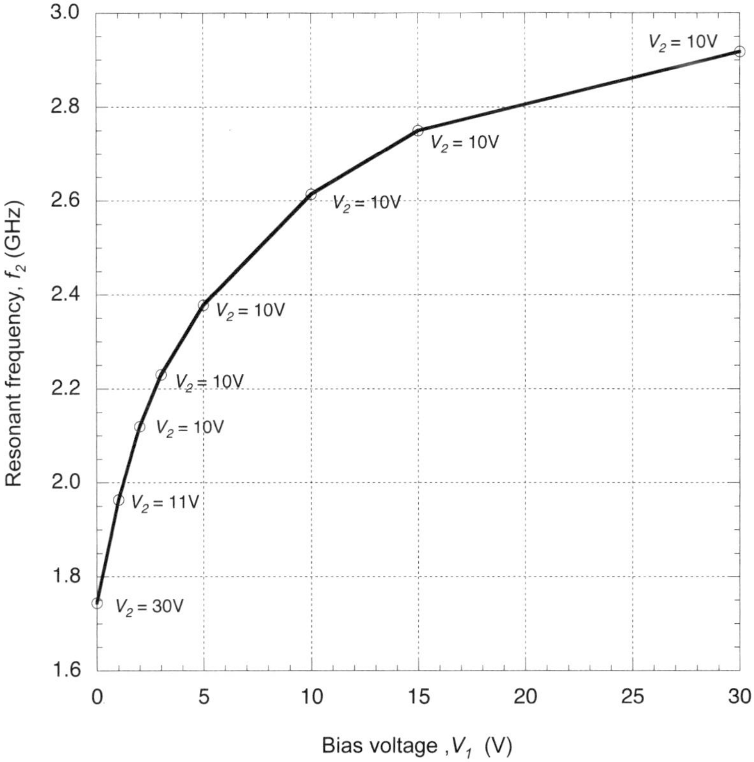

Figure 7.47 Tuning performance of f_2 for the varactor-loaded slot antenna. (From [36].)

DC- and RF-shorted to the ground plane. By varying the DC voltage from 0 to 30V, the resonant frequency of the slot shifted from 7.3 to 8.1 GHz (TR = 10%), as shown in Figure 7.49. Information on gain or radiation efficiency for this antenna was not provided.

7.4.7.6 Cavity-Backed Rectangular Slot Antenna

The previous examples of slot antennas all produced bidirectional radiation patterns, making them unsuitable for conformal mounting on a ground plane or other scattering objects. By adding a cavity backing to the slot antenna, the backlobe can be suppressed, resulting in a radition pattern with a single main beam. The final example of a microstrip-fed varactor-loaded slot antenna is a cavity-backed rectangular slot, shown in Figure 7.50 [39]. By connecting a varactor across the slot, with a capacitance ranging from 2.5 to 0.45 pF for an applied DC bias voltage of 0–20V, the resonant frequency of the antenna was

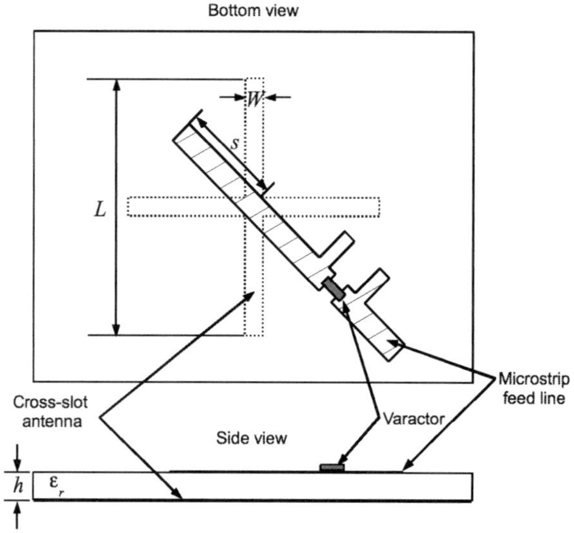

Figure 7.48 Varactor-loaded cross-slot antenna (L = 30.2 mm, W = 1.22 mm, s = 10 mm, h = 0.508 mm, and ε_r = 2.2). (After [38].)

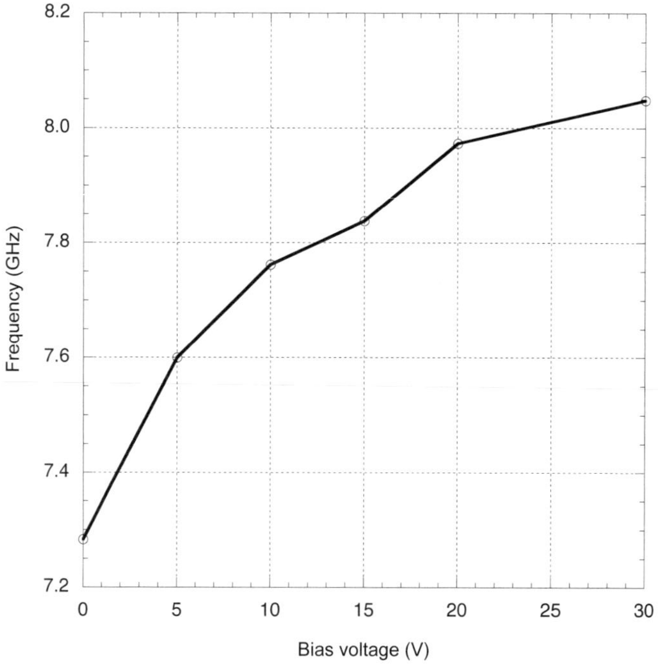

Figure 7.49 Tuning performance of the varactor-loaded cross-slot antenna. (From [38].)

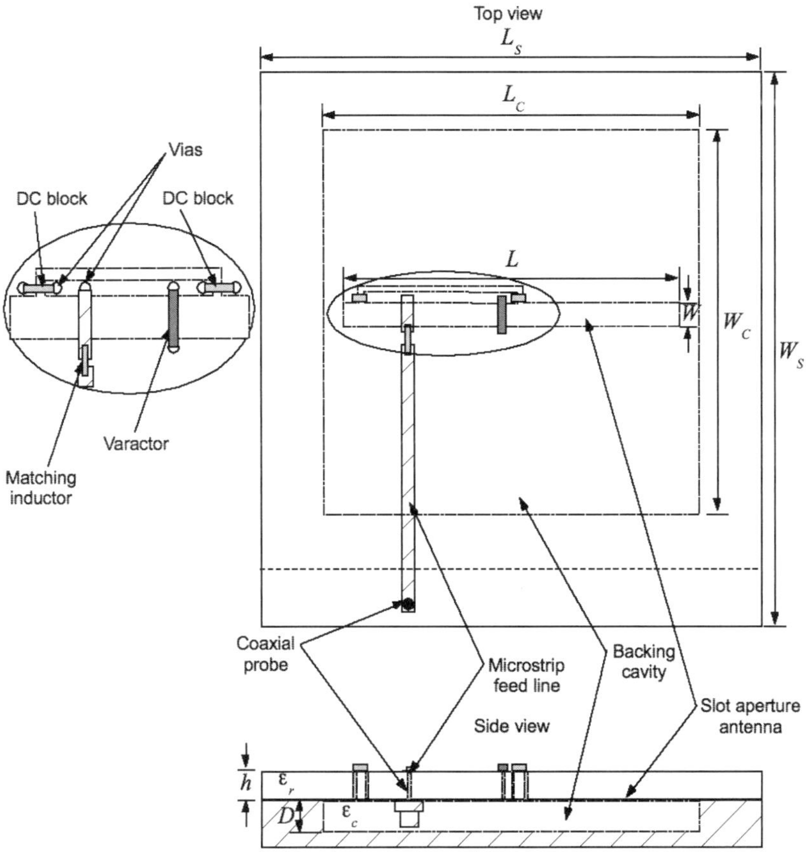

Figure 7.50 Varactor-loaded cavity-backed slot aperture antenna (L = 60 mm, W = 3 mm, h = 0.79 mm, ε_r = 2.2, L_c= W_c = 72 mm, D = 3.18 mm, ε_c = 2.1, L_s = 92 mm, and W_s = 102 mm). (After [39].)

tuned from 1 to 1.9 GHz (TR = 62%), as shown in Figure 7.51. The antenna attained a total spectrum of TS ~ 63%. Since the slot was etched on the bottom side of the dielectric and the components (varactor, 22-pF DC blocks, and a 7.5-nH RF choke) were mounted on the opposite side, connections were made using via holes, as shown in the inset of Figure 7.50. The measured gain of the antenna varied from -2 to 5.0 dBi with a corresponding radiation efficiency ranging from 18% to 76%. The largest contribution to the loss was the series resistance of the varactor diode, followed by the metal/dielectric losses. The 1-dB compression point (P_{1dB}) of the antenna was also measured to get a sense of its power-handling capabilities. The value of P_{1dB} ranged from -10 dBm at 1 GHz to 10 dBm at 1.9 GHz. Other examples of varactor-loaded slot antennas can be found in [40–42].

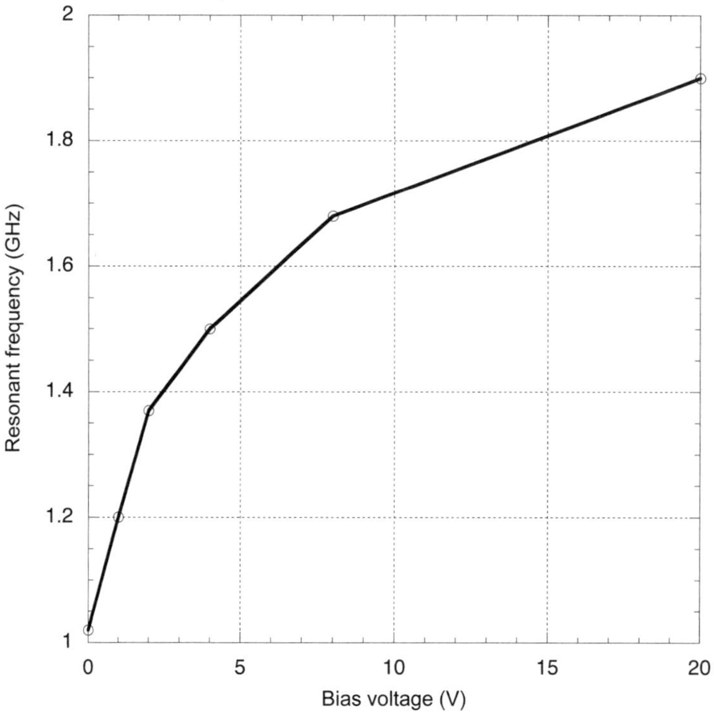

Figure 7.51 Tuning performance of the varactor-loaded cavity-backed slot aperture antenna. (From [39].)

7.4.8 DRAs

An example of a tunable DRA is illustrated in Section 6.1.3, where the DRA is fabricated from a ferrite material. Frequency tuning was achieved by altering the permeability of the ferrite with an applied DC magnetic field. This section considers the use of varactor loading to achieve frequency agility. This method is based on the fact that placing metal walls on certain faces of a rectangular DRA results in a reduction in resonant frequency. This behavior can be explained with the aid of Figure 7.52. Since there is symmetry about the vertical centerline of the E-field distribution for the $TE^{x}_{\delta 11}$ mode, shown in the aperture-coupled DRA in Figure 7.52(a), halving the dimension *W*, and placing a grounded metallic plate along the line of symmetry does not perturb the fields. As a result, the resonant frequency remains unchanged. Alternatively, if a shorting plate is added to the original DRA of dimension *W*, as shown in Figure 7.52(b), its resonant frequency

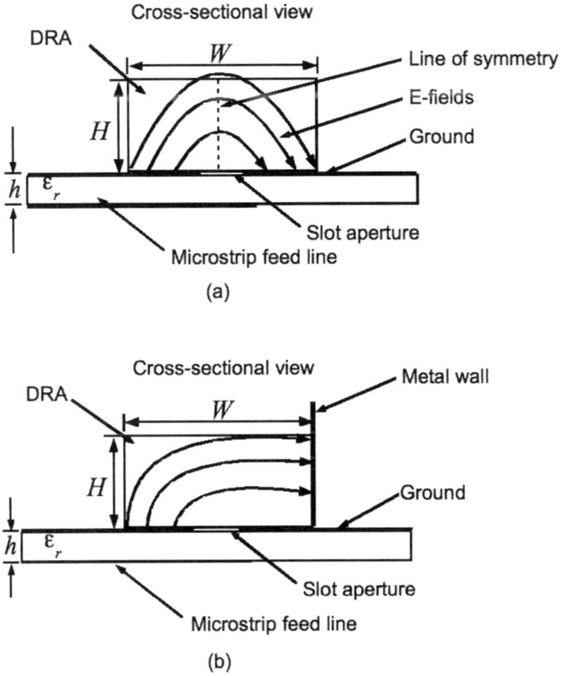

Figure 7.52 Effect of a metallic wall on the $TE^{x}_{\delta 11}$ mode E-field configuration of a rectangular DRA.

will shift downward, since its effective dimension *W* is doubled due to the image from the shorting plate. If one could dynamically control the dimensions of the plate, then the DRA would achieve frequency agility. This was demonstrated passively using a combination of small shorting tabs in [43] and will be discussed in more detail in Section 8.1.6 on discrete tuning using PIN diodes.

Here, continuous frequency agility in the DRA is achieved by introducing a pair of varactors to opposite sides of a differentially fed slot-coupled DRA, as shown in Figure 7.53 [44]. The two opposite walls of the DRA are partially covered with metal plates, and the varactors are connected across a small gap in the thin tab between the metal walls and the ground plane. By applying a DC bias of 2–8V, the capacitance of the varactor varies from approximately 1.9 to 0.6 pF, with a resultant change in resonant frequency from 2.6 to 2.76 GHz (TR = 6%), as shown in Figure 7.54. The antenna achieved a total spectrum of TS ~ 10. Gain and radiation efficiency measurements with the varactors were not reported.

A similar, but more complex, example of a varactor-loaded DRA is shown in Figure 7.55, where three pairs of varactors were integrated along the side walls of an aperture-fed DRA [45]. Each pair of varactors was controlled by an

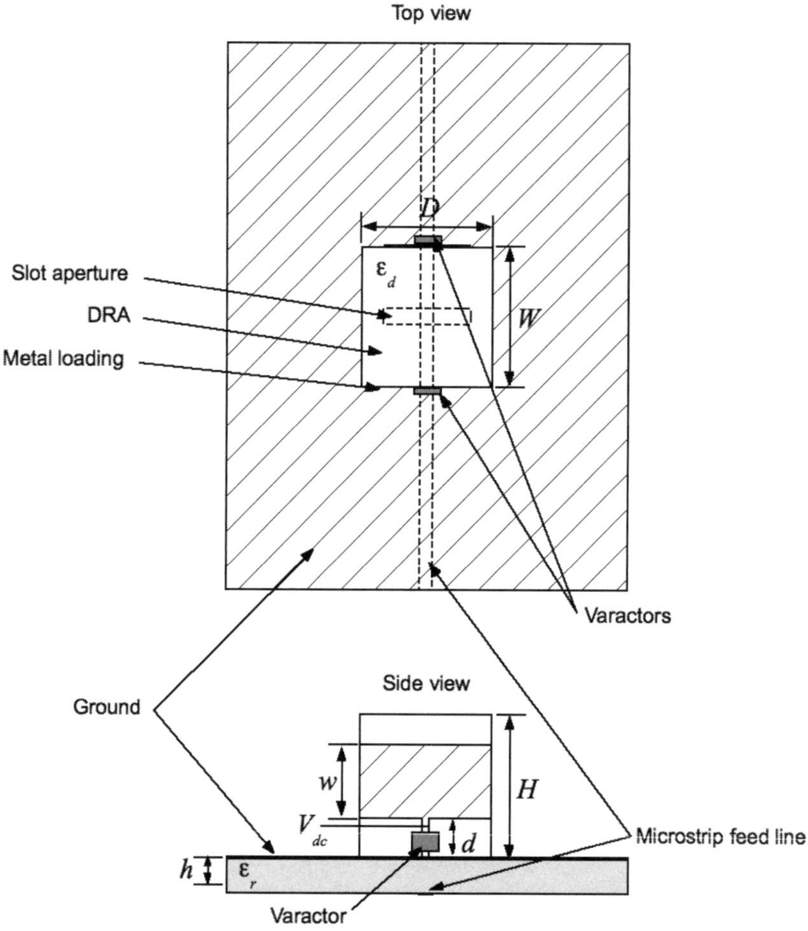

Figure 7.53 Varactor-loaded DRA ($W = D = 20$ mm, $H = 15$ mm, $\varepsilon_d = 9.4$, $w = 4$ mm, $d = 1.8$ mm, $h = 1.57$ mm, and $\varepsilon_r = 2.33$). (After [44].)

independent DC bias voltage that was applied to the metal side walls via RF chokes. Small slits ($s = 0.3$ mm) were cut in the side walls in order to separate the DC voltage signals. By varying the three DC bias voltages in various combinations between 0 and 19V, the resonant frequency of the DRA was tuned

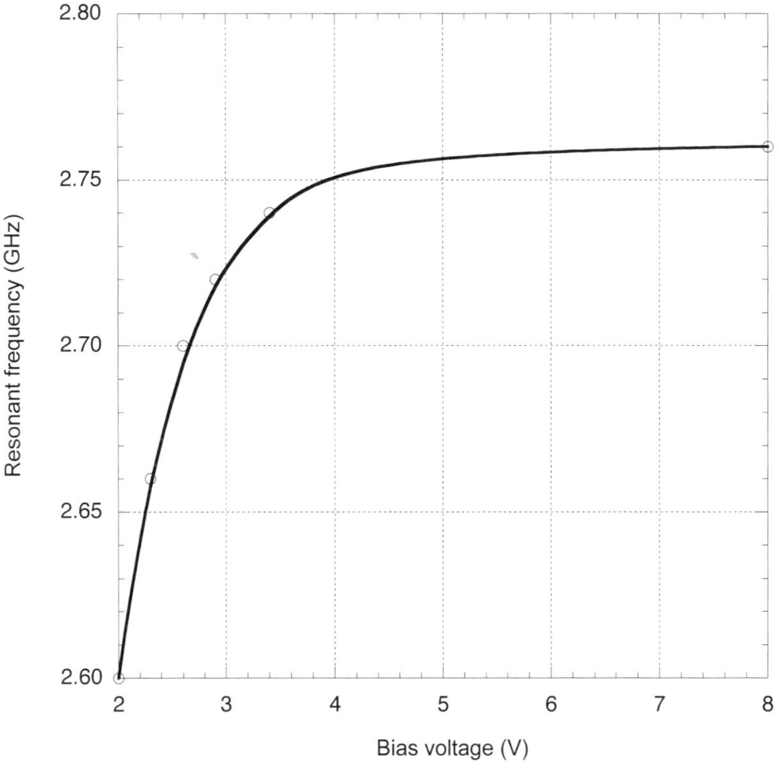

Figure 7.54 Tuning performance of the varactor-loaded DRA. (From [44].)

between 3.2 to 5.7 GHz (TR = 55%), with a total spectrum of TS = 58%. Measured gain peaks varied between 2.7 and 3.7 dBi, approximately 3 dB lower than a corresponding passive DRA. Cross-polarization levels also increased significantly compared to the passive DRA, due to the physical presence of the varactors and the accompanying bias wires in the radiating aperture. By somewhat altering the design and moving the varactors below the ground plane, the negative effects on the radiation patterns can be removed, as was demonstrated in [46].

7.4.9 Microstrip Patch Antennas

This final section examines microstrip patch antennas loaded with one or more varactor diodes. The microstrip patch antenna is the most common radiator in which varactors have been integrated to achieve frequency agility, and several examples will be presented in this section to highlight the diverse range of

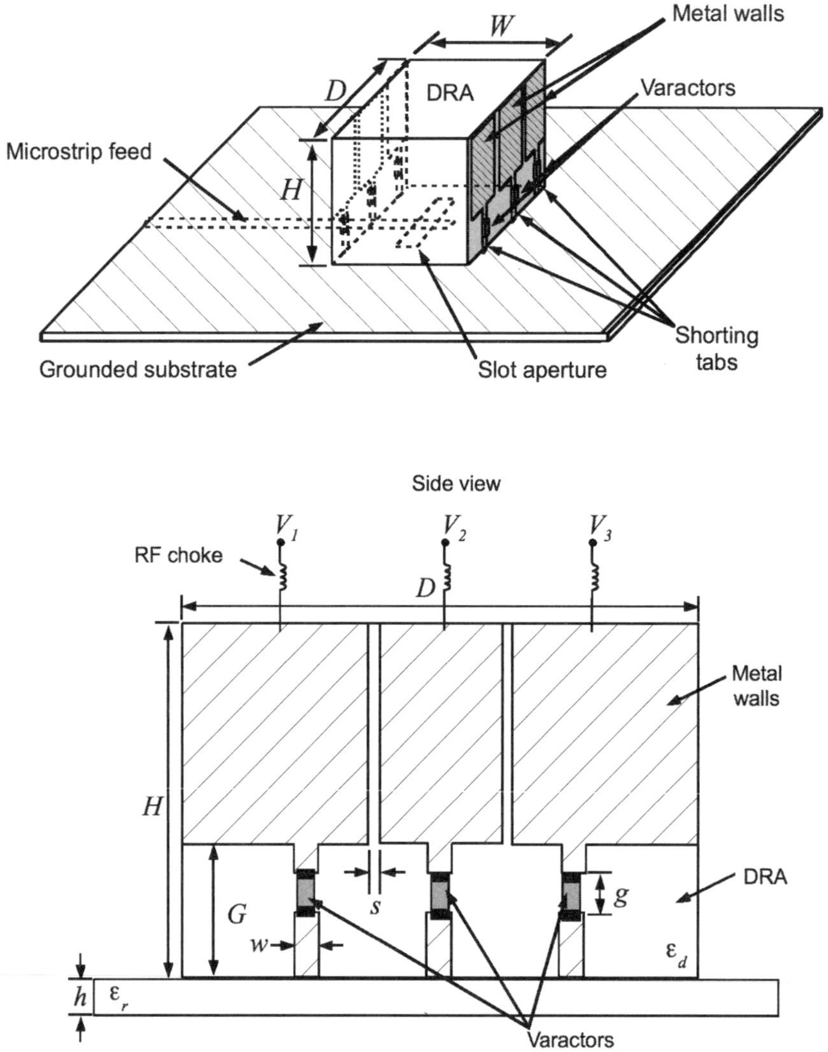

Figure 7.55 Varactor-loaded DRA ($W = D = 12$ mm, $H = 8$ mm, $\varepsilon_d = 10$, $h = 0.508$ mm, $\varepsilon_r = 3.38$, $G = 3$ mm, $s = 0.3$ mm, $g = 1$ mm, and $w = 0.5$ mm). (After [45].)

possibilities. The various designs are grouped by the type of feed used to excite the microstrip patch. The feed types include probes, slot apertures, direct microstrip line feeds, and CPW feeds.

7.4.9.1 Probe-Fed Microstrip Patch Antennas

The use of probes is one of the most common methods for exciting microstrip patch antennas, especially at lower frequencies. The probe can be implemented by simply extending the center conductor of a coaxial cable to connect to the patch while the outer conductor is connected to the ground plane. Often, a commercially available coaxial connector with a long center pin is used. The advantage of the probe lies in its simplicity and the fact that it does not require any additional matching, since it can be offset from the radiating edge of the patch at a location where the input impedance of the patch matches that of the coaxial line. However, probes do require drilling into the substrate and soldering to the patch in order to ensure a reliable connection. At higher frequencies, where the size of the patch becomes quite small, inserting and soldering a probe may be challenging and other feeding techniques, such as aperture-coupling, are more attractive.

The first example of a frequency-agile microstrip patch antenna loaded with varactor diodes was published in 1982 [47]. Two varactors were connected to opposite radiating edges of a probe-fed rectangular patch antenna, as shown in Figure 7.56. As mentioned in Section 3.4.1, this antenna can be modeled by an ideal lossless transmission line loaded at each end. For the varactor-loaded microstrip patch, the model is modified as shown in Figure 7.57. The conductance G and susceptance B account for the radiation and reactance of the radiating edges of the antenna, while suspectance B_v accounts for the reactance of the varactors. Based on this model, the resonant frequency of the varactor-loaded microstrip patch antenna can be estimated using [47]

$$f_o = \frac{c}{2\pi L\sqrt{\varepsilon_{eff}}}\tan^{-1}\left[\frac{2Y_o(B+B_v)}{(B+B_v)^2+G-Y_o^2}\right] \tag{7.8}$$

where $B = \frac{k_o \Delta L\sqrt{\varepsilon_{eff}}}{Z_o}$, $B_v = 2\pi f C_v$, G was given in (3.21), ε_{eff} in (3.9), Z_o in (3.11), and C_v is the variable capacitance of the varactor. Using two varactors whose capacitance ranged from 2.4 to 0.4 pF when the DC bias voltage was altered from 0 to 30V, the resonant frequency of the antenna was shifted between 1.42 and 1.82 GHz (TR = 21.8%), as shown in Figure 7.58.

A varactor-loaded circular microstrip patch antenna (with a radius of a = 20 mm) was also investigated in [47], achieving a frequency range from 1.85 to 2.45 GHz (TR = 28%) for a DC bias voltage range of 0 to 30V. The authors reported no significant degradation in gain performance for either antenna, with measured gains between 4.5 and 5.5 dBi.

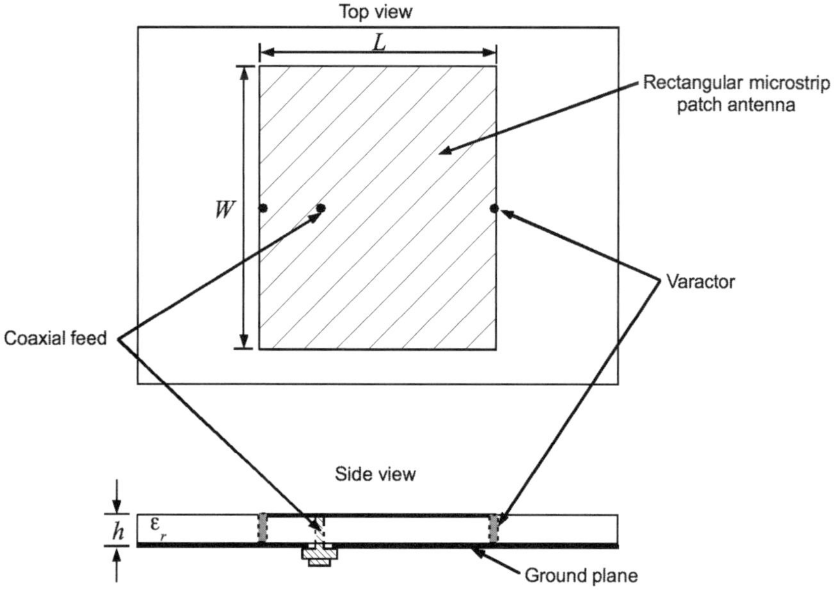

Figure 7.56 Varactor-loaded probe-fed microstrip antenna (L = 46.5 mm, W = 30 mm, h = 1.59 mm, and ε_r = 2.32). (After [47].)

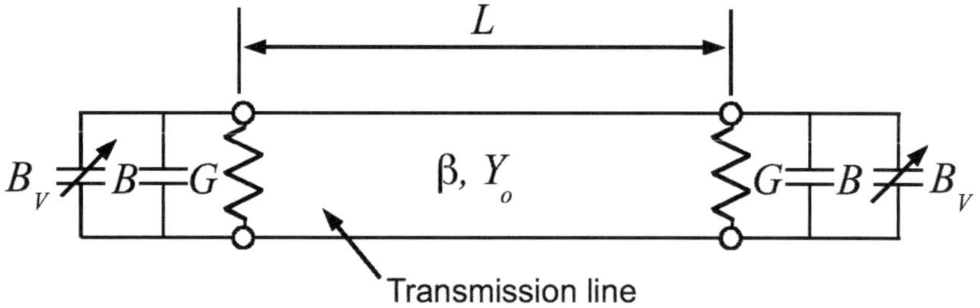

Figure 7.57 Transmission line model of the varactor-loaded microstrip antenna. (After [47].)

A similar design of a varactor-loaded rectangular microstrip patch antenna is shown in Figure 7.59 [48]. In this example, however, only a single varactor was introduced. The value of the varactor capacitance varied from 6 to 0.6 pF, causing a corresponding shift in frequency from 1.75 to 2.77 GHz (TR = 45%) and a total spectrum of TS = 49% when the DC bias voltage was increased from 0 to 10V, as shown in Figure 7.60 The larger capacitance ratio (TX = 10) of the varactor used in [48] compared to that used in [47] (TX = 6) resulted in a corresponding

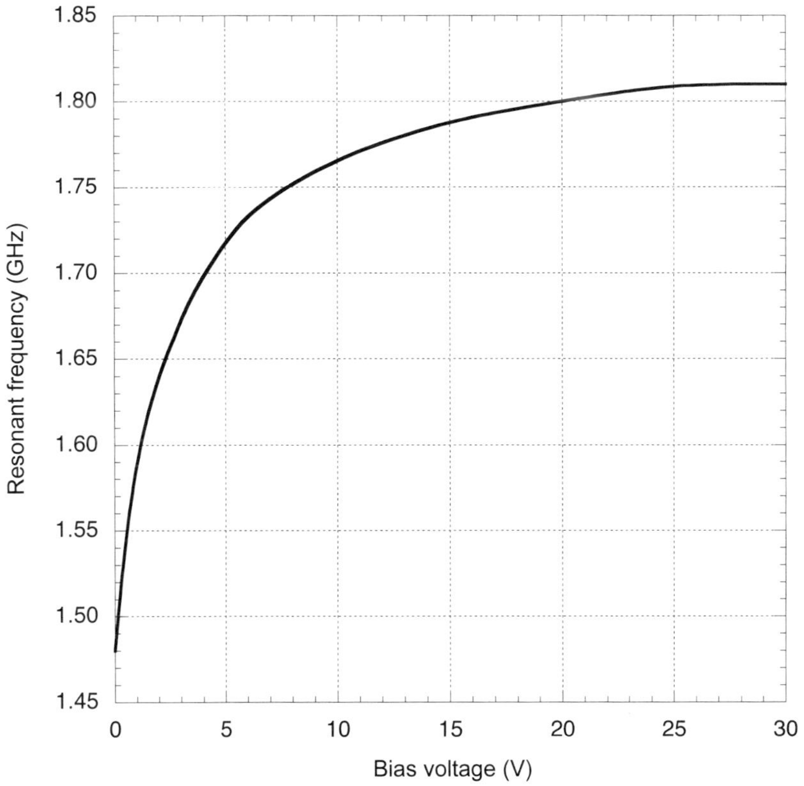

Figure 7.58 Tuning performance of the varactor-loaded microstrip antenna. (From [47].)

increase in tuning range. No mention of the gain or radiation efficiency of the antenna was provided.

The addition of more varactors along the radiating edges of microstrip patches does not result in an increase in tuning range. However, the added capacitance is used to miniaturize the size of the patch. Figure 7.61 shows an example of an L-band rectangular microstrip patch with four varactors on each radiating edge [49]. By varying the DC bias voltage from 0.2 to 20V, the resonant frequency of the antenna shifted from 1.43 to 1.8 GHz (TR = 23%), as shown in Figure 7.62, with a total spectrum of TS = 24%. The measured gain of the antenna varied from 4.8 to 7.5 dBi as the bias voltage was increased to 20V.

The designs presented so far all required the varactor to be connected between the microstrip patch and the ground plane, necessitating drilling large enough holes into the dielectric substrate for the varactor to be inserted. Each end of the varactor must then be soldered to ensure a reliable metal contact. This adds to the complexity and cost of fabrication and can be avoided by designs using surface-mount varactors. The example shown in Figure 7.63 uses six surface-mount

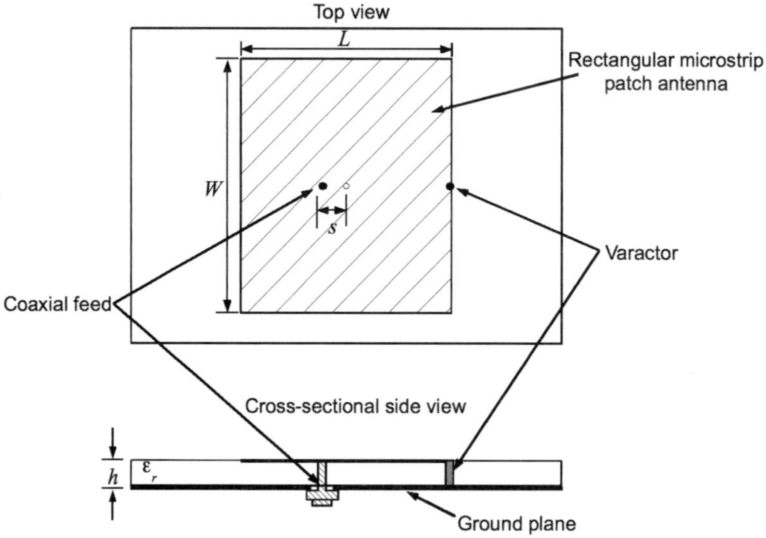

Figure 7.59 Varactor-loaded probe-fed rectangular microstrip patch antenna (L = 32.85 mm, W = 39.50 mm, s = 5.48 mm, h = 1.59 mm, and ε_r = 2.2). (After [48].)

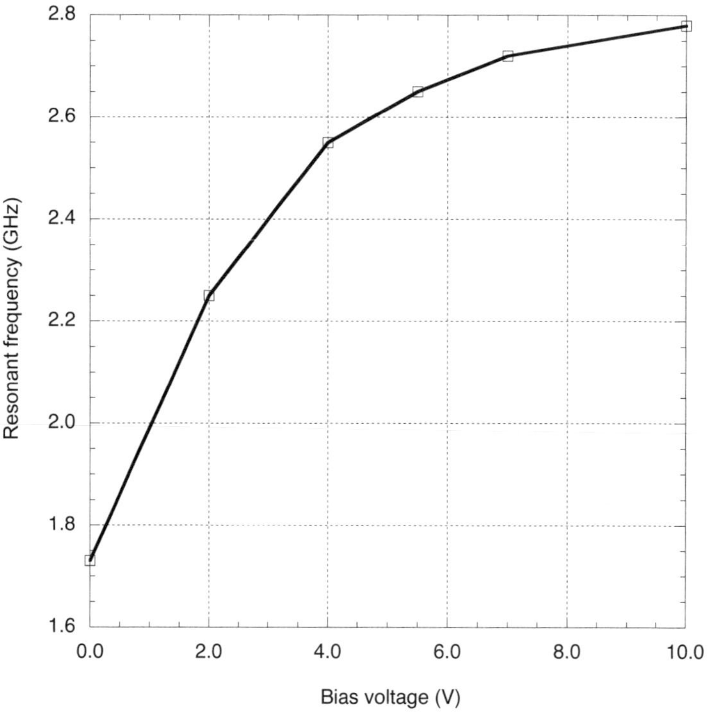

Figure 7.60 Tuning performance of the varactor-loaded microstrip patch antenna. (From [48].)

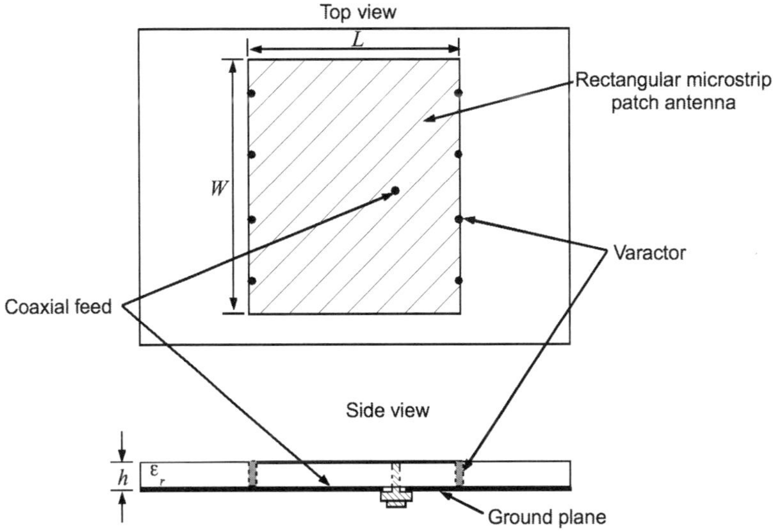

Figure 7.61 Varactor-loaded probe-fed rectangular microstrip patch antenna (L = 50 mm, W = 80 mm, h = 1.6 mm, and ε_r = 2.2). (After [49].)

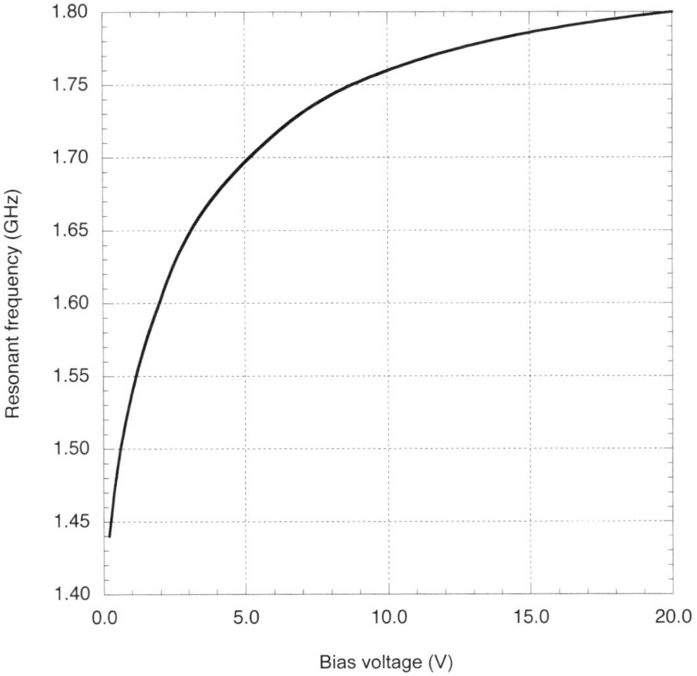

Figure 7.62 Tuning performance of the varactor-loaded microstrip patch antenna. (From [49].)

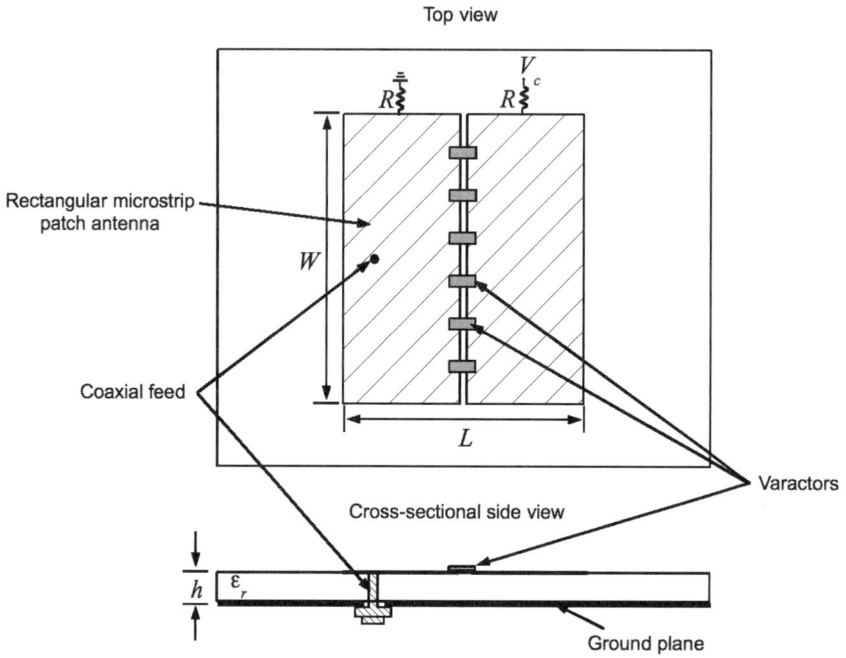

Figure 7.63 Varactor-loaded probe-fed rectangular microstrip patch antenna ($L = W = 60$ mm, $h = 0.52$ mm, $\varepsilon_r = 4.8$, and $R = 100$ kΩ). (After [50].)

varactors that were connected across a thin gap in the microstrip patch antenna [50]. This configuration eliminates the requirement for drilling holes to accommodate the varactors. The patch on the left side of the gap was connected to DC ground while the bias voltage was applied to the patch on the right side of the gap. High-value resistors (100 kΩ) were used to isolate the RF signal from the DC bias lines, instead of RF chokes, since the varactors drew negligible bias current. Applying a DC bias between 1 and 15V resulted in a frequency shift from 1.4 to 1.92 GHz (TR = 31%), as shown in Figure 7.64. Measured gain and radiation efficiency of this design were not reported.

The antenna shown in Figure 7.65 consists of a square annular microstrip patch designed for dual-band operation, achieved by the addition of a small square patch in one corner [51]. Connecting a varactor diode at the corner of the small square patch resulted in a large shift in the resonant frequency of one mode, while hardly affecting the resonant frequency of the second mode. Using a varactor whose capacitance varied from 2.2 to 0.5 pF for a DC bias voltage of

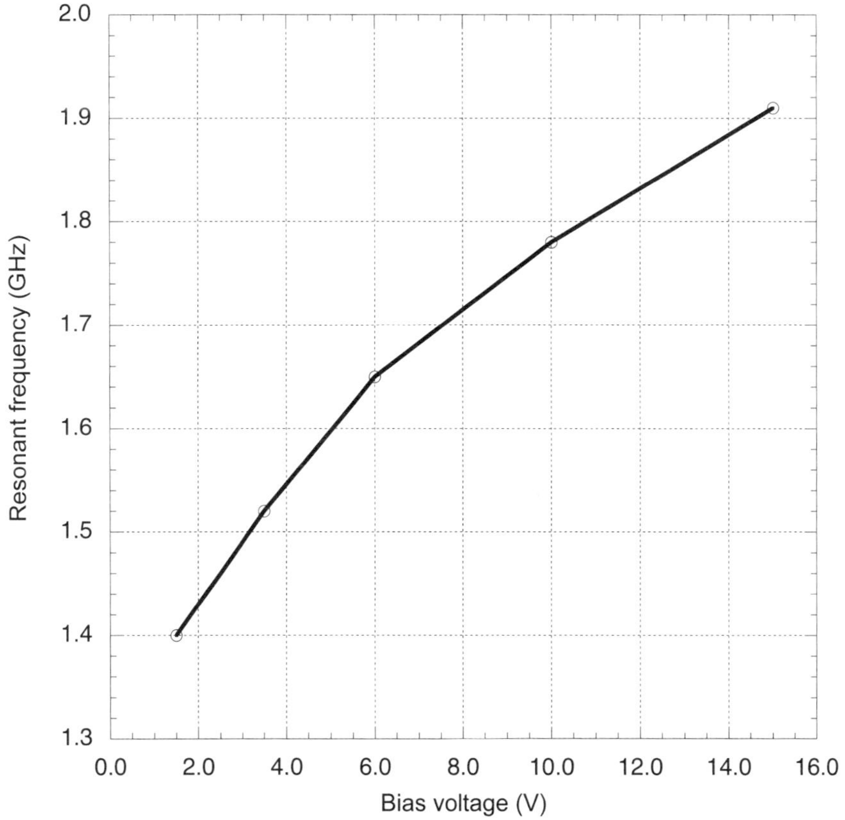

Figure 7.64 Tuning performance of the varactor-loaded microstrip patch antenna. (From [50].)

0–30V, the resonant frequency of the first mode of the antenna was shifted from 1.37 to 1.68 GHz (TR = 20%), as shown in Figure 7.66. The antenna achieved a total spectrum of TS = 26%. The second frequency varied very little around a value of 1.93 GHz as the bias voltage was altered. The measured gain and radiation efficiency were not reported.

An example of a probe-fed triangular-shaped microstrip patch is shown in Figure 7.67 [52]. In this design, two varactors were connected across thin F-shaped slots cut along two of the three sides of the equilateral triangle, with appropriate DC blocks and RF chokes to isolate the DC and RF signals. The varactors, whose capacitance varied from 2.67 to 0.64 pF for a DC bias of 0–30V, caused a shift in the resonant frequency of the antenna from 1.95 to 2.16 GHz (TR = 10%), as shown in Figure 7.68, with a total spectrum of TS = 13%. The measured gain remained fairly constant with applied DC bias, varying from 2.87 to 3.1 dBi.

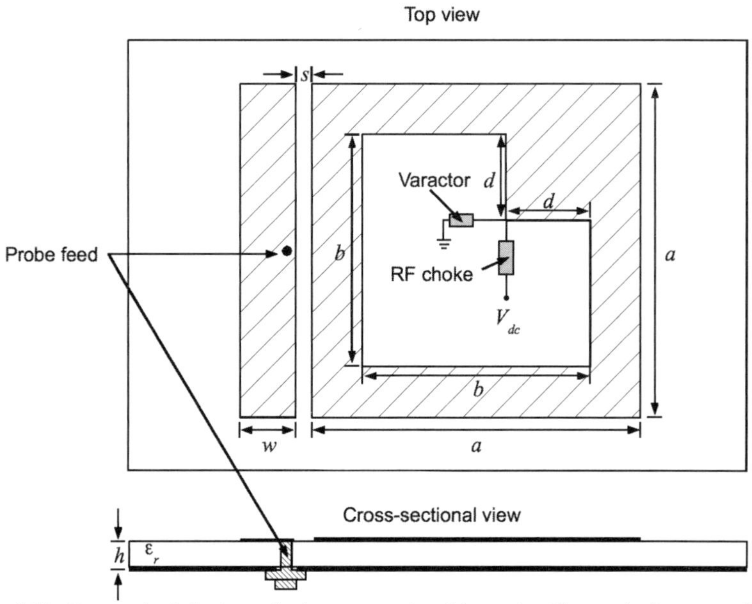

Figure 7.65 Varactor-loaded microstrip ring antenna (a = 26 mm, b = 20 mm, d = 3 mm, w = 3 mm, s = 0.2 mm, h = 1.5 mm, and ε_r = 4.5). (After [51].)

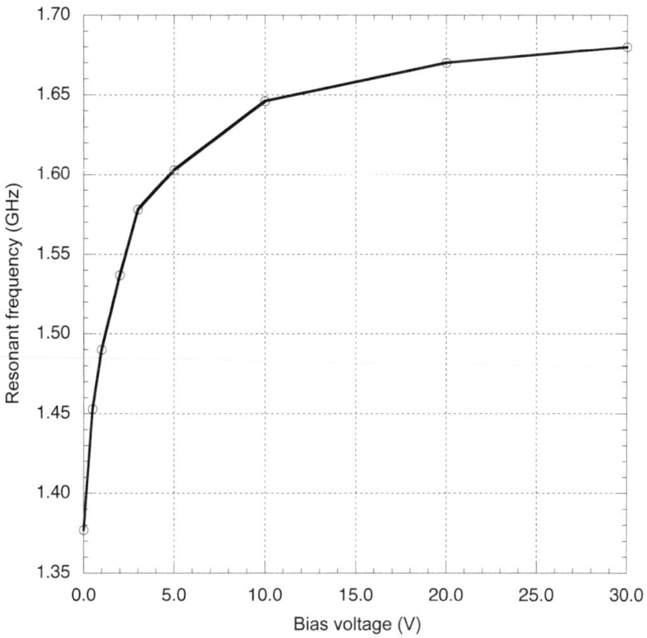

Figure 7.66 Tuning performance of the varactor-loaded microstrip ring antenna. (From [51].)

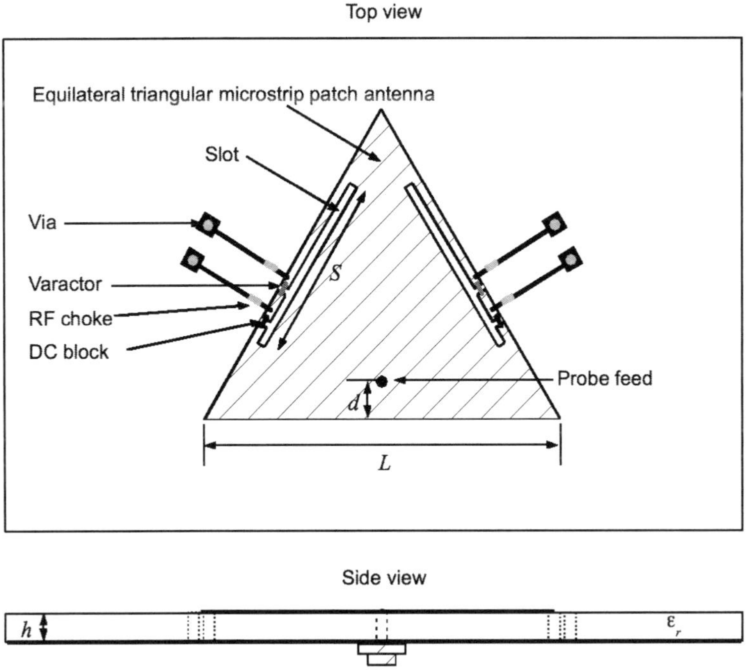

Figure 7.67 Varactor-loaded probe-fed triangular microstrip antenna (L = 50 mm, s = 26 mm, d = 4 mm, h = 1.6 mm, and ε_r = 4.4). (After [52].)

The next example, shown in Figure 7.69, is a probe-fed compact dual-band microstrip patch antenna designed for GSM900 and DCS1800 applications [53]. The microstrip patch consists of two resonant sections, each loaded with a varactor diode used to separately tune each resonant frequency. The varactors used had a junction capacitance that varied from 2.2 to 1.1 pF when a DC bias of 0–3V was applied. When the bias voltage of the first varactor was varied between 0 and 3V, while the bias of the second varactor was kept fixed at 3V, the first resonant frequency of the antenna was altered from f_1 = 814 to 973 MHz (TR_1 = 17.8%) with a total spectrum of TS_1 = 14%. The second resonant frequency remained fairly constant (f_2 = 1,876 ± 9 MHz). When the first varactor was held at a constant 3-V bias, while the bias voltage of the second varactor was varied between 0 and 3V, the second resonance shifted from f_2 = 1,432 to 1,885 MHz (TR_2 = 27.3%) with a total spectrum of TS_2 = 18.9%. The first resonant frequency was nearly constant at f_1 = 970 ± 3 MHz. The resonant frequencies as a function of the corresponding varactor biases are overlaid in the plot of Figure 7.70. Measured gain or radiation patterns were not reported, but simulated gains were estimated to be 0.1 dB for the GSM900 band and 1.3 dB for the DCS1800 band.

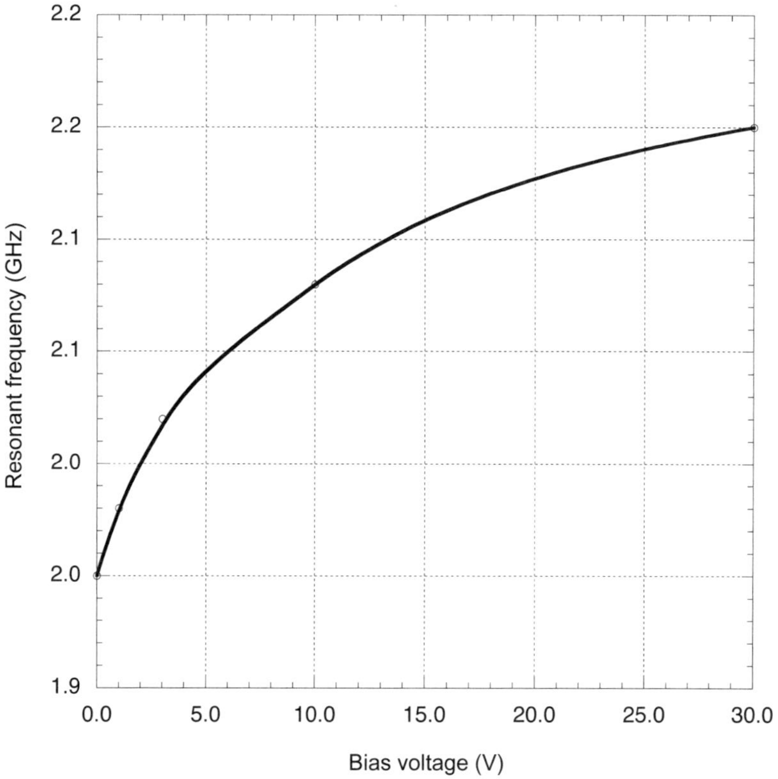

Figure 7.68 Tuning performance of the varactor-loaded triangular microstrip antenna. (From [52].)

The final example of a probe-fed varactor-loaded microstrip patch antenna is shown in Figure 7.71 [54]. This antenna contains two substrate layers and is commonly referred to as a stacked patch. The probe-fed rectangular microstrip patch antenna was etched on the top of the lower substrate while a second rectangular patch was etched on the bottom of the upper substrate and was suspended at a height *g* above the surface of the lower substrate. The upper patch was electromagnetically coupled to the lower patch and was designed to resonate at a somewhat different frequency than the lower patch, the combination of which resulted in a wideband response. Two varactors made a connection between each of the nonradiating sides of the lower patch and ground through a via. The capacitance value of the varactors varied from 1 to 0.3 pF as the DC bias voltage was increased 4 to 20V, resulting in a change in the resonant frequency of the antenna from 2.6 to 5.5 GHz (TR = 72%). Measured gains were not provided but simulations estimated the gain to vary between -5 and 5 dBi over the tuning range.

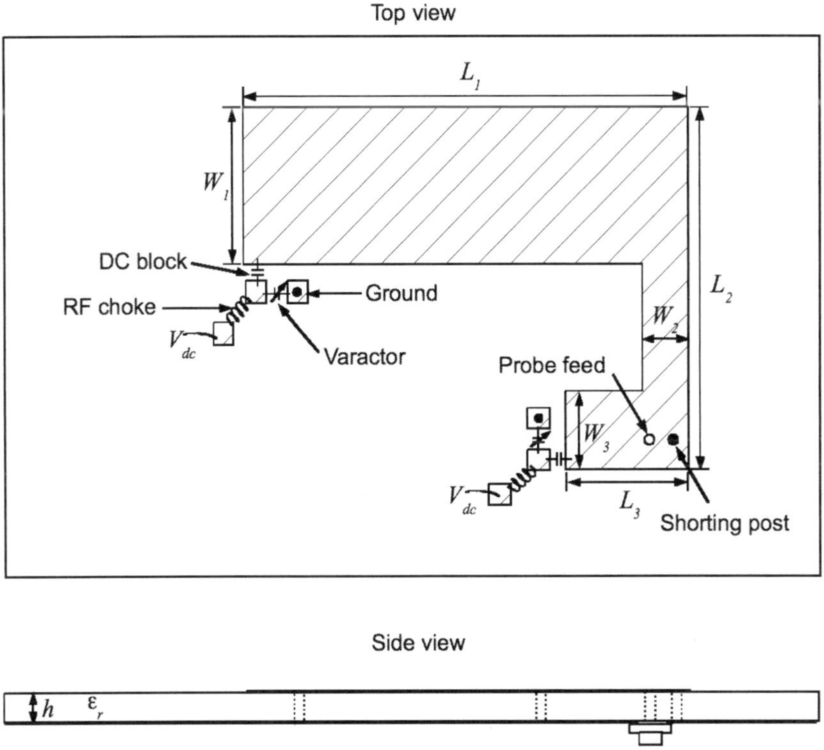

Figure 7.69 Varactor-loaded dual-band probe-fed microstrip patch antenna (L_1 = 29 mm, W_1 = 10 mm, L_2 = 23 mm, W_2 = 3 mm, L_3 = 8 mm, W_3 = 5 mm, ε_r = 1.2, and h = 2.44 mm). (After [53].)

A dual-polarized version of this design (with a total of eight varactors) has also been developed [55]. As the DC bias to the varactors was varied between 4 to 20V, the frequency of the microstrip patch could be tuned between 2.1 to 4.1 GHz (TR = 65%). The gain of the antenna varied between -4 and 4 dBi over the tuning range.

7.4.9.2 Aperture-Coupled Microstrip Patch Antennas

This section examines varactor-loaded microstrip patch antennas that are excited using aperture coupling. Aperture coupling, another common method for exciting microstrip patch antennas, is usually implemented using a thin rectangular slot etched in the ground plane beneath the patch. The slot is fed by a microstrip line, which shares a common ground plane with the patch antenna.

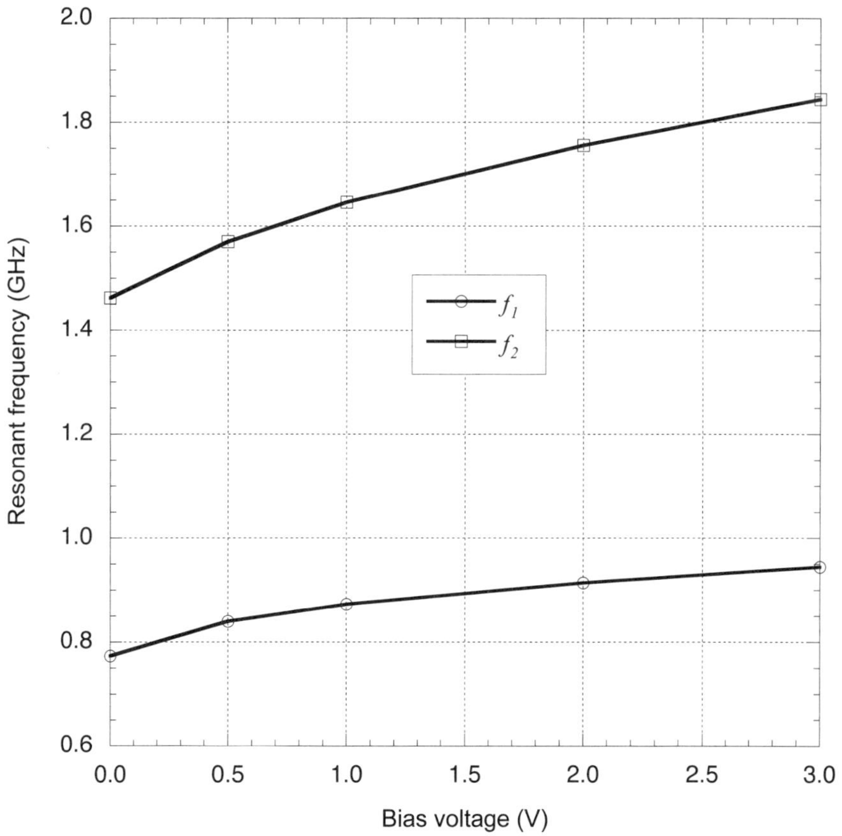

Figure 7.70 Tuning performance of the varactor-loaded dual-band microstrip antenna. (From [53].)

The signal from the microstrip line is electromagnetically coupled through the slot to excite the patch antenna. The advantage of aperture coupling is that there is no connection required between the aperture and the patch, eliminating the need for soldering or drilling through the substrate, which was required for the probe feed. This is attractive for high-frequency designs where the size of the patch can be as small as a few millimeters. The drawback to aperture coupling is that it requires a second substrate for the microstrip feed line, which needs to be bonded to the ground plane of the patch substrate, thus increasing the overall thickness and material cost. Care must also be taken in selecting the length of the slot to ensure that the radiation from the slot itself remains low and does not significantly contribute to the back lobe of the microstrip patch antenna pattern.

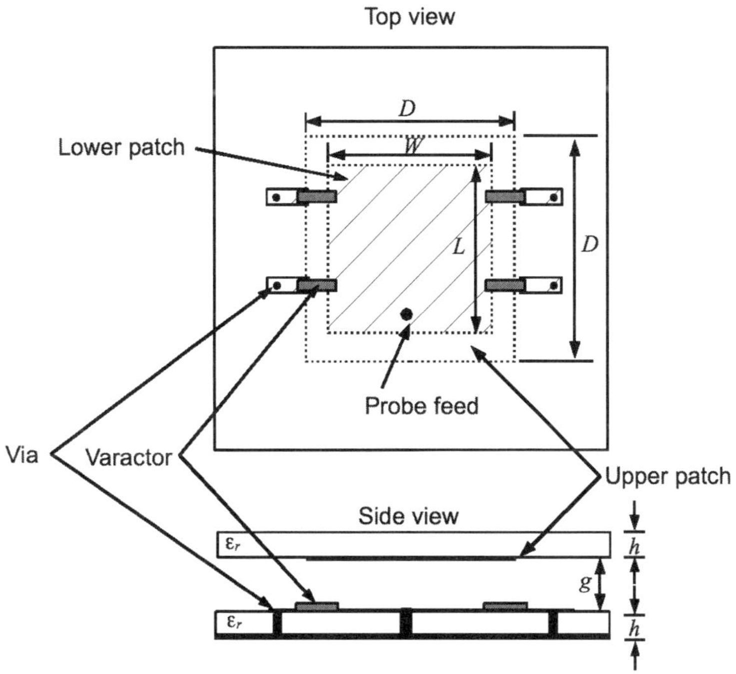

Figure 7.71 Varactor-loaded stacked patch microstrip patch antenna (D = 36 mm, $L = W$ = 15 mm, g = 8 mm, h = 1.57 mm, and ε_r = 2.33). (After [54].)

The first example of a varactor-loaded microstrip patch excited by a slot aperture is shown in Figure 7.72 [56]. Two varactor diodes were located at each of the radiating edges of the rectangular microstrip patch, connecting the patch to the ground plane, which was sandwiched between the two dielectric layers. The microstrip patch was excited by the rectangular slot aperture etched into the ground plane and fed by a microstrip line located on the bottom of the lower dielectric layer. The DC bias voltage was provided by a bias line also located on the bottom layer through a via connection to the microstrip patch.

This antenna was modeled with the circuit in Figure 7.73, based on which the input impedance Z_{in} was determined using the following equations [57, 58]

$$Z_{in} = \frac{N_2^2}{N_1^2 Y_{patch} + Y_a} - jZ_{om}\cot(\beta_m L_s) \tag{7.9}$$

where

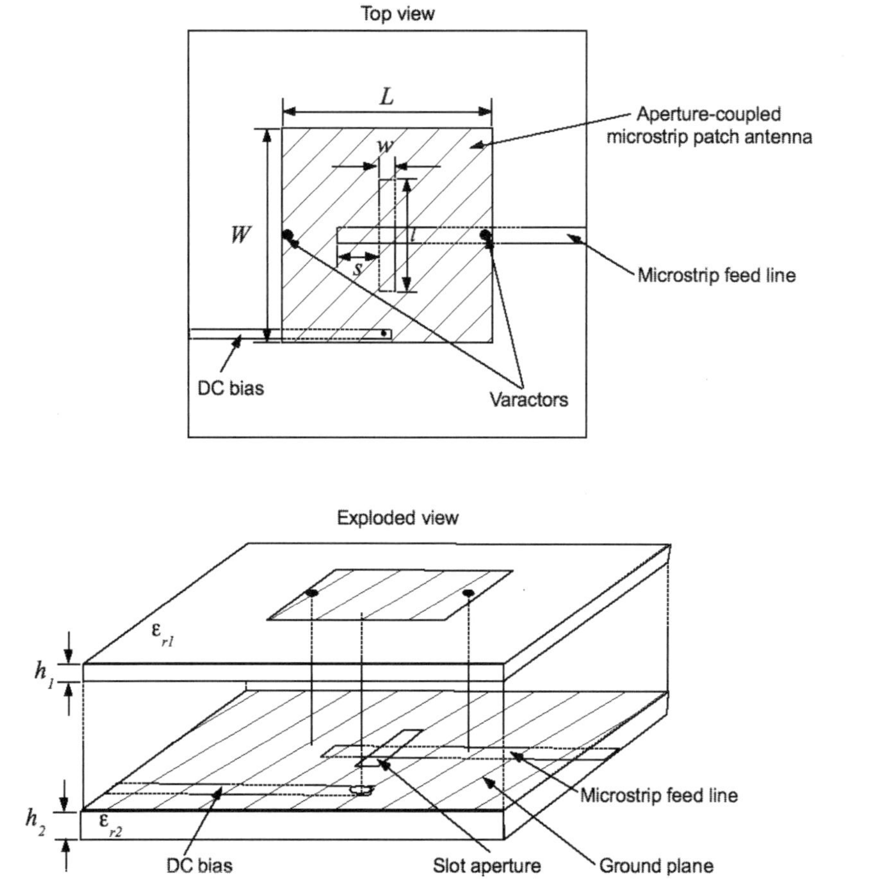

Figure 7.72 Varactor-loaded slot-fed microstrip antenna ($L = W = 47.7$ mm, $\varepsilon_{r1} = 2.2$, $h_2 = 0.786$ mm, $\varepsilon_{r2} = 2.55$, $h_1 = 1.587$ mm, $l = 14.4$ mm, $w = 2.1$ mm, and $s = 23$ mm). (After [56].)

$$Z_{patch} = \frac{1}{Y_1} + \frac{1}{Y_2} \tag{7.10}$$

$$Y_1 = Y_o \frac{G + j\left[B + B_v + Y_o \tan(\beta L_1)\right]}{Y_o + j\left[G + j(B + B_v)\tan(\beta L_1)\right]} \tag{7.11}$$

$$Y_2 = Y_o \frac{G + j\left[B + B_v + Y_o \tan(\beta L_2)\right]}{Y_o + j\left[G + j(B + B_v)\tan(\beta L_2)\right]} \tag{7.12}$$

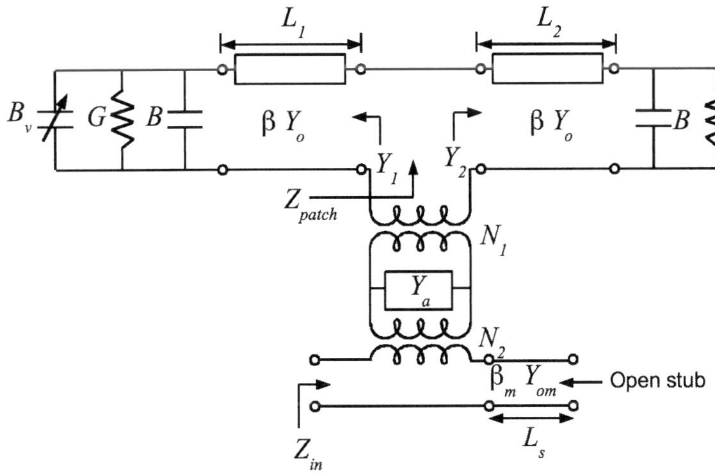

Figure 7.73 Circuit model for the varactor-loaded aperture-coupled microstrip patch antenna. (After [56].)

$$Y_a \approx -j2Y_{os}\cot(\beta_s l/2) \tag{7.13}$$

$$N_1 = \frac{l}{W} \tag{7.14}$$

$$N_2 = \sum_{n=1,3,\ldots}^{\infty} \frac{\sqrt{2\pi}}{aR} \frac{F_n \cos\left(\frac{n\pi}{a}\delta\right)\beta_m\left[\cos\left(\frac{\beta_m l}{2}\right) - \cos\left(\frac{n\pi}{a}x\frac{l}{2}\right)\right]}{\left(\frac{n\pi}{a}\right)^2 - \beta_m^2} \tag{7.15}$$

$$R^2 = \sum_{n=1,3,\ldots}^{\infty} \frac{1}{n} F_n^2 \left[\frac{1}{2}\sinh\left(\frac{2\pi n}{a}h_2\right) + \sinh^2(\pi n)\right] \tag{7.16}$$

$$F_n = \frac{J_o\left(\frac{n\pi l}{2a}\right)\sin\left(\frac{n\pi}{2}\right)}{\varepsilon_{r2}\cosh\left(\frac{n\pi}{a}h_2\right) + \sinh\left(\frac{n\pi}{a}h_2\right)} \tag{7.17}$$

$$\beta_m = \frac{2\pi\sqrt{\varepsilon_{2eff}}}{\lambda_o} \tag{7.18}$$

and $J_o()$ is the Bessel function of order zero. Altering the bias voltage on the varactors caused a change in the susceptance B_v in (7.11) and (7.12), resulting in a shift of the resonant frequency. The varactors used had a capacitance range of 2–40 pF, and when the applied DC bias voltage was varied between 0 and 22V, the resonant frequency of the antenna shifted from 1.62 to 2.57 GHz (TR = 45.3%), as shown in Figure 7.74. The achieved total spectrum for this design was TS = 47%. There was no mention of the gain or radiation efficiency of this antenna.

A second example of a similar design is shown in Figure 7.75, where only a single varactor diode was used [59]. In this instance, the slot was not centered under the patch, but located closer to one of the radiating edges, while the diode was located on the radiating edge opposite to the slot. This simplified design resulted in a lower tuning range, with a variation in DC bias from 3.5 to 30V causing a frequency shift from 770 to 870 MHz (TR = 12.2%), as shown in Figure 7.76. The total spectrum attained was TS = 14%. The gain and radiation efficiency of this antenna were not published.

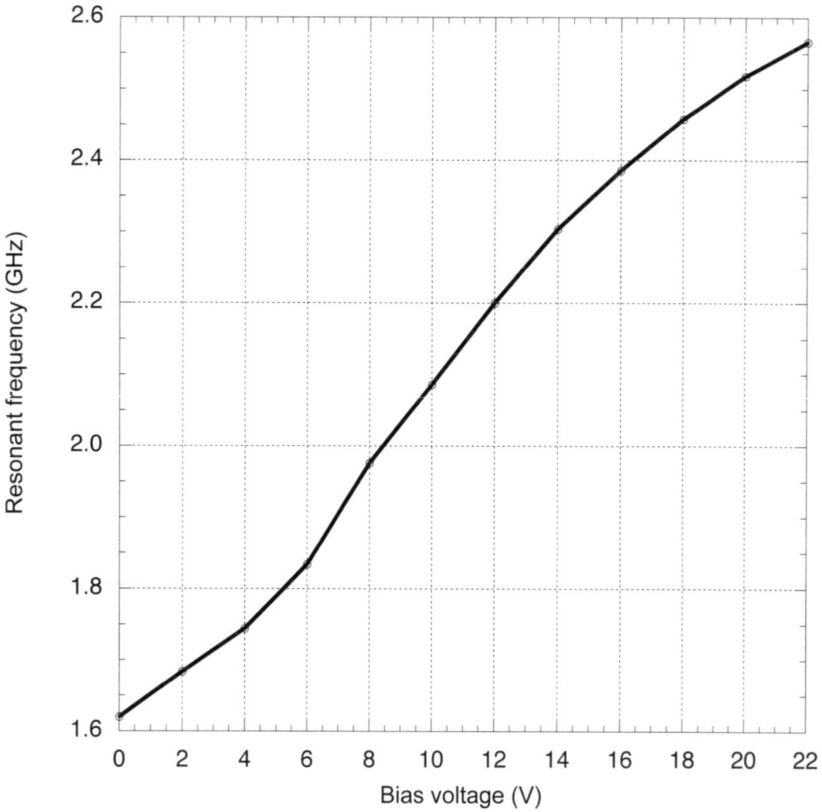

Figure 7.74 Tuning performance of the varactor-loaded microstrip antenna. (From [56].)

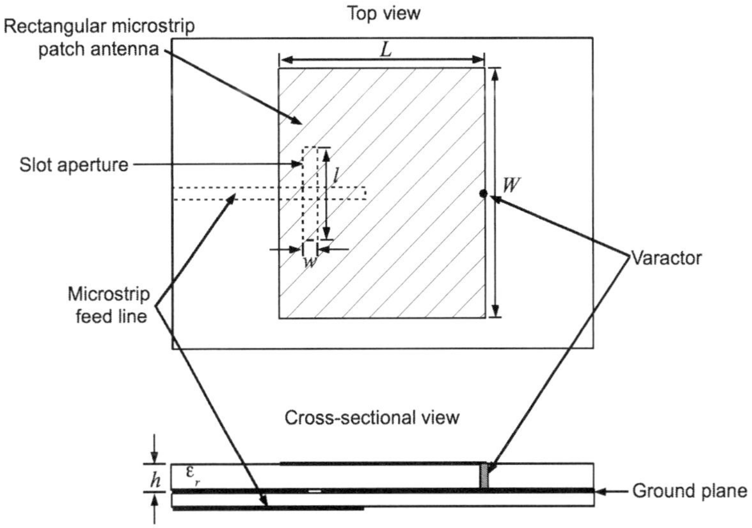

Figure 7.75 Varactor-loaded slot-fed rectangular microstrip patch antenna ($L = W = 152.5$ mm, $h = 2.59$ mm, $\varepsilon_r = 1.1$, $l = 60$ mm, and $w = 4$ mm). (After [59].)

7.4.9.3 Microstrip Line-Coupled Microstrip Patch Antennas

Another method to excite a microstrip patch antenna is by placing the patch over an open-circuited microstrip line. The fringing fields from the microstrip line will electromagnetically couple to the microstrip patch, leading to radiation. The advantage of this approach over the aperture-coupled approach is that there will not be any back radiation (assuming a large enough ground plane). As was the case for the aperture-coupled patch, the microstrip line and patch are each etched on different layers, so that the dielectric constant and thickness of the respective layers can be independently selected for optimized performance.

An example of a varactor-loaded microstrip patch antenna fed by a coupled microstrip line is shown in Figure 7.77 [60]. In this design, six varactors were used: two on each radiating edge, connecting the microstrip patch to ground, and two across a thin central gap, connecting the two halves of the patch. A technique known as differential feeding was used in this design, where the patch was fed on opposite sides by two microstrip lines with signals of equal magnitude but opposite phase. Differential feeding offers intrinsic broadband performance and can accommodate differential transceivers. The resonant frequency of the microstrip patch varied from 1.35 to 3.17 GHz (TR = 80.5%) with a total spectrum of TS ~ 60% as the DC bias voltage to the varactors was varied from 2 to 24V. Figure 7.78 plots the resonant frequency versus the corresponding

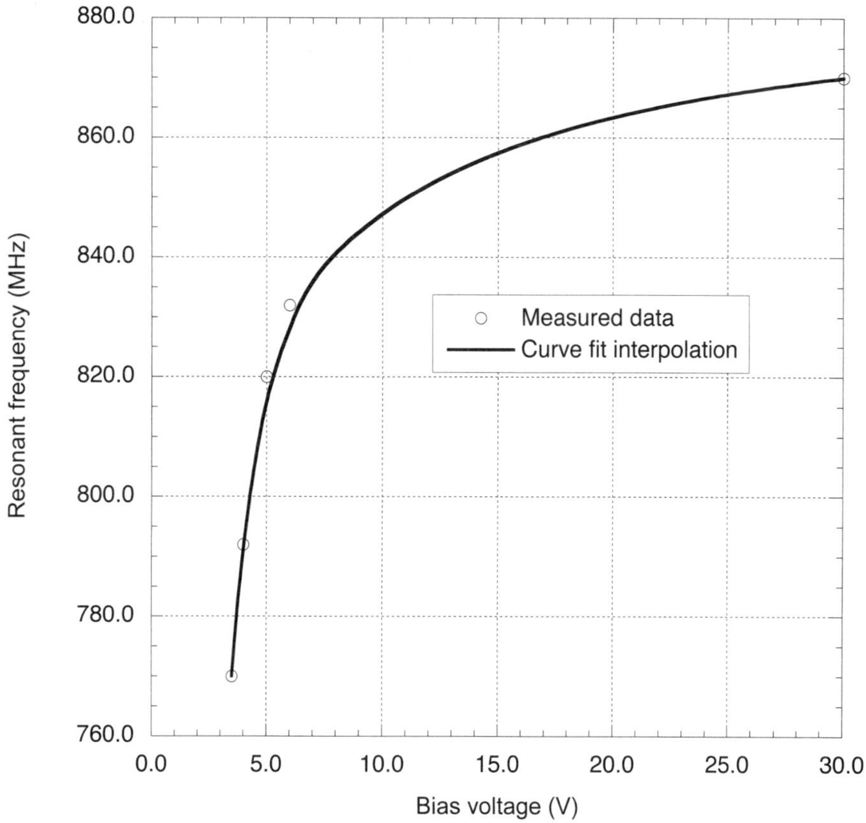

Figure 7.76 Tuning performance of the varactor-loaded microstrip patch antenna. (From [59].)

varactor capacitance values. The radiation efficiency of the antenna was determined to vary between 10% and 90% over the tuning range, corresponding to gain values between about -6 and 6 dBi.

A second example of a microstrip-line-coupled patch antenna is shown in Figure 7.79 [61, 62]. A hexagonal aperture with three rectangular arm extensions was etched out of the rectangular microstrip patch, with a varactor connected across each of the three extensions. The aperture in the rectangular patch resulted in dual-band, dual-polarized operation, and by varying the DC bias voltage across all the varactors from 0 to 30V, the two resonant frequencies shifted from f_1 = 1.272–1.536 GHz (TR_1 = 18.8%) and f_2 = 1.616–1.848 GHz (TR_2 = 13.4%), as shown in Figure 7.80. The gain and radiation efficiency of the antenna were not reported.

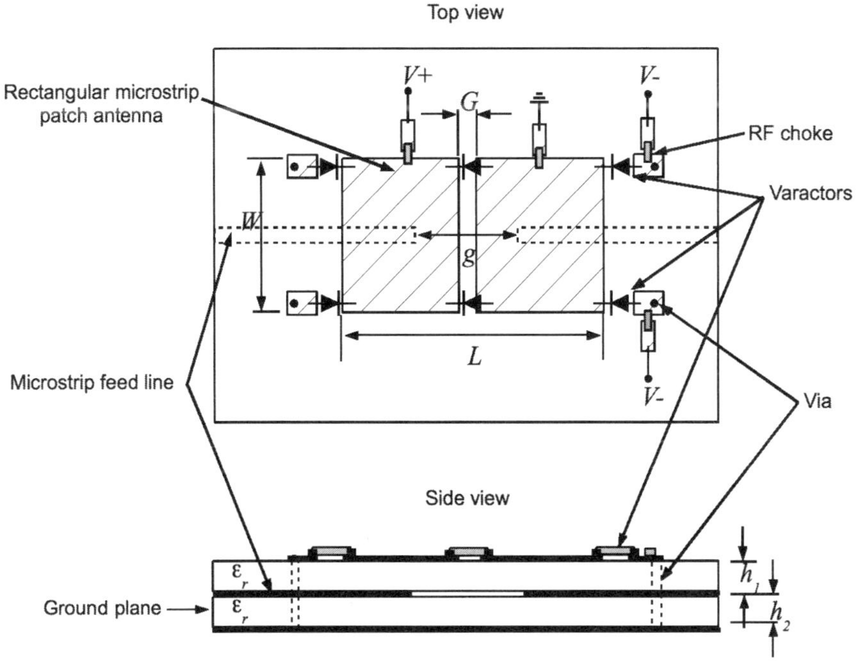

Figure 7.77 Varactor-loaded differentially fed microstrip patch antenna (L = 32 mm, W = 9 mm, G = 6 mm, g = 1 mm, h_1 = 1.542 mm, h_2 = 0.813 mm, and ε_r = 3.02). (After [60].)

7.4.9.4 Microstrip Line-Fed Microstrip Patch Antennas

One of the simplest methods for exciting a microstrip patch is to connect a microstrip line directly to one of the radiating edges of the patch. The microstrip line and patch can be etched in a single process, and only one dielectric sheet is required, unlike aperture coupling or microstrip line coupling where two sheets are needed. The main drawback of this approach is that a compromise is often required when choosing the dielectric constant and thickness of the substrate since the values that are optimal for the patch are not necessarily the best choice for the microstrip line.

An example of a varactor-loaded rectangular microstrip patch excited directly by a microstrip line is shown in Figure 7.81 [63]. The two varactors were each offset from one of the two radiating edges and connected the patch to the ground plane. By biasing the varactors between 2 and 16V, the resonant frequency of the microstrip patch antenna shifted from 367 to 650 MHz (TR = 55.6%), as shown in Figure 7.82. No information was given on the radiation efficiency or gain of this antenna.

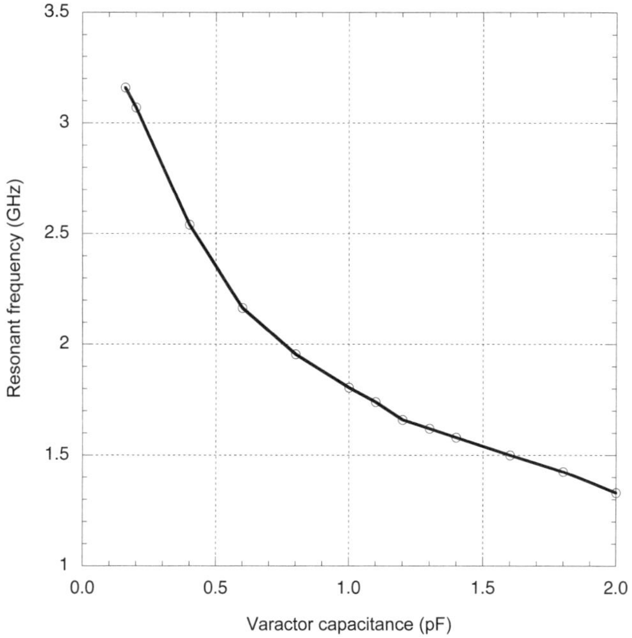

Figure 7.78 Tuning performance of the varactor-loaded differentially fed microstrip patch antenna. (From [60].)

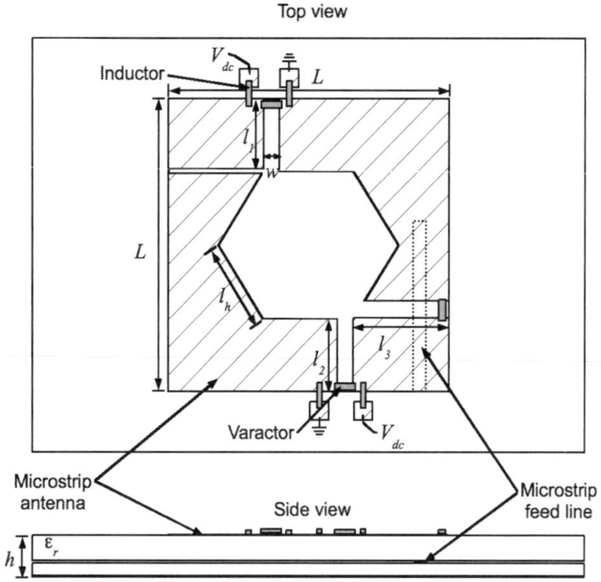

Figure 7.79 Varactor-loaded dual-band microstrip-fed hexagonal microstrip patch antenna (L = 40 mm, l_h = 8 mm, $l_1 = l_2$ = 11 mm, l_3 = 15 mm, w = 1 mm, h = 1.6 mm, and ε_r = 3.98). (After [61, 62].)

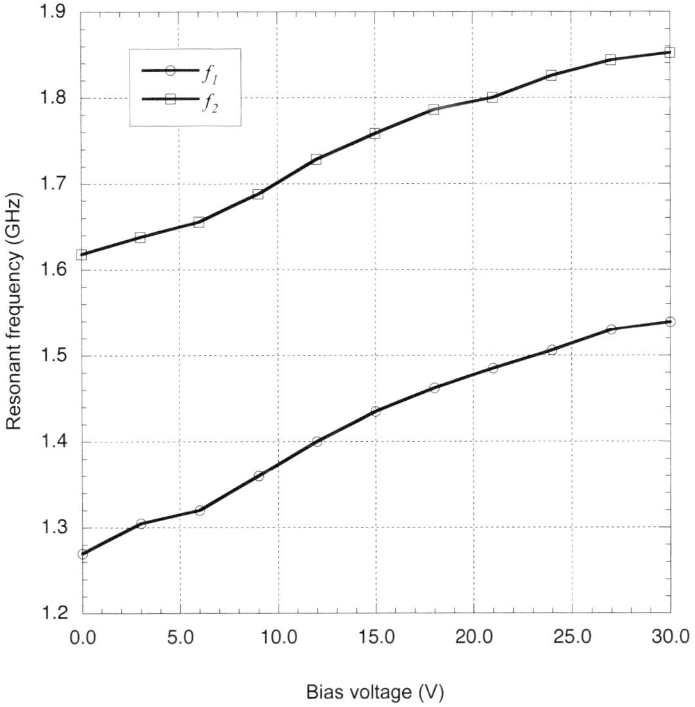

Figure 7.80 Tuning performance of the varactor-loaded dual-band microstrip antenna. (From [61].)

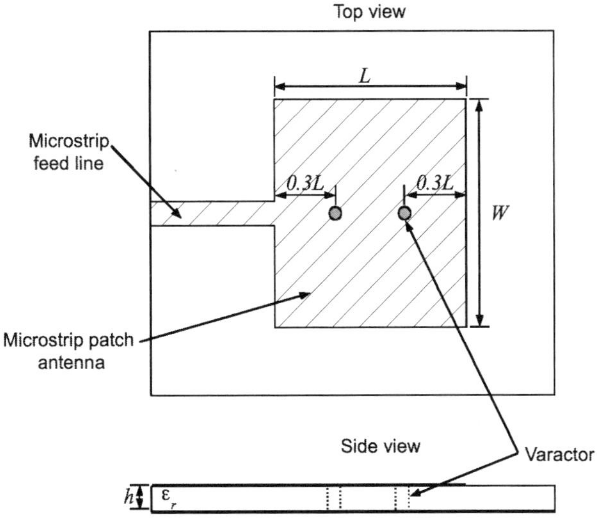

Figure 7.81 Varactor-loaded microstrip-fed microstrip patch antenna (L = 45 mm, W = 53 mm, h = 0.787 mm, and ε_r = 2.17). (After [63].)

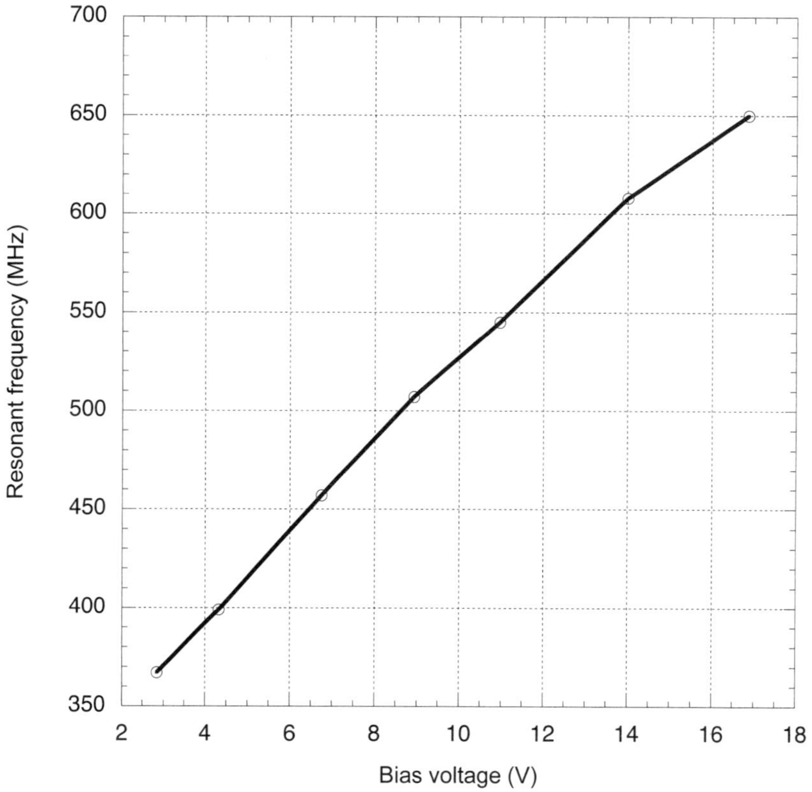

Figure 7.82 Tuning performance of the varactor-loaded shorted microstrip antenna. (From [63].)

7.4.9.5 CPW-Fed Microstrip Patch Antennas

A CPW can also be used to excite a microstrip patch antenna by electromagnetic coupling, similar to the microstrip line. With CPW, the signal trace and ground plane are located on the same layer, which may offer advantages in certain applications. Also, since the characteristic impedance of the CPW is a function of both the width of the signal line and the gaps between the line and the ground plane, various combinations of these dimensions can result in the same value of characteristic impedance for a given substrate thickness and dielectric constant. This degree of freedom in choosing dimensions is not available with microstrip lines, where the width of the line corresponds to a unique characteristic impedance for a given substrate thickness and dielectric constant. The CPW thus has more design flexibility than the microstrip line.

An example of a varactor-loaded microstrip patch antenna excited by a CPW line is shown in Figure 7.83 [64]. The patch was made compact by using the common technique of short-circuiting one of the radiating edges. A varactor was located near the remaining radiating edge, connecting the microstrip patch to the signal trace of the CPW line. A bias-T was used to provide the DC bias to the varactor. Varying the bias between 0 and 25V resulted in a frequency shift from 4.26 to 4.58 GHz (TR = 7.2%), as shown in Figure 7.84. The total spectrum for the antenna was TS = 9%. The gain and radiation efficiency of the antenna were not reported, but as is common with shorted patches, the cross-polarization levels of this antenna were high.

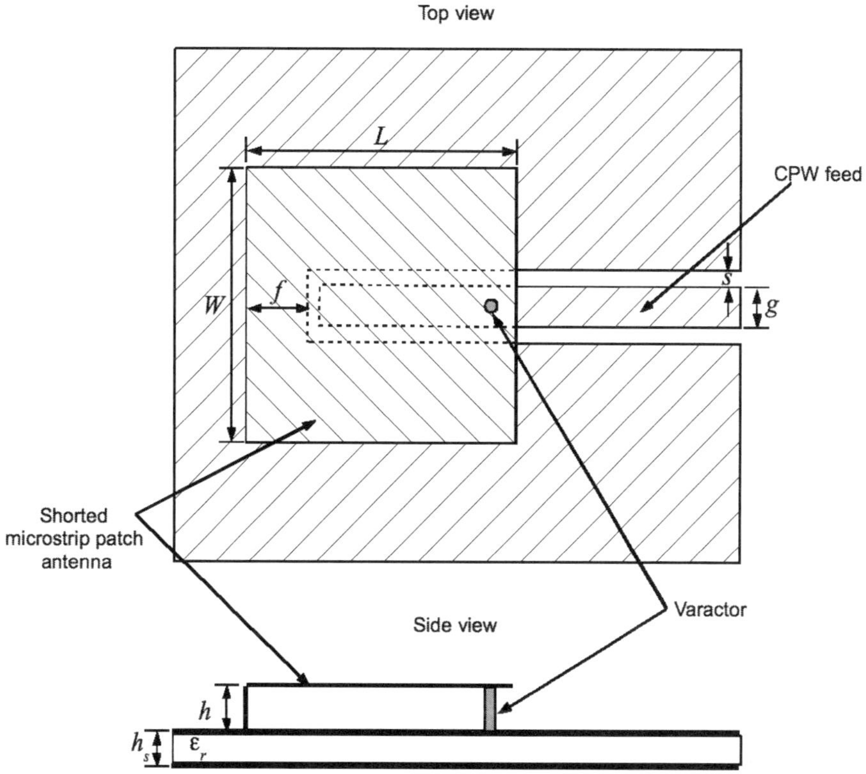

Figure 7.83 Varactor-loaded shorted CPW-fed microstrip patch antenna (L = 30 mm, W = 15 mm, s = 0.5 mm, g = 4 mm, f = 3 mm, h = 1 mm, h_s = 1.524 mm, and ε_r = 2.2). (After [64].)

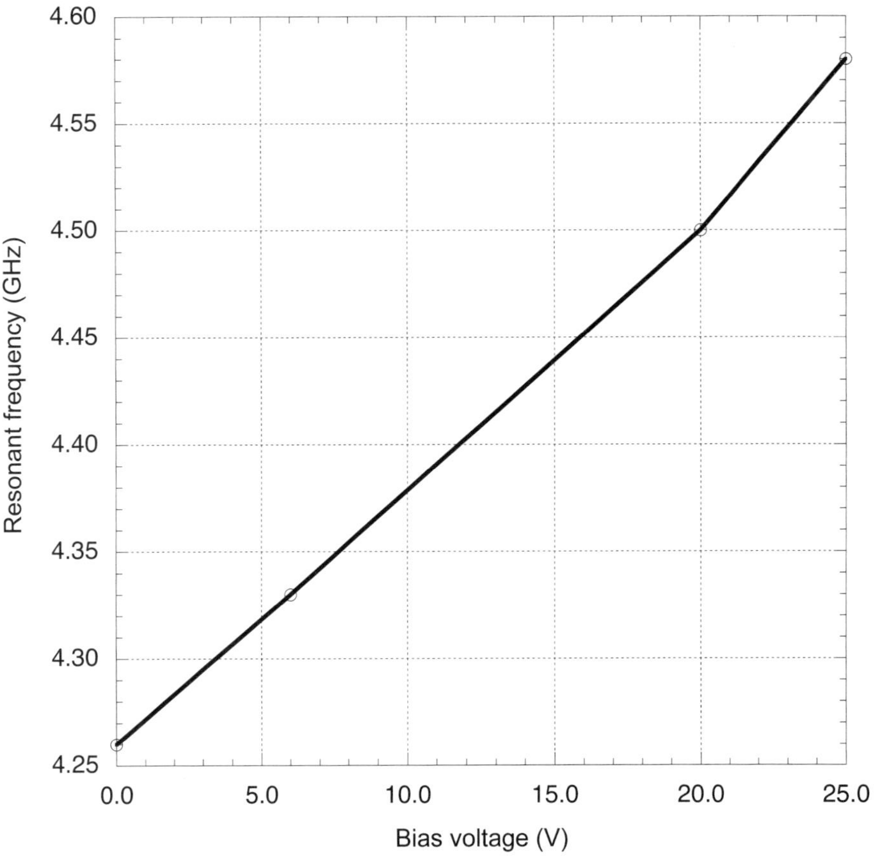

Figure 7.84 Tuning performance of the varactor-loaded CPW-fed microstrip antenna. (From [64].)

7.5 SUMMARY

This chapter examined antenna configurations where electronic devices were used to enable continuous frequency-tuning capability. The majority of the cases were based on varactor diodes, with a few examples of transistors, photodiodes, and MEMS devices. Most of the antenna configurations presented were designed for terrestrial wireless applications in frequency bands within the 1–5 GHz range. Achievable tuning ranges varied from a few percent to over 90%, with typical maximum bias voltages of 30V or less. The main features of the selected examples examined in this chapter are summarized in Table 7.1. Other examples of varactor-loaded microstrip patch antennas using various feed excitations are found in [65–75].

Table 7.1

Summary of FAA Continuous Tuning Examples

Tuning Technique	Antenna	Bias	Frequency Range (GHz)	TR (%)	TS (%)	Reference
Shottky diode	Microstrip	0–30V	2.19–2.28	4.4	N/A	[1]
Photo diode	Slot	0–88 mA	2.08–2.85	31.2	>33	[4]
MESFET	Slot	0–0.6V	9.22–10.22	10.3	20	[5]
	Yagi slot	0–1.5V	10.1–11.0	8.5	12	[6]
MEMS	Ring slot	0–25V	7.7–8.7	12	10	[8]
	Microstrip	0–11.9V	15.75–16.05	1.9	N/A	[9]
Varactor	Loop	0–5V	0.47–0.675	35.8	39	[10]
		N/A	0.3–1.585	135	N/A	[11]
		2–22V	0.585–1.376	80.7	48	[12]
	Monopole	1.2–23.7V	1.4–2.23	45.7	34	[13]
	Dipole	10–29V	2.78–3.45	21.5	23	[15]
		0–30V	2.065–2.365	13.5	25	[16]
	Yagi	0–16.6V	1.8–2.45	30.6	40	[17]
	Helix	0.3–2.7V	0.474–0.702	38.8	7	[18]
		0–3.5V	0.178–0.210	16.5	18	[19]
	IFA	0–9.5V	1.6–2.3	35.9	36	[20]
		0–22V	0.48–0.85	55.6	58	[21]
	PIFA	0–18.4V	1.65–2.4	37	41	[22]
		0–5V	0.41–0.77	61	15	[23]
	Slot	0–15V	1.92–2.55	28.2	30	[32]
		0.5–20V	4.92–5.4	9.7	14	[33]
		0–20V	0.93–1.72	59.6	65	[34]
		0–20V	0.78–2.09	91.3	91	[35]
		0–30V	1.1–1.45, 1.75–2.95	27.5 50	30 54	[36]
		0–30V	7.3–8.1	10	N/A	[38]
		0–20V	1.0–1.9	62	63	[39]
	DRA	2–8V	2.6–2.76	6	10	[44]
		0–19V	3.2–5.7	55	58	[45]
	Microstrip	0–30V	1.42–1.82	21	N/A	[46]
		0–30V	1.85–2.45	28	N/A	[47]
		0–10V	1.75–2.77	45	49	[48]
		0.2–20V	1.43–1.8	23	24	[49]
		1–15V	1.4–1.92	31	N/A	[50]
		0–30V	1.37–1.68	20	26	[51]
		0–30V	1.95–2.16	10	12.6	[52]
		0–3V	0.814–0.973, 1.432–1.885	17.8 27.3	13.7 18.9	[53]
		4–20V	2.6–5.5	72	N/A	[54]
		4–20V	2.1–4.1	65	N/A	[55]
		0–22V	1.62–2.57	45.3	47	[56]
		3.5–30V	0.77–0.87	12.2	14	[59]
		2–24V	1.35–3.17	80.5	60	[60]
		0–30V	1.272–1.536, 1.616–1.848	18.8 13.4	N/A N/A	[61]
		2–16V	0.367–0.65	55.6	N/A	[63]
		0–25V	4.26–4.58	7.2	9	[64]
		4–20V	2.17–2.65	19.9	25	[65]

References

[1] Haskins, P. M., P. S. Hall, and J. S. Dahele, "Active Patch Antenna Element with Tuning Diode," *Electronics Letters,* Vol. 27, No. 20, September 1991, pp. 1846-1848.

[2] Srivastava, S., B. R. Vishvakarma, and J. A. Ansari, "Tunnel Diode-Loaded Rectangular Microstrip Antenna for Millimeter Range," *IEEE Transactions on Antennas and Propagation,* Vol. 51, No. 4, April 2003, pp. 750-755.

[3] Gautam, A. K., and B. R. Vishvakarma, "Frequency Agile Microstrip Antenna Symmetrically Loaded With Tunnel Diodes," *Microwave and Optical Technology Letters,* Vol. 48, No. 9, September 2006, pp. 1807-1810.

[4] Carraquillo-Rivera, I., Z. Popovic, and R. A. Rodriquez Solis, "Tunable Slot Antenna Using Varactors and Photodiodes," *Proc. IEEE International Symposium on Antennas and Propagation,* Columbus, OH, June 22-27, 2003, pp. 532-535.

[5] Kawasaki, S., and T. Itoh, "A Slot Antenna With Electronically Tunable Length," *Proc. IEEE International Symposium on Antennas and Propagation,* Vol. 1, London, Canada, June 24-28, 1991, pp. 130–133.

[6] Ke, B. S., Y. Qian, and T. Itoh, "Two-Element Yagi-Uda Array Using Tunable Slot Antenna," *Proc. Asia Pacific Microwave Conference,* Hong Kong, December 2-5, 1997, pp. 437-440.

[7] Erdil, E., et al., "Reconfigurable CPW-Fed Dual-Frequency Rectangular Slot Antenna Using MEMS Technology," *Proc. IEEE International Symposium on Antennas and Propagation,* Washington, D.C., July 3-8, 2005, pp. 392-395.

[8] Erdil, E., et al., "Reconfigurable Slot Antenna With Fixed-Fixed Beam Type RF MEMS Capacitors," *Proc. European Conference on Antennas and Propagation EuCAP,* Nice, France, November 6-10, 2006.

[9] Erdil, E., et al., "Frequency Tunable Microstrip Patch Antenna Using RF MEMS Technology," *IEEE Transactions on Antennas and Propagation,* Vol. 55, No. 4, April 2007, pp. 1193-1196.

[10] Abdallah, M., et al., "Frequency Tunable Monopole Coupled Loop Antena With Broadside Radiation Pattern," *Electronics Letters,* Vol. 45, No. 23, November, 2009, pp. 1149-1151.

[11] Hallbjorner, P., "Electrically Small Unbalanced Four-Arm Wire Antenna," *IEEE Transactions on Antennas and Propagation,* Vol. 52, No. 6, June 2004, pp. 1424-1428.

[12] Herraiz-Martinez, F. J., et al., "Left-Handed Wire Antennas Over Ground Plane With Wideband Tuning," *IEEE Transactions on Antennas and Propagation,* Vol. 59, No. 5, May 2011, pp. 1460-1471.

[13] Mirzaei, H., and G. V. Eleftheriades, "A Compact Frequency-Reconfigurable Metamaterial-Inspired Antenna," *IEEE Antennas and Wireless Propagation Letters,* Vol. 10, 2011, pp. 1154-1157.

[14] Zhu, J., and G. V. Eleftheriades, "Dual-Band Metamaterial-Inspired Small Monopole Antenna for WiFi Applications," *Electronics Letters,* Vol. 45, No. 22, October 2009, pp. 1104-1106.

[15] Carrere, J. M., R. Staraj, and G. Kossiavas, "Small Frequency Agile Antennas," *Electronics Letters,* Vol. 37, No. 12, June 2001, pp. 728-279.

[16] Kolsrud, A. T., M. Y. Li, and K. Chang, "Dual-Frequency Electronically Tunable CPW-Fed CPS Dipole Antenna," *Electronics Letters,* Vol. 32, No. 7, April 1998, pp. 609-611.

[17] Cai, Y., Y. J. Guo, and A. Weily, "A Frequency-Reconfigurable Quasi-Yagi Dipole Antenna," *IEEE Antennas and Wireless Propagation Letters,* Vol. 9, 2010, pp. 883-886.

[18] Kim, J., et al., "A Tunable Internal Antenna With an Epsilon Negative Zeroth Order Resonator for DVB-H Service," *IEEE Transactions on Antennas and Propagation,* Vol. 57, No. 12, December 2009, pp.4014-4017.

[19] Yoon, I. C., S. H. Park, and Y. E. Kim, "Frequency Tunable Antenna for Mobile TV Signal Reception," *Proc. IEEE International Symposium on Antennas and Propagation,* Hawaii, HI, June 10-15, 2007, pp. 5861-5864.

[20] Liang, J., and H. Y. D. Yang, "Frequency Reconfigurable Printed Inverted-F Antennas," *Proc. IEEE International Symposium on Antennas and Propagation,* San Diego, CA, July 5-11, 2008.

[21] Huitema, L., et al., "A Compact and Reconfigurable DVB-H Antenna for Mobile Handheld Devices," in *Proc. 5th European Conference on Antennas and Propagation EuCAP 2011,* Rome Italy, April 11-15, 2011.

[22] Nguyen, V. A., R. A. Bhatti, and S. O. Park, "A Simple PIFA-Based Tunable Internal Antenna for Personal Communication Handsets," *IEEE Antennas and Wireless Propagation Letters,* Vol. 7, 2008, pp. 130-133.

[23] Mi, H., and Z. Feng, "A Novel Compact Varactor-Tunable PIFA Antenna for ISDB-T Application," *Proc. Asia-Pacific Microwave Conference APMC,* Bangkok, Thailand, December 11-14, 2007.

[24] Y. Tsutsumi, et al., "Low Profile Double Resonance Frequency Tunable Antenna Using RF MEMS Variable Capacitor for Digital Terrestrial Broadcasting Reception," *Proc. IEEE International Symposium Antennas and Propagation,* Charleston, SC, June 1-5, 2009.

[25] Panayi, P. K., M. O. Al-Nuaimi, and L. P. Ivrissimtzis, "Tuning Techniques for Planar Inverted-F Antenna," *Electronics Letters,* Vol. 37, No. 16, August 2001, pp. 1003-1004.

[26] Berg, M., et al., "Frequency-Tunable DVB-H Antenna for Mobile Terminals," *Proc. IEEE International Symposium on Antennas and Propagation,* Honolulu, HI, June 10-15, 2007.

[27] Oh, S. K., H. S. Yoon, and S. O. Park, "A PIFA-Type Varactor-Tunable Slim Antenna With a PIL Patch Feed for Mulitband Applications," *IEEE Antennas and Wireless Propagation Letters,* Vol. 6, 2007, pp. 103-105.

[28] Lim, J. H., et al., "A Reconfigurable PIFA Using a Switchable PIN-Diode and a Fine-Tuning Varactor for USPCS/WCDMA/m-WiMAX/WLAN," *IEEE Transactions on Antennas and Propagation,* Vol. 58, No. 7, July, 2010, pp. 2404-2411.

[29] Milosavljevic, Z. D., "Varactor-Tuned DVB-H Antenna," *Proc. International Workshop on Antenna Technology IWAT,* March 21-23, 2007, Cambridge, United Kingdom, pp. 124-127.

[30] Zheng, Y., et al., "Suppression of Harmonic Radiation of Tunable Planar Inverted-F Antenna by Ferroelectric Varactor Loading," *Proc. IEEE International Microwave Symposium,* Atlanta, GA, June 15-20, 2008, pp. 959-962.

[31] Canneva, F., et al., "Reconfigurable Miniature Antenna for DVB-H Standard," *Proc. IEEE International Symposium on Antennas and Propagation,* Toronto, Canada, July 11-17, 2010.

[32] Urwin-Wright, et al., "A Reconfigurable Electrically-Small Antenna Operating in the 'DC' Mode," *Proc. IEEE Vehicular Technology Conference,* Dresden, Germany, April 22-25, 2003, pp. 857-861.

[33] Holland, B. R., et al., "Tunable Coplanar Patch Antenna Using Varactor," *Electronics Letters,* Vol. 42, No. 6, March 2006, pp. 319-321.

[34] White, C. R., and G. M. Rebeiz, "Single- and Dual-Polarized Tunable Slot-Ring Antennas," *IEEE Transactions on Antennas and Propagation,* Vol. 57, No. 1, January 2009, pp. 19-26.

[35] Chen, S. Y., et al., "Frequency-Agile, Miniaturized Slot Antenna for Hand-Held Devices,"

Proc. URSI International Symposium on Electromagnetic Theory EMTS 2010, Berlin, Germany, August 16-19, 2010, pp. 377-380.

[36] Behdad, N., and K. Sarabandi, "Dual-Band Reconfigurable Antenna With a Very Wide Tunability Range," *IEEE Transactions on Antennas and Propagation,* Vol. 54, No. 2, February 2006, pp. 409-416.

[37] Janaswamy, R., and D. H. Schaubert, "Characteristic Impedance of a Wide Slotline on Low-Permittivity Substrates," *IEEE Transactions on Microwave Theory and Techniques,* Vol. 34, No. 8, August 1986, pp. 900-902.

[38] Forman, M. A., and Z. B. Popovic, "A Tunable Second-Resonance Cross-Slot Antenna," *Proc. IEEE International Symposium on Antennas and Propagation,* Vol. 1, Montreal, Canada, July 13-18, 1997, pp. 18-21.

[39] White, C. R., and G.M. Rebeiz, "A Shallow Varactor-Tuned Cavity-Backed Slot Antenna with a 1.9:1 Tuning Range," *IEEE Transactions on Antennas and Propagation,* Vol. 58, No. 3, March, 2010, pp. 633-639.

[40] Carrasquillo-Rivera, I., R. A. Rodriguez Solis, and J. G. Colom-Ustariz, "Tunable and Dual-Band Rectangular Slot-Ring Antenna," *Proc. International Symposium on Antennas and Propagation,* Monterey, CA, June 20-25, 2004, pp. 4308-4311.

[41] Erdil, E., et al., "Reconfigurable Slot Antenna With Fixed-Fixed Beam Type RF MEMS Capacitors," *Proc. European Conference on Antennas and Propagation EUCAP,* Nice, France, November 6-10, 2006.

[42] Sim, C. Y. D., T. Y. Lin, and J. S. Row, "Aperture-Coupled Antenna with Switchable Polarisation and Frequency Agility," *Electronics Letters,* Vol. 45, No. 17, August 2009, pp. 865-866.

[43] Petosa, A., S. Thirakoune, and A. Ittipiboon, "Frequency Agile DRA," *Proc. IEEE International Symposium on Antennas and Propagation,* Charleston, SC, June 1-5, 2009.

[44] Hao, C. X., et al., "Frequency-Tunable Differentially Fed Rectangular Dielectric Resonator Antenna," *IEEE Antennas and Wireless Propagation Letters,* Vol. 10, 2011, pp. 884-887.

[45] Desjardins, J., et al., "Electronically Frequency-Reconfigurable Rectangular Dielectric Resonator Antenna," *IEEE Transactions on Antennas and Propagation,* Vol. 60, No. 6, June 2012, pp. 2997-3002.

[46] Petosa, A., and S. Thirakoune, "Frequency-Agile Dielectric Resonator Antenna," *Proc. IEEE International Symposium on Antennas and Propagation,* Chicago, IL, July 8-14, 2012.

[47] Bhartia, P., and I. J. Bahl, "Frequency Agile Microstrip Antennas," *Microwave Journal,* Vol. 25, No. 10, October 1982, pp. 67-70.

[48] Al-Charchafchi, S. H., and M. Frances, "Electronically Tunable Microstrip Patch Antennas," *Proc. IEEE International Symposium on Antennas and Propagation,* Vol. 1, Atlanta, GA, June 21-26, 1998, pp. 304-307.

[49] Jin, Z., and A. Mortazawi, "An L-Band Tunable Microstrip Antenna Using Multiple Varactors," *Proc. IEEE International Symposium on Antennas and Propagation,* Columbus, OH, June 22-27, 2003, pp. 524-527.

[50] Fayyaz, N., et al., "A Novel Electronically Tunable Rectangular Patch Antenna With One Octave Bandwidth," *Proc. IEEE Canadian Conference on Electrical and Computer Engineering,* Vol. 1, Kitchener-Waterloo, Canada, May 25-28, 1998, pp. 25-28.

[51] Alkanhal, M. A., and A. F. Sheta, "A Novel Dual-Band Reconfigurable Square-Ring Microstrip Antenna," *Progress in Electromagnetics Research PIER,* Vol. 70, 2007,

pp. 337-349.

[52] Lee, K. M., et al., "A Tunable Triangular Microstrip Patch Antenna for Multiband Applications," *Microwave and Optical Technology Letters,* Vol. 51, No. 7, July 2009, pp. 1788-1790.

[53] Sheta, A. F., and M. A. Alkanhal, "Compact Dual-Band Tunable Microstrip Antenna for GSM/DCS-1800 Applications," *IET Microwaves, Antennas and Propagation,* Vol. 2, No. 3, 2008, pp. 274-280.

[54] Nishiyama, E., and T. Itoh, "Widely Tunable Stacked Microstrip Antenna Using Varactor Diodes," *Proc. Asia Pacific Microwave Conference,* Hong Kong, December 16-20, 2008.

[55] Nishiyama, E., and T. Itoh, "Dual Polarized Widely Tunable Stacked Microstrip Antenna Using Varactor Diodes," *Proc. International Conference on Antenna Technology IWAT,* Santa Monica, CA, March 2-4, 2009.

[56] Le Ray, G., M. Himdi, and J. Daniel, "Frequency Agile Slot-Fed Patch Antenna," *Electronics Letters,* Vol. 32, No. 1, January 1996, pp. 2-3.

[57] Das, B. N., and K.K. Joshi, "Impedance of a Radiating Slot in the Ground Plane of a Microstrip Line," *IEEE Transactions on Antennas and Propagation,* Vol. 30, No. 5, September 1982, pp. 922-926.

[58] Himdi, M., J. P. Daniel, and C. Terret, "Transmission Line Analysis of Aperture-Coupled Microstrip Antenna," *Electronics Letters,* Vol. 25, No. 18, 1989, pp. 1229-1230.

[59] Garcia, J. A., L. Cabria, and A. Mediavilla, "A Frequency Agile Phase-Conjugating Active Antenna for Full-Duplex Retrodirective Arrays," *Proc. IEEE International Microwave Symposium,* San Francisco, CA, June 11-16, 2006, pp. 622-625.

[60] Hum, S. V., and H. Y. Xiong, "Analysis and Design of a Differentially-Fed Frequency Agile Microstrip Patch Antenna," *IEEE Transactions Antennas and Propagation,* Vol. 58, No. 10, October 2010, pp. 3122-3130.

[61] Shynu, S. V., et al., "Triple Slot Arm Loaded Reconfigurable Dual Frequency Microstrip Antenna Using Varactors," *Proc. IEEE International Symposium on Antennas and Propagation,* Washington D.C., July 3-8, 2005, pp. 609-612.

[62] Shynu, S. V., et al., "A Reconfigurable Dual-Frequency Slot-Loaded Microstrip Antenna Controlled by PIN Diodes," *Microwave and Optical Technology Letters,* Vol. 44, No. 4, February 2005, pp. 374-376.

[63] Waterhouse, R. B., and N. V. Shuley, "Frequency Agile Microstrip Rectangular Patches Using Varactor Diodes," *Proc. IEEE International Symposium on Antennas and Propagation,* Chicago, IL, July 20-24, 1992, pp. 2188-2191.

[64] Turki, Y., and R. Staraj, "CPW-Fed Frequency-Agile Shorted Patch," *Microwave and Optical Technology Letters,* Vol. 25, No. 5, July 2000, pp. 291-294.

[65] Aberle, J. T., M. Chu, and C. R. Birtcher, "Scattering and Radiation Properties of Varactor-Tuned Microstrip Antennas," *Proc. International Symposium on Antennas and Propagation,* Chicago, IL, July 20-24, 1992, pp. 2229-2232.

[66] Waterhouse, R. B., and N. V. Shuley, "Full Characterisation of Varactor-Loaded, Probe-Fed, Rectangular, Microstrip Patch Antennas, *IEE Proceedings on Microwaves, Antennas and Propagation*, Vol. 141, No. 5, October 1994, pp. 367-373.

[67] Rostbakken, O., G. S. Hilton, and C. J. Railton, "An Adaptive Microstrip Patch Antenna for Use in Portable Tranceivers," *Proc. IEEE Vehicular Technology Conference,* Atlanta, GA, April 28-May 1, 1996, pp. 339-343.

[68] Dauget, S., et al., "Global Electromagnetic Analysis of Microstrip Agile Antenna," *Electronics Letters,* Vol. 33, No. 13, June 1997, pp. 1111-1112.

[69] Gautam, A. K., and B. R. Vishvakarma, "Frequency Agile Microstrip Antenna Symmetrically Loaded With Tunnel Diodes," *Microwave and Optical Technology Letters,* Vol. 48, No. 9, September 2006, pp. 1807-1810.

[70] Garcia, J. A., L. Cabria, and A. Mediavilla, "A Frequency Agile Phase-Conjugating Active Antenna for Full-Duplex Retrodirective Arrays," *Proc. IEEE International Microwave Symposium*, San Francisco, CA, June 11-16, 2006, pp. 622-625.

[71] Liu, J., J. Zhang, W. Wang, and D. Wang, "Compact Reconfigurable Microstrip Antenna for Multi-Band Wireless Applications," *Proc. International Symposium on Microwave, Antenna, Propagation and EMC Technologies for Wireless Communications*, Hangzhou, China, August 14-16, 2007, pp. 608-611.

[72] Weily, A. R., T. S. Bird, Y. J. Guo, "A Reconfigurable High-Gain Partially Reflecting Surface Antenna," *IEEE Transactions on Antennas and Propagation*, Vol. 56, No. 11, November 2008, pp. 3382-3390.

[73] Costa, F., et al., "An Active High-Impedance Surface for Low-Profile Tunable and Steerable Antennas," *IEEE Antennas and Wireless Propagation Letters*, Vol. 7, 2008, pp. 676-680.

[74] AbuTarboush, H. F., et al., "Reconfigurable Wideband Patch Antenna for Cognitive Radio," *Proc. Loughborough Antennas & Propagation Conference LAPC*, Loughborough, United Kingdom, November 16-17, 2009, pp. 141-144.

[75] Sufyar, S., C. Delaveaud, and R. Staraj, "A Frequency Agility Technique on a Miniature Omnidirectional Antenna," *Proc. European Conference on Antennas and Propagation EUCAP*, Barcelona, Spain, April 12-16, 2010.

Appendix A

Characteristic Impdedance of a Wide Slotline on Low-Permittivity Substrates (from [37])

Case 1. $2.22 \le \varepsilon_r \le 3.8,\ \ 0.015 \le W/\lambda_o \le 0.075$

$$\frac{\lambda_g}{\lambda_o} = 1.045 - 0.365 \ln \varepsilon_r + \frac{6.3(W/h)\varepsilon_r^{0.945}}{238.64 + 100W/h} - \left[0.148 - \frac{8.81(\varepsilon_r + 0.95)}{100\varepsilon_r}\right] \ln(h/\lambda_o)$$

q

$$Z_o = 60 + 3.69 \sin\left[\frac{(\varepsilon_r - 2.22)\pi}{2.36}\right] + 133.5 \ln(10\varepsilon_r)\sqrt{W/\lambda_o}$$
$$+\ 2.81\left[1 - 0.11\varepsilon_r(4.48 + \ln \varepsilon_r)\right](W/h)\ln(100h/\lambda_o)$$
$$+131.1(1.028 - \ln \varepsilon_r)\sqrt{h/\lambda} + 12.48(1 + 0.18 \ln \varepsilon_r)\frac{W/h}{\sqrt{\varepsilon_r - 2.06 + 0.85(W/h)^2}}$$

Case 2. $2.22 \le \varepsilon_r \le 3.8,\ \ 0.075 \le W/\lambda_o \le 1.0$

$$\frac{\lambda_g}{\lambda_o} = 1.194 - 0.24 \ln \varepsilon_r - \frac{0.621(W/\lambda_o)^{0.48}\varepsilon_r^{0.835}}{1.334 + W/h} - 0.0617\left[1.91 - \frac{\varepsilon_r + 2}{\varepsilon_r}\right]\ln(h/\lambda_o)$$

$$Z_o = 133 + 10.34\ (\varepsilon_r - 1.8)^2 + 2.87\left[2.96 + (\varepsilon_r - 1.582)^2\right]$$
$$\cdot \left\{\left[W/h + 2.32\varepsilon_r - 0.56\right]\left[(32.5 - 6.67\varepsilon_r)\ (100h/\lambda_o)^2 - 1\right]\right\}^{0.5}$$
$$-\ (684.45h/\lambda_o)\ (\varepsilon_r + 1.35)^2 + 13.23\left[(\varepsilon_r - 1.722)\ W/\lambda_o\right]^2$$

Case 3. $3.8 \le \varepsilon_r \le 9.8,\ \ 0.015 \le W/\lambda_o \le 0.075$

$$\frac{\lambda_g}{\lambda_o} = 0.9217 - 0.277 \ln \varepsilon_r + 0.0322(W/h)\left[\frac{\varepsilon_r}{W/h + 0.435}\right]^{0.5}$$
$$-\ 0.01 \ln(h/\lambda_o)\left[4.6 - \frac{3.65}{\varepsilon_r^2\sqrt{W/\lambda_o}(9.06 - 100W/\lambda_o)}\right]$$

$$Z_o = 73.6 - 21.5\varepsilon_r + (638.9 - 31.37\varepsilon_r)(W/\lambda_o)^{0.6} + \left(36.23\sqrt{\varepsilon_r^2 + 41} - 225\right)\frac{W/h}{W/h + 0.876\varepsilon_r - 2}$$
$$+\ 0.51(\varepsilon_r + 2.12)(W/h)\ln(100h/\lambda_o) - 0.753\varepsilon_r(h/\lambda_o)\sqrt{W/\lambda_o}$$

Case 4. $3.8 \le \varepsilon_r \le 9.8,\ \ 0.075 \le W/\lambda_o \le 1.0$

$$\frac{\lambda_g}{\lambda_o} = 1.05 - 0.04\varepsilon_r + 0.01411\,(\varepsilon_r - 1.421)\ln\left[W/h - 2.012\,(1 - 0.146\varepsilon_r)\right]$$

$$+\ 0.111(1 - 0.366\varepsilon_r)\,\sqrt{W/\lambda_o} +\ 0.139\,(1 + 0.52\varepsilon_r\,\ln(14.7 - \varepsilon_r))\,(W/h)\ \ln(h/\lambda_o)$$

$$Z_o = 120.75 - 3.74\varepsilon_r +\ 14.21(1 - 0.458\varepsilon_r)\left[100\frac{h}{\lambda_o} + 5.1\ln\varepsilon_r - 13.1\right]\left[\frac{W}{h} + 0.33\right]^2$$

$$+\ 50\,\left[\tan^{-1}(2\varepsilon_r) - 0.8\right]\left[\frac{W}{h}\right]^{1.11 + \left[0.132(\varepsilon_r - 27.7)/(100h/\lambda_o + 5)\right]}\ln\left[100\frac{h}{\lambda_o} + \sqrt{\left[100\frac{h}{\lambda_o}\right]^2 + 1}\right]$$

Chapter 8

FAAs With Integrated Devices and Discrete Tuning

This chapter examines a cross-section of antennas with discrete frequency tuning capabilities. Discrete frequency tuning is realized by integrating various electronic components with the antennas. The majority of these electronic devices can be grouped into one of four categories: diodes, transistors, MEMS, and optical switches. These electronic devices have been incorporated into various antenna elements including slot antennas, PIFAs, CPW antennas, and DRAs and a variety of printed antennas including loops, dipoles, patches, rings, and monopoles. Representative examples of the various configurations are illustrated in this chapter to provide an overview of the wide variety of designs and an indication of achievable performance. The TR and, where available, the tunable impedance bandwidth (BW_{TZ}) and the TS of these designs are provided, both based on a 10-dB return loss, along with bias conditions and antenna performance.

8.1. PIN-DIODE CONTROL

This section examines a selection of antennas loaded with PIN diodes to achieve discrete frequency tuning. The operation, performance, advantages, and disadvantages of PIN diodes are discussed in Section 4.3.1. PIN diodes are readily available commercially, are inexpensive, come in a variety of packaged sizes, and require low bias voltages. They do, however, draw current in their On state, which will increase DC power consumption, and their series resistance in the On state will contribute to some degradation in the antenna's radiation efficiency. The use of PIN diodes is one of the most common techniques for achieving discrete frequency tuning, and examples of PIN-diode-loaded antenna elements in this section include microstrip patch antennas, IFAs, PIFAs, monopoles, dipoles, meander-line antennas, slot antennas, and DRAs.

8.1.1 Microstrip Patch Antennas

A selection of examples illustrating the various ways in which PIN diodes are used to achieve discrete frequency tuning in microstrip patch antennas is presented in this section. Figure 8.1 shows a microstrip patch fed by a rectangular slot aperture, excited using a CPW line [1]. Two PIN diodes are connected across the slot and are activated by a DC bias voltage across two thin isolation slots. The width (W_s = 0.5 mm) of the slot was chosen so that the beam-lead PIN diodes

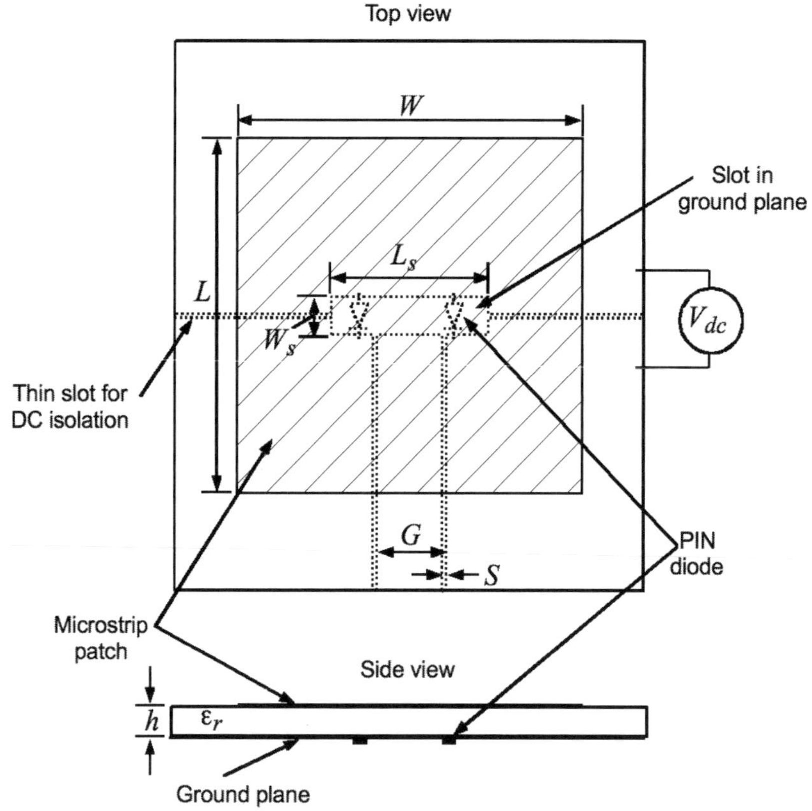

Figure 8.1 PIN-diode-loaded microstrip patch ($L = W = 20$ mm, $h = 2.3$ mm, $\varepsilon_r = 2.2$, $L_s = 18$ mm, $W_s = 0.5$ mm, $G = 4.4$ mm, and $S = 0.2$ mm). (After [1].)

(having a total length of 0.8 mm) could easily be soldered. When the diodes were in the Off state, the entire length of the slot excited the microstrip patch antenna, which resonated at 3.51 GHz. When the diodes were in the On state, they drew 30 mA of DC current and essentially acted as short circuits to shorten the effective length of the slot. This in turn shifted the resonant frequency up to 4.28 GHz, resulting in a tuning range of TR = 19.8%. The total spectrum for this antenna was TS = 23%, and the tunable impedance bandwidth was BW_{TZ} = 6%. The gain or radiation efficiency of the antenna was not mentioned in [1]; however this configuration was then further analyzed by different investigators and reported in [2]. Their analysis related the gain of the antenna to the series resistance of the diodes in the On state. For a typical series resistance of 5Ω, the gain was found to be about 5.7 dBi, compared to a gain of 6.4 dBi for an equivalent passive antenna, corresponding to a 15% reduction in radiation efficiency.

A similar idea was proposed using a single diode in [3]; however, that configuration required a specific slot length in order to achieve a good impedance match for both states of the PIN diode, whereas this is not a restriction in the configuration proposed in [1].

A more complex design is shown in Figure 8.2 [4]. In this case the microstrip patch antenna is fed using a coaxial probe, and the slot beneath the patch contains seven PIN-diode switches (S_1 to S_7), each of which is independently controlled by bias voltages (V_1 to V_7). By turning on or off the PIN diodes in the combinations shown in Table 8.1, the resonant frequency of the antenna shifted from 1.87 to 3.19 GHz (TR = 52.2%). The antenna achieved a total spectrum of TS = 56% and a tunable impedance bandwidth of BW_{TZ} = 24%. A 60-mA bias current was supplied to each PIN-diode switch in the On state. The slot was designed to resonate (and thus radiate) at the lower resonant frequencies, while the microstrip patch antenna resonated at the highest resonant frequencies. The radiation pattern is thus nearly omnidirectional at the lower frequencies and directional at the higher frequencies. Measured gain values were not reported but simulated values ranged from 4.3 to 6.4 dBi.

Table 8.1

Resonant Frequency for the Various PIN-Diode States for the Microstrip-Slot Antenna

State	S_1	S_2	S_3	S_4	S_5	S_6	S_7	Frequency (GHz)
1	Off	Off	Off	Off	Off	Off	Off	1.87
2	On	Off	Off	Off	Off	Off	Off	1.96
3	On	Off	Off	Off	Off	Off	On	2.08
4	Off	Off	Off	Off	On	On	Off	2.20
5	On	On	Off	Off	Off	On	On	2.41
6	On	On	On	Off	On	On	On	2.74
7	On	On	On	On	On	On	On	3.19

(From [4])

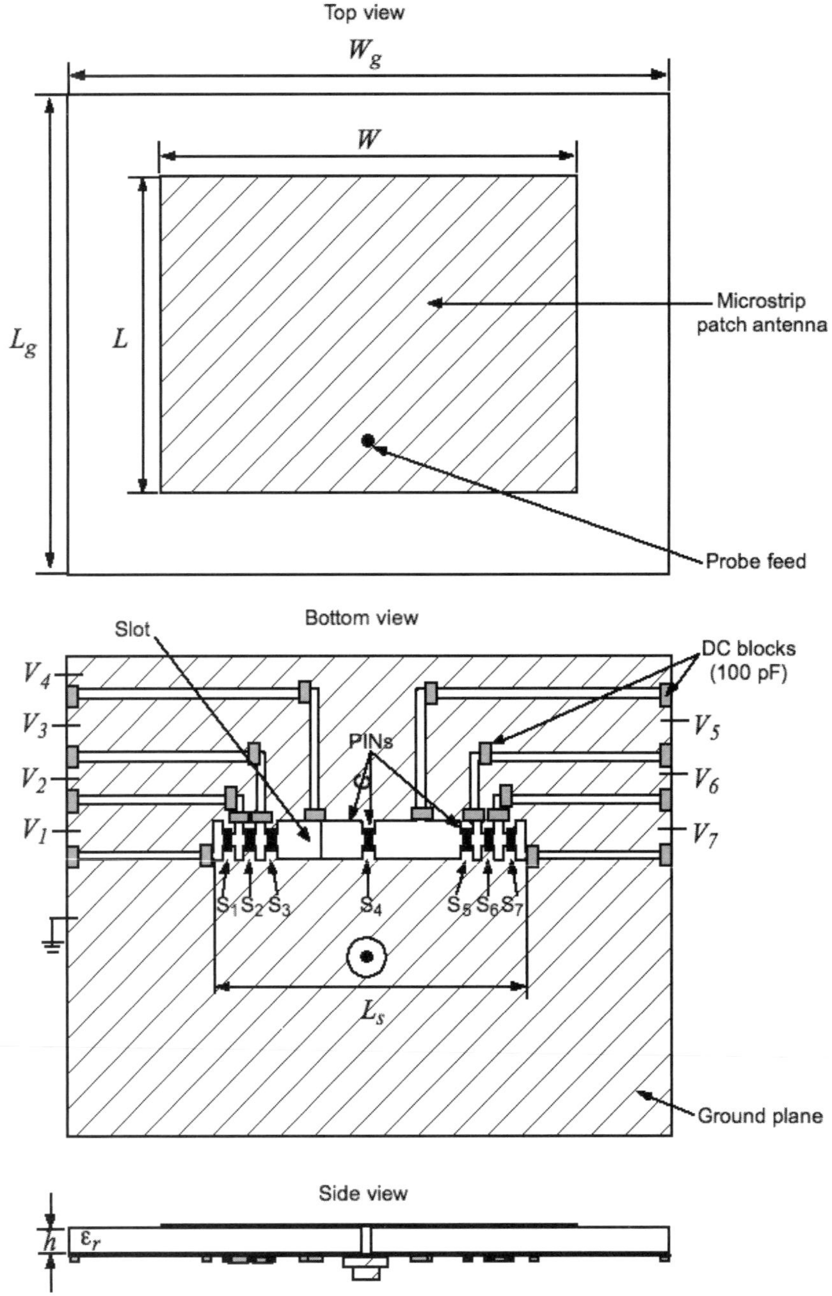

Figure 8.2 PIN-diode-loaded microstrip patch (L_g = 41.5 mm, W_g = 50 mm, L = 21.5 mm, W = 30 mm, h = 1.58 mm, ε_r = 3.5, and L_s = 25.8 mm). (After [4].)

In the next example, shown in Figure 8.3, a probe-fed microstrip patch was loaded with a rectangular slot, across which a PIN diode was connected [5]. The DC bias voltage (of 2V) was provided via a bias-T through the probe feed along with the RF signal. Two capacitors were connected across the two ends of the slot to provide an RF signal path but block the DC component. A high-impedance quarter-wave microstrip line acted as the RF choke, preventing the RF signal from flowing through the DC ground. When the PIN diode was in the Off state, the RF current generated by the probe was forced to flow around the slot through the two capacitors, increasing the path length and thus lowering the resonant frequency of the patch. When the PIN diode was in the On state, the RF current could follow the shorter path through the diode, thus increasing the resonant frequency. For the patch dimensions shown in Figure 8.3, the resonant frequency shifted from 4.38 GHz in the Off state to 4.82 GHz (TR = 9.6%) in the On state. The antenna also achieved a total spectrum of TS = 16% and a tunable impedance bandwidth of BW_{TZ} = 14%. The gain of the patch dropped from about 6 dBi for the Off state to 5 dBi in the On state, due to the series resistance of the PIN diode. Similar examples of PIN diodes connected across slot-loaded microstrip patches can be found in [6–8].

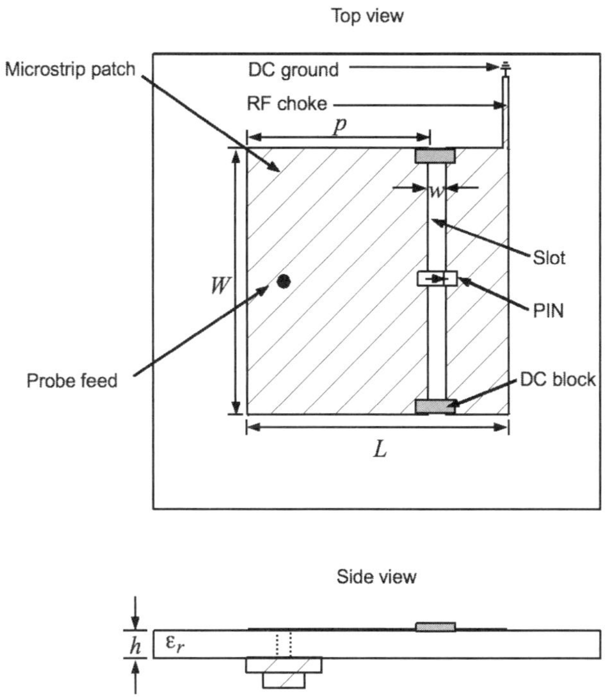

Figure 8.3 Microstrip patch loaded with PIN-diode switches ($L = W = 18$ mm, $h = 3$ mm, $\varepsilon_r = 2.2$, $w = 1$ mm, and $p = 13$ mm). (After [5].)

A more complicated design is shown in Figure 8.4 where four pairs of PIN diodes (D_1 to D_4) are integrated with an aperture-coupled microstrip patch antenna [9]. The required DC bias voltage to the PIN diodes was supplied through wideband RF chokes located either on the top surface or the bottom surface, due to

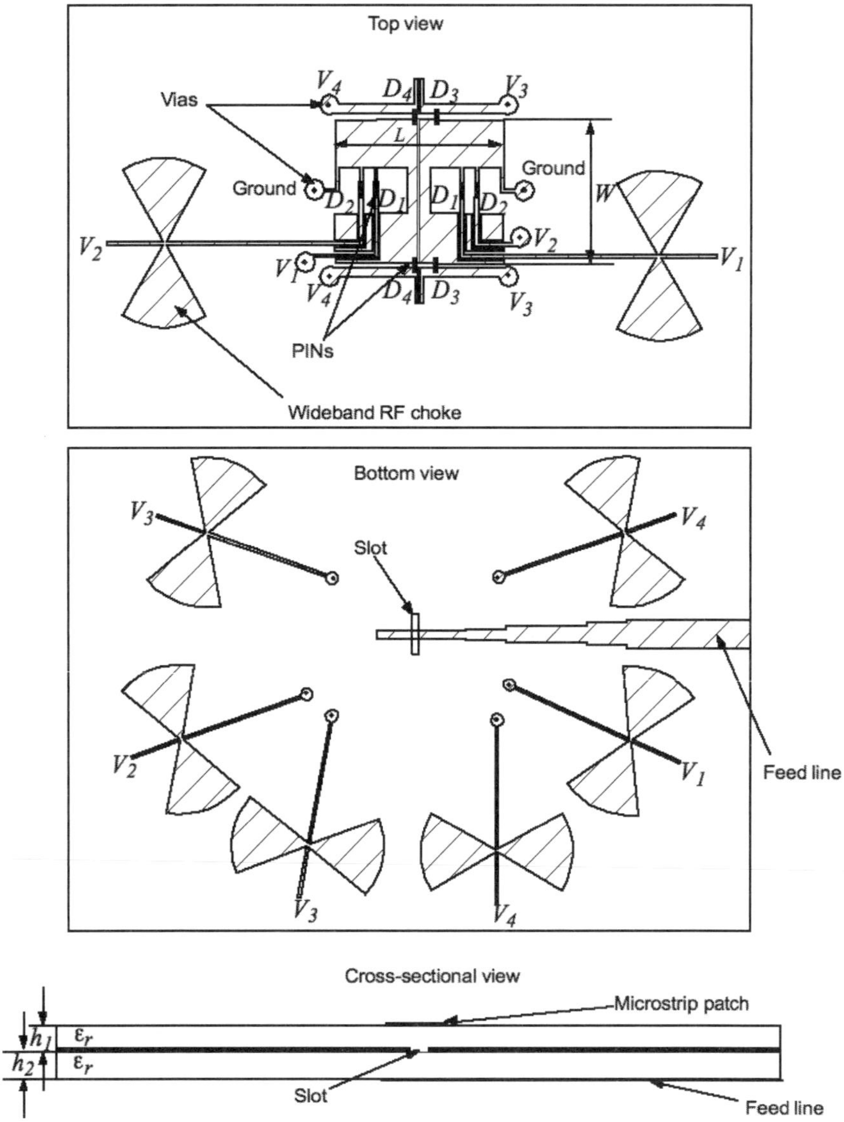

Figure 8.4 Microstrip patch loaded with PIN-diode switches (L = 2.6 mm, W = 2.4 mm, $h_1 = h_2$ = 0.254 mm, and ε_r = 2.2). (After [9].)

lack of space to fit them all on one surface. By turning on or off the diodes in the combinations shown in Table 8.2, the resonant frequency of the patch was switched between one of five values, ranging from 10.2 to 21.3 GHz (TR = 70.5%), achieving a total spectrum of TS = 72%, and a tunable impedance bandwidth of BW_{TZ} = 16%. The gain and radiation efficiency performance of the antenna were not reported.

Table 8.2

Resonant Frequency for Different PIN-Diode State Combinations

D1	D2	D3	D4	Frequency (GHz)
Off	Off	Off	Off	10.2
On	Off	Off	Off	15.5
On	On	Off	Off	16.8
On	On	On	Off	18.3
On	On	On	On	21.3

(From [9])

The final example of a PIN-diode-loaded microstrip patch antenna is illustrated in Figure 8.5 [10]. The antenna actually consisted of two radiators of slightly different dimensions, combined by a simple microstrip T-branch. A PIN diode was connected across a small gap in each of the branches of the microstrip combiner. By turning on or off the diodes, the RF signal could be directed to either radiator or to both radiators (if both diodes were on). Introducing the slots in each radiator added a second resonant frequency. Table 8.3 shows the resonant frequencies (f_1 and f_2) and the corresponding 10-dB return loss fractional bandwidths (BW_1 and BW_2) for the On-Off and Off-On states of the two PIN diodes. The corresponding tuning ranges for f_1 and f_2 are TR_1 = 11.3% and TR_2 = 8.4%. The corresponding total spectrums and tunable impedance bandwidths are $TS_1 = BW_{TZ1}$ = 24% and $TS_2 = BW_{TZ2}$ = 38%. When both diodes were on, the antenna exhibited a wideband frequency response, with a 10-dB return loss from 4.99 to 7 GHz. The radiation efficiency of the antenna was determined to be about 80% for all states.

Table 8.3

Resonant Frequencies and Fractional Bandwidths for the PIN-Diode States

Diodes States	f_1 (GHz)	f_2 (GHz)	BW_1	BW_2
On-Off	5	5.7	5%	25%
Off-On	5.6	6.2	22%	33%

(From [10])

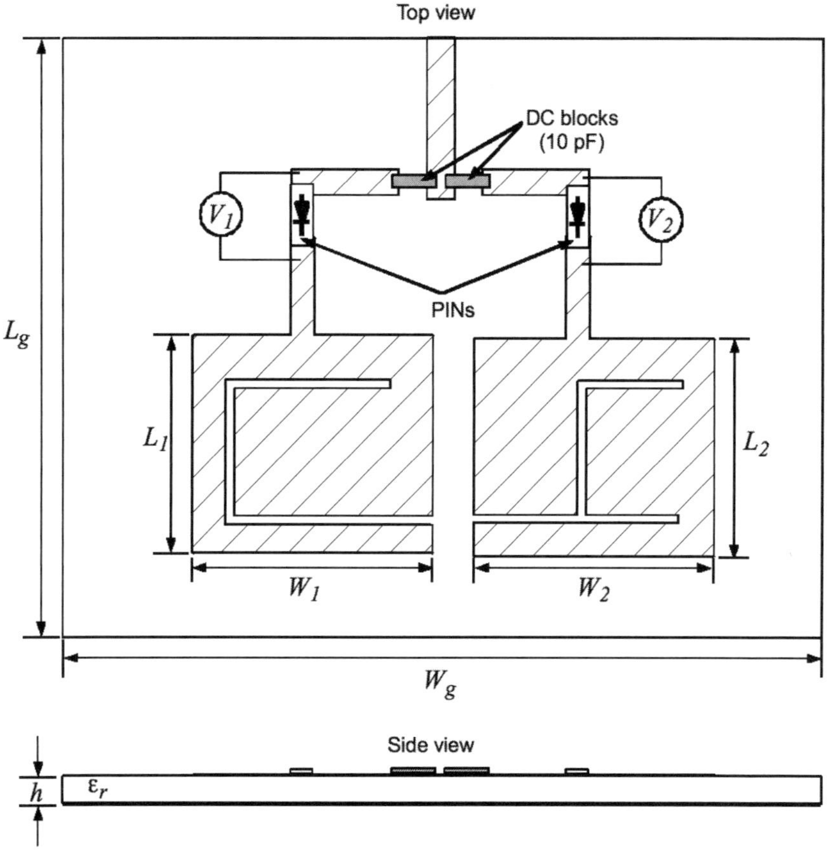

Figure 8.5 Dual patches with PIN-diode switches ($L_1 = L_2 = 24$ mm, $W_1 = 3.5$ mm, $W_1 = 20$ mm, $W_2 = 22$ mm, $L_g = 36$ mm, $W_g = 46$ mm, $h = 1.57$ mm, and $\varepsilon_r = 4.4$). (After [10].)

8.1.2 IFAs

This section examines several designs of IFAs or IFA-like antennas with integrated PIN diodes for achieving discrete frequency tuning. The first example is shown in Figure 8.6, where the antenna was etched onto a small (W_g x H_g x h) substrate mounted orthogonally to the larger grounded substrate (W_g x L_g x h) used to represent the circuit board of a handheld phone [11]. The PIN diode was mounted on the large ground plane and connected to the antenna via a short microstrip line. With the addition of the DC blocking capacitor and an RF choke inductor, the DC bias could be injected on the same line as the RF signal. In the On state, the PIN diode made a connection across the microstrip line to effectively lengthen the antenna, lowering the resonant frequency. The antenna was designed to provide dual-band coverage for GSM900/GSM1800 when the PIN diode was in the On state and GSM850/GSM1800 and DCS1900 when the PIN diode was in the Off state. Table 8.4 lists the resonant frequencies ($f_1 + f_2$) and the 10-dB return loss bandwidth (BW_1 and BW_2) for the two diode state achieved by the applying a DC bias voltage of ±1V. The tuning range for the frequency shifts in f_1 and f_2 were TR_1 = 10.8% and TR_2 = 2.2%. The combined response of the two states results in total spectrums of TS_1 = 15% and TS_2 = 8%, and tunable impedance bandwidths of BW_{TZ1} = 9% and BW_{TZ2} = 8%. The maximum gain for the antenna in the Off state was measured to be 2.2 dBi, while in the On state the maximum gain dropped to 0.9 dBi, attributed to the series resistance of the PIN diode when in the On state.

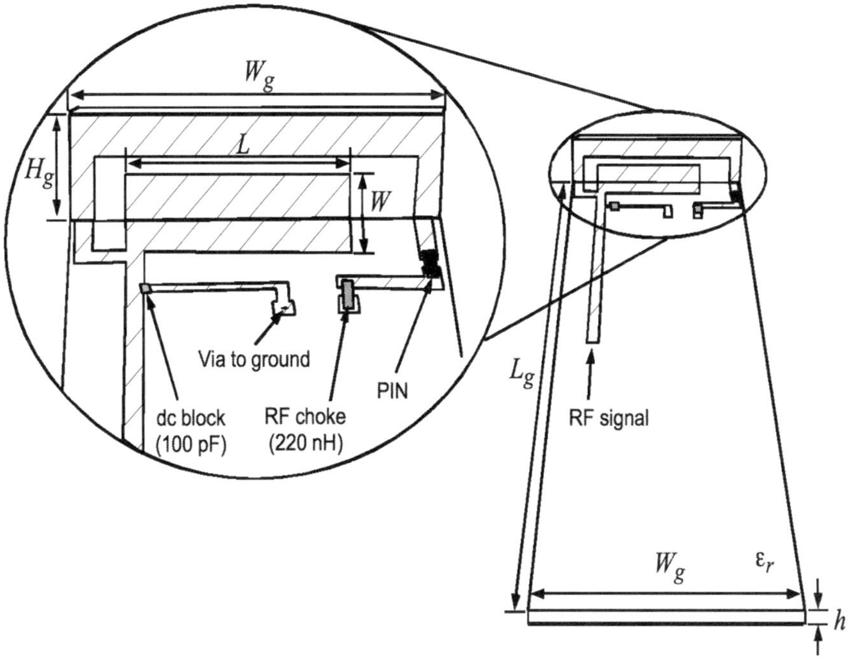

Figure 8.6 PIN-diode-loaded IFA (L_g = 100 mm, W_g = 40 mm, H_g = 9.2 mm, L = 24 mm, W = 8 mm, h = 0.8 mm, and ε_r = 3.38). (After [11].)

Table 8.4

Resonant Frequencies and Factional Bandwidths for the PIN-Diode States

Diode State	f_1 (MHz)	f_2 (MHz)	BW_1 (MHz)	BW_2 (MHz)
On	858	1,820	847–873	1,775–1,870
Off	956	1,860	934–983	1,795–1,915

(From [11])

A second example of a dual-band IFA is shown in Figure 8.7(a) and was again designed to operate in two sets of dual-band wireless services GSM850/personal communications services (PCS) and wireless local area network (WLAN)/worldwide interoperability for microwave access (WiMax) [12]. The antenna was composed of two IFAs placed side-by-side that share the same RF signal feed point. A PIN diode was connected to each IFA which determined the location of the short circuit when in the On state. The effective antenna geometries when the PINs were oppositely biased (with a DC bias voltage of ±1V) are shown in Figure 8.7(b,c). The antennas were etched on a thin layer (0.05 mm) of Kapton and placed atop a 32 cm x 20 cm rectangular plate, used to simulate the lid and body of a laptop. The resonant frequencies and bandwidths for the two diode states are listed in Table 8.5. The antennas had a tuning range of TR_1 = 96% and TR_2 = 56% for the f_1 and f_2 frequencies, respectively. The total spectrum covered

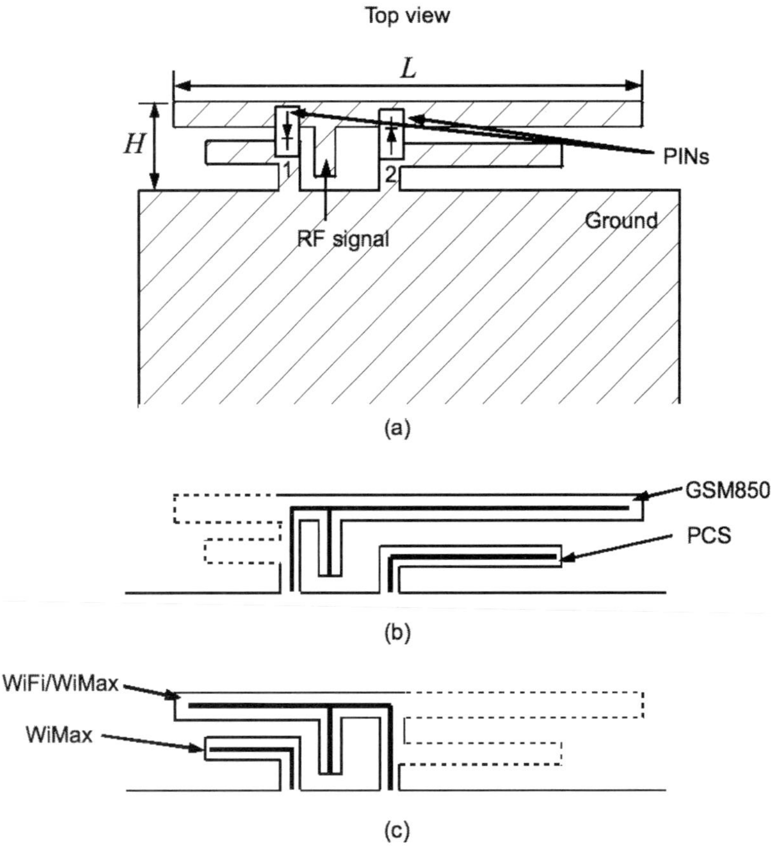

Figure 8.7 PIN-diode-loaded IFA: (a) multiband antenna geometry, (b) PIN 1: On, PIN 2: Off, and (c) PIN 1: Off, PIN 2: On (L_g =90 mm, and H = 10 mm). (After [12].)

by the antenna was TS_1 = 108% and TS_2 = 70% with a tunable impedance bandwidth of BW_{TZ1} = 39% and BW_{TZ2} = 23%. Peak measured gains at various frequencies for the different diode states ranged from approximately 0 to 2 dBi.

A similar example is shown in Figure 8.8 where, in this instance, the two multiband IFAs were etched on the same side of a thin (0.05 mm) Kapton substrate, which was then folded so that the antennas were back-to-back, separated by a 3-mm foam spacer [13]. The antenna dimensions were suitable for integration into a laptop case. PIN diodes were then integrated with the two antennas, as sketched in Figure 8.8. The DC bias signal was injected with the RF signal through a bias-T, and the IFAs resonated at different sets of frequencies for the different states of the PIN diodes. Table 8.6 shows the first five resonant frequencies and bandwidths attained for the two states, where in State 1 the bottom two switches were on and the top one was off (referring to Figure 8.8), while for State 2, the bottom two switches were off and the back switch was on. The antenna achieved a total spectrum of TS = 142% and a tunable impedance bandwidth of BW_{TZ} = 43% and covered various of the terrestrial wireless services' frequency bands. Each diode required about 100 mA of DC current when operating in the On state to keep the gain of the antennas to within 0.9 dB of the passive versions, which achieved gains of between 2 and 4 dBi.

The final example of an IFA-like antenna is shown in Figure 8.9, where the antenna was located at the top of the dielectric sheet, whose dimensions were similar to those of a cellular phone [14]. The PIN diode was connected across a small gap in the antenna line and lengthened or shortened the effective length when turned on or off, respectively. With a DC bias voltage of ±1V, the resonant frequency of the antenna shifted from 2.39 GHz in the Off state to 1.8 GHz in the On state, for a tuning range of TR = 28.2%. With 10-dB return loss bandwidths of 210 MHz for the Off state and 220 MHz for the On state, the antenna achieved a total spectrum of TS = 40% and a tunable impedance bandwidth of BW_{TZ} = 23%. The peak radiation efficiency for the antenna was 68% in the Off state and 63% in the On state, compared to 85% and 80% for the corresponding passive versions, thus resulting in a tuning loss of about 20%.

Table 8.5

Resonant Frequencies and Fractional Bandwidths for the PIN-Diode States

Diode States	f_1 (MHz)	f_2 (MHz)	BW_1 (MHz)	BW_2 (MHz)
On-Off	879	1,915	864-890	1,871-1,963
Off-On	2,500	3,500	2,195-2,900	3,232-3,798

(From [12])

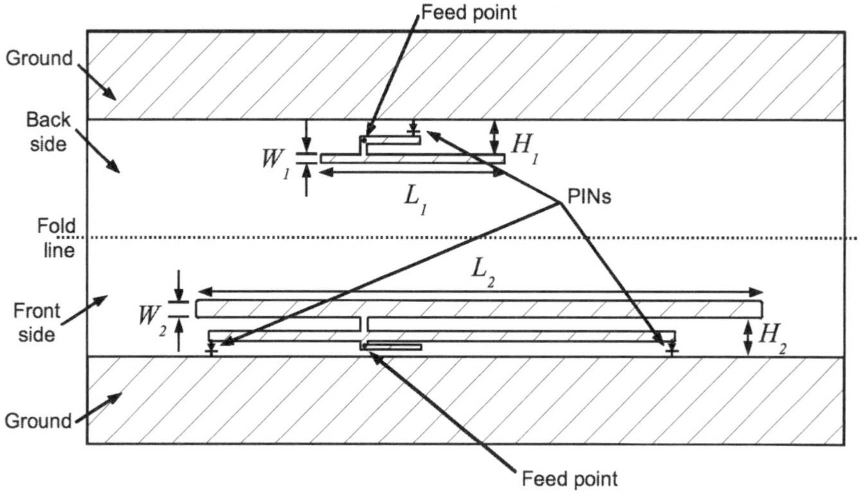

Figure 8.8 Multi-branch printed monopole (L_1 = 31 mm, H_1 = 3.5 mm, W_1 = 1.5 mm, L_2 = 95 mm, H_2 = 6.5 mm, and W_2 = 2 mm). (After [13].)

Table 8.6

Resonant Frequencies for the PIN-Diode-Loaded IFAs

State 1		State 2	
f_1: 828 MHz	BW_1: 780–875 MHz	f_1: 840 MHz	BW_1: 820–840 MHz
f_2: 1,925 MHz	BW_2: 1,880–1,970 MHz	f_2: 1,585 MHz	BW_2: 1,530–1,630 MHz
f_3: 2,685 MHz	BW_3: 2,600–2,770 MHz	f_3: 2,395 MHz	BW_3: 2,340–2,450 MHz
f_4: 4,035 MHz	BW_4: 3,950–4,120 MHz	f_4: 3,605 MHz	BW_4: 3,570–3,640 MHz
f_5: 5,335 MHz	BW_5: 5,260–5,410 MHz	f_5: 4,485 MHz	BW_5: 4,380–4,590 MHz

(From [13])

8.1.3 PIFAs

This section examines four examples of PIFA antennas loaded with one or more PIN diodes for achieving discrete frequency agility. The first example is shown in Figure 8.10, where the PIN diode is used to extend the length of a tuning stub connecting the short post of the PIFA to ground [15]. The tuning stub, PIN diode,

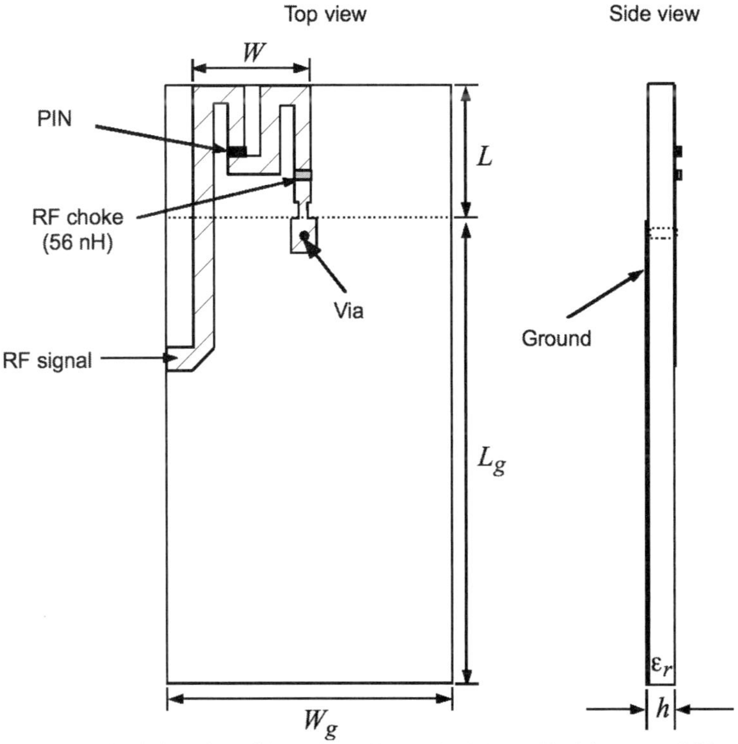

Figure 8.9 PIN-diode-loaded IFA (L_g =65 mm, W_g = 40 mm, L = 15 mm, W= 14 mm, ε_r = 3.38, and h = 0.8 mm). (After [14].)

and accompanying DC bias circuitry were all located on a thin dielectric sheet placed adjacent to the PIFA. When the PIN diode was switched from the Off state to the On state, the resonant frequency of the PIFA shifted from 941 to 898 MHz (TR = 4.7%), with a total spectrum of TS = 9% and a tunable impedance bandwidth of BW_{TZ} = 7%. When the DC bias for the On state was set to 10 mA and the DC bias for the Off state set to -20V, the radiation efficiency for both states was measured to be 95%. Reducing the DC bias to 0V for the Off state, resulted in a decrease in radiation efficiency to 92%, while reducing the bias current for the On state to 1 mA resulted in a decrease in radiation efficiency to 88%. There is thus a trade-off between DC power consumption and radiation efficiency.

The second example, shown in Figure 8.11, consists of two nested PIFAs, stacked one above the other for dual-band operation [16]. The lower PIFA is directly excited by a coaxial probe and resonated at the higher frequency, while the upper PIFA was electromagnetically coupled and resonated at the lower frequency. Each PIFA included six shorting straps, which were each connected to ground

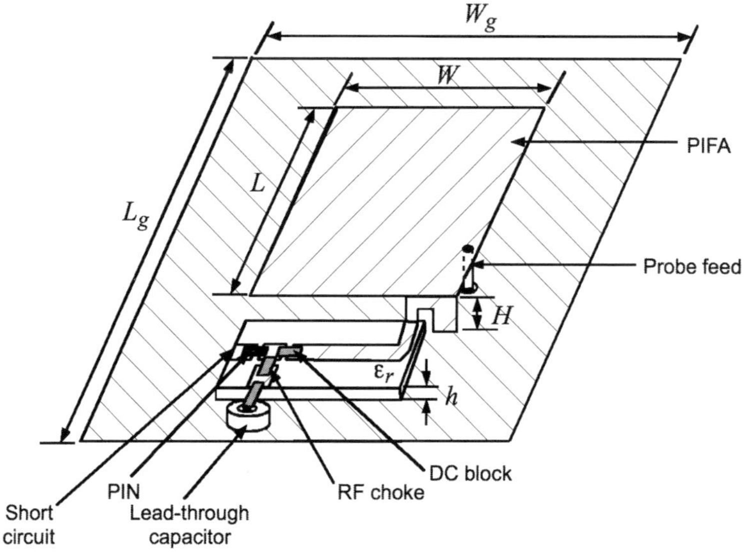

Figure 8.10 PIN-diode-loaded PIFA ($L_g = W_g = 300$ mm, $L = 46.6$ mm, $W = 46.7$ mm, $H = 12$ mm, $\varepsilon_r = 2.33$, and $h = 0.79$ mm). (After [15].)

through a PIN-diode switch. The ground connection was made when the PIN diode was in the On state and broken when the diode was in the Off state. A DC bias of ±5V was used to turn on or off each diode, and the resistor limited the current to a maximum of 10 mA in the On state. For the twelve PIN diodes, there were 720 possible tuning sequences for resonant frequencies; however, only a few combinations were investigated in [16]. When the first shorting strap was connected to ground (PIN diode in the On state) of the lower PIFA, and all shorting straps of the upper patch were open, the two resonant frequencies were at $f_1 = 806$ MHz and $f_2 = 1{,}240$ MHz. As the shorting straps for the upper PIFA were successively connected to ground, the lower resonant frequency shifted down to $f_1 = 730$ MHz ($TR_1 = 10\%$, $TS_1 = BW_{TZ1} = 6\%$), with almost no impact on the higher resonant frequency. When the first shorting strap of the upper PIFA was connected to ground, while the shorting straps of the lower PIFA were successively connected to ground, the upper resonant frequency shifted from $f_2 = 1{,}250$ MHz to $f_2 = 1{,}023$ MHz ($TR_2 = 20\%$, $TS_2 = 6\%$, $BW_{TZ2} = 4\%$). However, in this case there was an accompanying shift in the lower resonant frequency from about 775 MHz to 730 MHz. Measured gains for the antenna were -4.3 dBi at 745 MHz and 1.6 dBi at 1,240 MHz. This was considered low

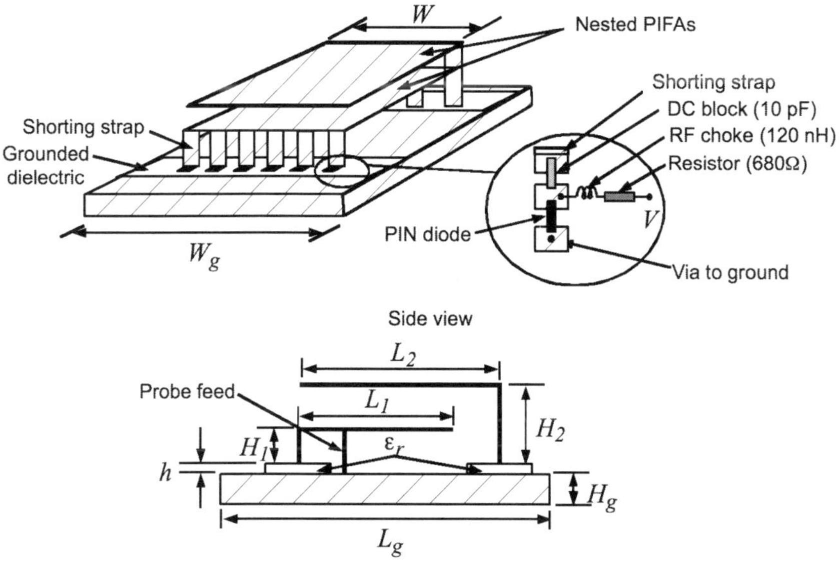

Figure 8.11 PIN-diode-loaded stacked PIFAs (L_g = 100 mm, W_g = 50 mm, H_g = 10 mm, L_1 = 56.5 mm, H_1 = 8 mm, L_2 = 51.5 mm, H_2 = 12 mm, W = 31 mm, h = 1.6 mm, and ε_r =4.2). (After [16].)

compared to a passive PIFA and was attributed to the series losses in the PIN diode in the On state and due to distortions in the pattern due to the proximity of the bias circuitry to the radiating aperture.

In the next example, illustrated in Figure 8.12, the PIN diode and biasing components (the DC blocks and the RF choke) were connected across slots in the top-surface metallization of the PIFA [17]. The PIFA was designed to operate in the GSM, PCS, and UMTS bands with frequencies ranging from 824 to 2,170 MHz. The selected PIN diode had a published insertion loss of 0.3 dB in the On state at a 10-mA forward bias current and an Off-state isolation of better than 20 dB up to 2 GHz. The diode was biased with a ±1V and drew 20 mA in the forward bias (On state). When the diode was switched on, the antenna resonated at f_1 = 845 MHz and f_2 = 1,770 MHz, while in the Off state, the antenna resonated at f_1 = 931 MHz and f_2 = 2,025 MHz, corresponding to tuning ranges of TR_1 = 9.7% and TR_2 = 13.4%, with accompanying total spectrums of TS_1 = 12% and TS_2 = 20% and tunable impedance bandwidths of BW_{TZ1} = 9% and BW_{TZ2} = 13%. In the On state, the peak radiation efficiency of the PIFA was 68% at f_1 and 88% at f_2, while in the Off state, the values for radiation efficiency were 65% at f_1 and

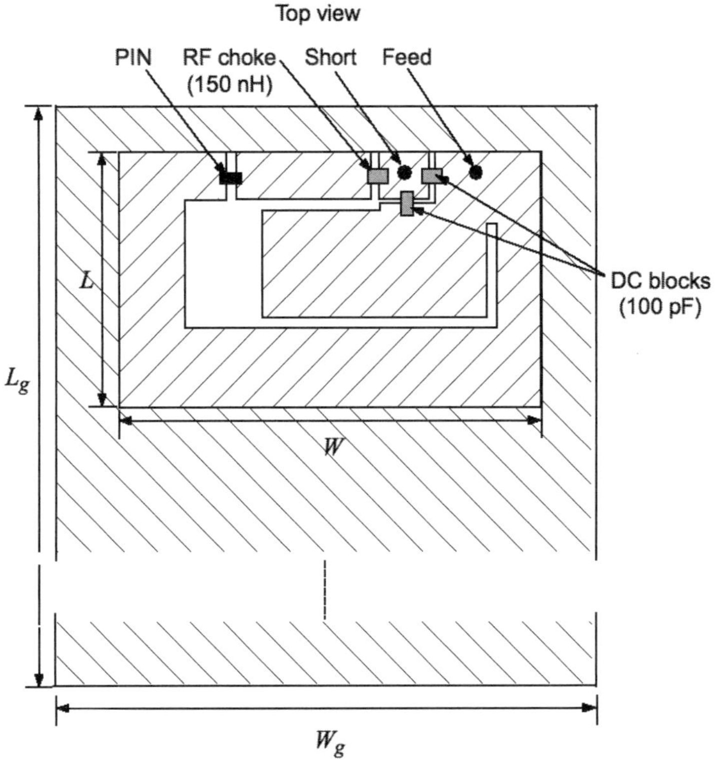

Figure 8.12 PIN-diode-loaded PIFA ($L_g = 110$ mm, $W_g = 45$ mm, $L = 21$ mm, $W = 39.5$ mm, $H = 9.2$ mm, $\varepsilon_r = 3.38$, and $h = 0.8$ mm). (After [17].)

85% at f_2. The minimum values for radiation efficiency ranged from 40% to 65% for the different bands. This represents a reduction of 20% from the radiation efficiency of the corresponding passive PIFA design. For the f_1 frequencies, the maximum gain was 1.8 dBi for the On state and 2.2 dBi for the Off state, while for the f_2 frequencies the maximum gain values were 4 dBi for the On state and 3.6 dBi for the Off state.

The final example is shown in Figure 8.13 and consisted of a PIFA combined with an IFA, each with its own PIN diode, for multiband operation [18]. The IFA was etched on the top side of the flat dielectric substrate by a microstrip line on the opposite side of the substrate. The PIN diode for the IFA was connected across a

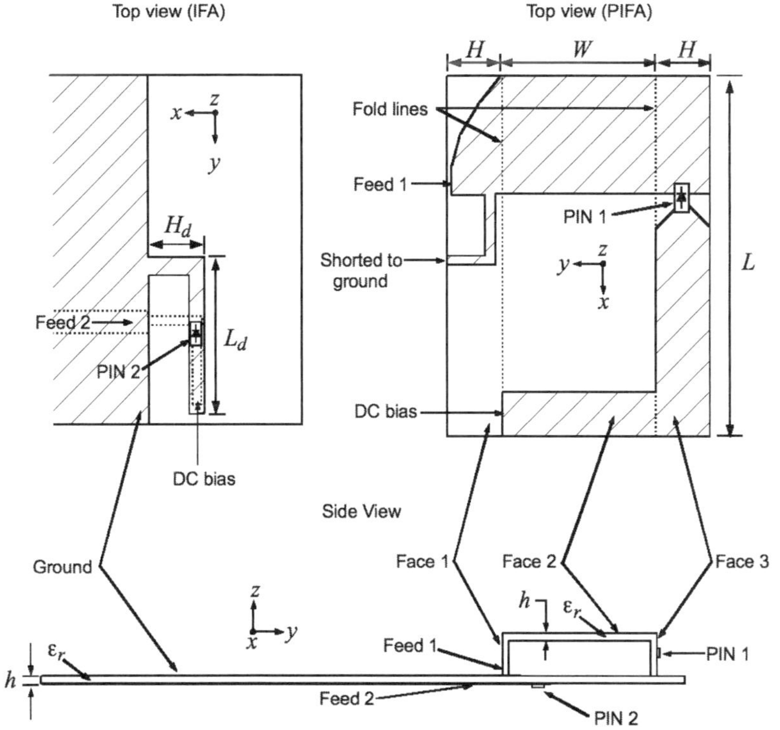

Figure 8.13 PIN-diode-loaded PIFA (L = 40 mm, W = 20 mm, H = 6 mm, L_d = 18 mm, H_d = 7 mm, ε_r = 4.4, and h = N/A). (After [18].)

small gap in the microstrip feed line. The feed for the PIFA was located on Face 1 (see Figure 8.13) while its PIN diode was connected across a small gap in the metallization on Face 3. The PIFA was designed to be switched between two sets of cellular bands: Global System for Mobile Communications (GSM)/code division multiple access (CDMA) and digital communication systems (DCS)/personal communications systems (PCS)/wideband code division multiple access (WCDMA) while the IFA could operate over three bands: wireless broadband internet (Wibro)/802.11b/g/n, mWiMax, and 802.11a/n. Table 8.7 lists the operating bands of the two antennas for the various combinations of the states of the PIN diodes. The resonant frequency of the PIFA shifted from 0.9 to 2.0 GHz when its PIN diode was switched from the On to Off state, achieving a tuning range of TR_1 = 75.9% with a total spectrum of TS_1 = 90% and a tunable impedance bandwidth of BW_{TZ1} = 41%. With its PIN diode in the On state, the IFA resonated at 3.5 GHz. When the PIN diode was switched to the Off state, the IFA resonated at both 2.4 and 5.2 GHz. The tuning range between the two lower

Table 8.7

Frequency Bands for the Different States of the PIN Diodes

	PIN 1: On PIN 2: On	PIN 1: Off PIN 2: On	PIN 1: On PIN 2: Off	PIN 1: Off PIN 2: Off
Port 1 (PIFA)	0.82–0.96	1.7–2.17	0.82–0.96	1.7–2.17
Port 2 (IFA)	3.4–3.6	3.4–3.6	2.3–2.5 5.15–5.35	2.3–2.5 5.15–5.35

(From [18])

frequencies was TR_2 = 37.3% with a total spectrum of TS_2 = 44% and a tunable impedance bandwidth of BW_{TZ2} = 14%. The gain and radiation efficiency of the antenna were not reported.

8.1.4 Printed Dipole Antenna

The diagram in Figure 8.14 is an example of a printed dipole antenna loaded with PIN diodes [19]. The dipole was etched on the top metal surface of a thin dielectric sheet and was excited by electromagnetic coupling from a microstrip line located on the bottom surface of the sheet. The diodes were connected across small gaps in the dipole arms, and their accompanying bias circuitry was located on the top surface of the dielectric sheet. When all the diodes were in the Off state, only the central portion (L_1) of the dipole was active, which had a resonant frequency of 5.9 GHz. When the inner pair of diodes were turned on, the effective length of the dipole increased to $L_1 + 2L_2$, and the dipole resonated at 5.8 GHz. When the outer pair of diodes were also turned on, the entire length of the dipole ($L_1 + 2L_2 + 2L_3$) was active and the dipole resonated at 5.6 GHz, resulting in a tuning range

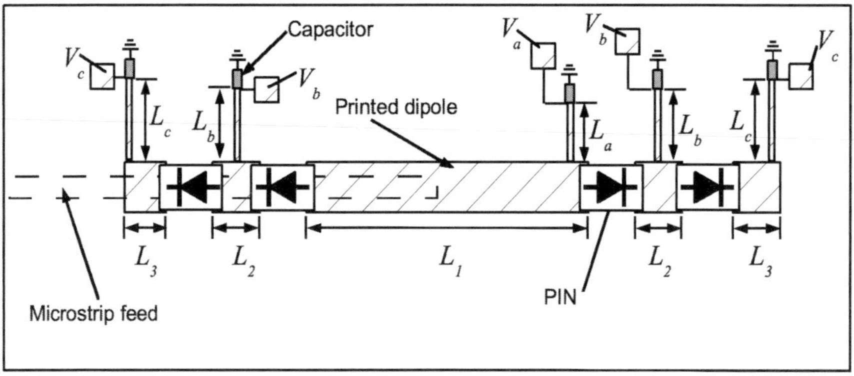

Figure 8.14 PIN-diode-loaded printed dipole ($L_1 = L_a = \lambda_g/4$ @ f_3, $L_1+2L_2 = L_b = \lambda_g/4$ @ f_2, and $L_1+2(L_2+L_3) = L_x = \lambda_g/4$ @ f_1). (After [19].)

of TR = 5.2%. Larger tuning ranges could be achieved by extending the lengths of the arms (L_2 and L_3) so that there is a greater range of values for the effective length of the dipole. The proximity coupled excitation of the dipole by the microstrip line led to adverse effects on the radiation pattern. This was resolved by using an aperture-coupled excitation. The gain or radiation efficiency of the antenna was not reported.

8.1.5 Slot Antennas

This section looks at three examples of slot antennas loaded with PIN diodes for discrete frequency tuning capabilities. The first example, shown in Figure 8.15 [20], uses the same approach as the printed dipole illustrated in Figure 8.14. In the case of the slot dipole, four pairs of PIN diodes were connected across the arms of the antenna. When all diodes were in the Off state, the entire length (L_S) of the antenna was excited, resonating at 2.96 GHz. As successive pairs of PIN diodes were switched on, starting from the most outer pair (S1) to the most inner pair (S4), the length of the dipole was progressively shortened, resulting in increasing resonant frequencies as listed in Table 8.8. The tuning range of the antenna was TR = 84.3%, and it achieved a total spectrum of TS = 102% and a tunable impedance bandwidth of BW_{TZ} = 60%. The measured gain for each resonant frequency, also listed in Table 8.8, varied from 3.5 to 5.2 dBi and it was reported that these gain values were little different from the gains of a passive version of the slot dipole, concluding that the resistance of the PIN diodes in the On state did not significantly alter the antenna gain.

The second example is an S-shaped slot, pictured in Figure 8.16 [21]. The slot was etched on the top metal surface of the dielectric sheet and was excited by a microstrip line located on the bottom surface. Four PIN diodes were connected at various locations across the slot. By applying individual DC bias voltages to each of the PIN diodes, they were independently switched between their On and Off states, altering the effective length of the slot and thereby shifting the resonant frequency. The four sets of applied voltages and the resultant resonant frequencies of the slot are listed in Table 8.9. The antenna achieved a tuning range of TR = 49.2%, a total spectrum of TS = 52%, and a tunable impedance bandwidth of BW_{TZ} = 12%. The measured gain of the antenna was -1.1 dBi, corresponding to a radiation efficiency of only 47%, attributed mainly to the loss in the PIN diodes.

The final example of a slot antenna loaded with PIN diodes is the bowtie-shaped slot depicted in Figure 8.17 [22]. The slot was etched on the top metal surface of a dielectric sheet and was excited by a microstrip line located on the bottom surface of the sheet. Two pairs of PIN diodes were connected across a short tuning stub connected to the bowtie antenna. For each pair of diodes, the positive and negative terminals were oriented in opposing directions, allowing the two pairs of diodes to be controlled by a single DC bias voltage. When no (0V) bias was applied, both pairs of diodes were off, and the bowtie resonated at

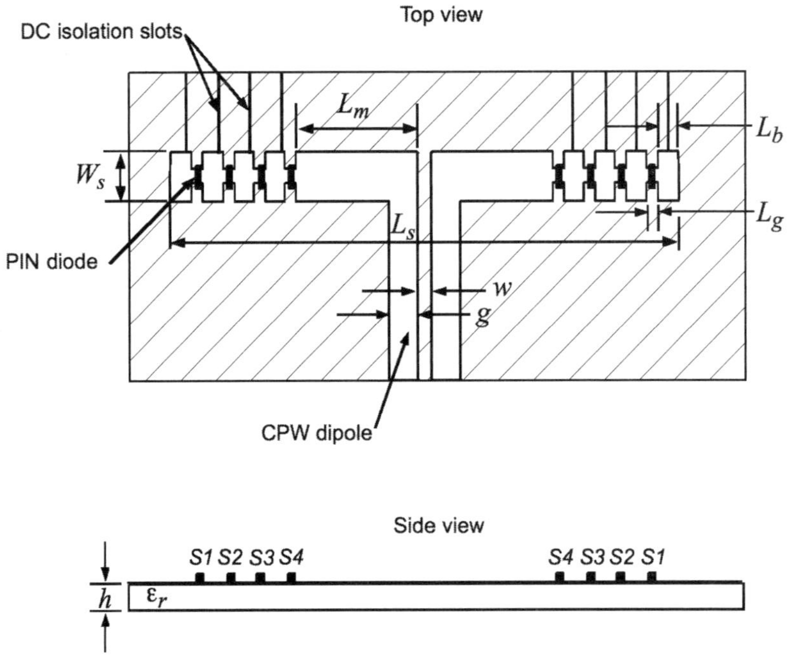

Figure 8.15 PIN-diode-loaded CPW dipole (L_s = 64.7 mm, L_m = 10 mm, L_b = 4.75 mm, L_g = 0.5 mm, W_s = 3 mm, w = 2.7 mm, g = 0.15 mm, h = 1.524 mm, and ε_r = 2.2). (After [20].)

Table 8.8

Frequency Bands for the Different States of the PIN Diodes

On PINs	None	S1	S1, S2	S1, S2, S3	S1, S2, S3, S4
Resonant frequency (GHz)	2.96	3.24	3.76	5.28	7.96
Fractional bandwidth (%)	18	17	12	21	8
Peak gain (dBi)	3.5	5.2	4.8	4.4	3.5

(From [20]).

3.45 GHz. When a positive (> 0.7-V) bias voltage was applied, the S1 diodes were on while the S2 diodes were off and the antenna resonated at 3.91 GHz. When a negative (< -0.7-V) bias voltage was applied, the S1 diodes were off and the S2 diodes were on, and the resonant frequency shifted to 4.42 GHz. The tuning range of the antenna was TR = 24.7% with a total spectrum of TS = 44% and a tunable impedance bandwidth of BW_{TZ} = 29%. The measured gain or radiation efficiency of the antenna was not reported.

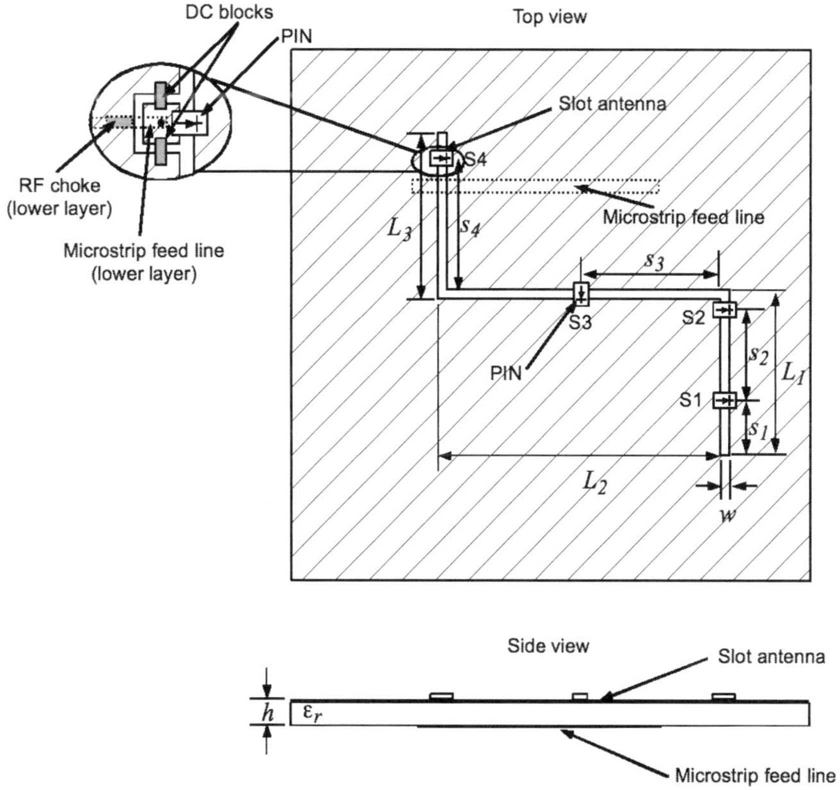

Figure 8.16 PIN-diode-loaded slot antenna (L_1 =37.5 mm, L_2 =64 mm, L_3 = 49.6 mm, w = 2 mm, s_1 = 14.9 mm, s_2 = 19 mm, s_3 = 31 mm, s_4 =35.2 mm, h = 2.54 mm, and ε_r =10.2). (After [21].)

Table 8.9
Frequency Bands for the Different States of the PIN Diodes

Bias Voltage (V)				Frequency (MHz)
S1	S2	S3	S4	
-20	-20	-20	1.1	537
1.1	-20	-20	1.1	603
0	1.1	-20	1.1	684
0	1.1	1.1	0.2	887

(From [21])

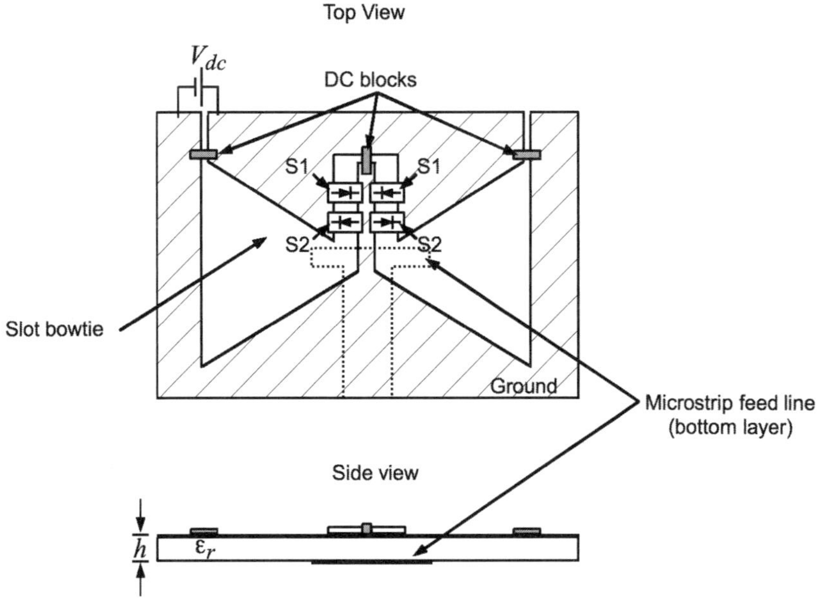

Figure 8.17 PIN-diode-loaded bowtie slot antenna (h = 1.6 mm and ε_r = 4.5). (After [22].)

8.1.6 DRA

The final example of the use of PIN diodes for discrete frequency tuning is of a DRA, shown in Figure 8.18 [23]. The dimensions of the DRA and the biasing circuit for the PIN diodes were identical to those shown in Figure 7.54; the only difference being that PIN diodes were used instead of varactors. Each of the three

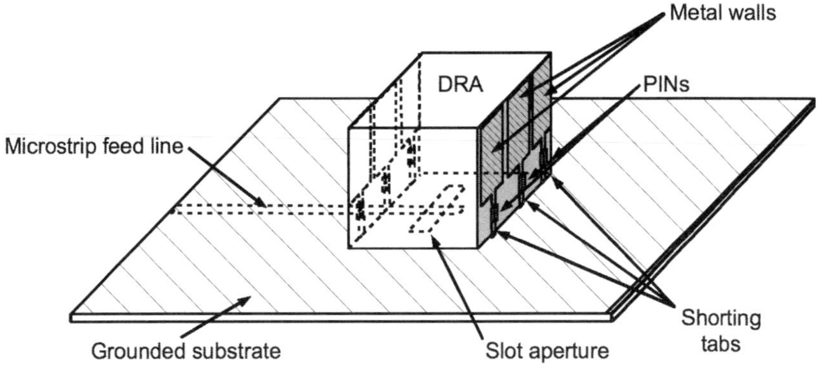

Figure 8.18 PIN-diode-loaded DRA. (After [23].)

diodes on the DRA face was paired with the diode directly opposite on the other DRA face and biased in the same state. This ensured symmetric current distributions on the metallic walls on the opposite faces, which helped to maintain pattern symmetry. With this biasing arrangement, there were six distinct On/Off state combinations for the three pairs of diodes, with resonant frequencies ranging from 3 to 8 GHz, corresponding to a tuning range of TR = 91%. The antenna had a total spectrum of TS = 95% and a tunable impedance bandwidth of BW_{TZ} = 32%. The maximum bias current was 43 mA. Measured peak gains ranged from 2.6 to 4.6 dBi, these values being between 0.6 and 3 dB lower than a corresponding version of the DRA using ideal switches (where a metallic tab was used for the On state and an open circuit for the Off state). The measured radiation efficiency of the DRA was approximately 54%, compared to a simulated efficiency of 93% for an ideal lossless switch and 79% using a PIN diode circuit model (that did not account for the biasing circuitry). Finally, the high cross-polarization levels observed at some frequencies can be reduced by relocating the PIN diodes and accompanying bias circuitry below the ground plane, as demonstrated in [24].

8.2 TRANSISTOR SWITCH CONTROL

Transistor switches can also be integrated with antennas to allow for discrete frequency tuning capability. Figure 8.19 shows a PIFA antenna with an adjustable tuning stub, controlled by a FET switch [25]. As discussed in Section 4.3.3, the FET switch draws practically no current and thus has a lower DC power consumption than the PIN diode. In this example, the switch allows one of two microstrip tuning stubs of different lengths (L_2 and L_3), located on the circuit board, to be connected to the PIFA, thus shifting the resonant frequency from 0.85 to 1.0 GHz, for a tuning range of TR = 16.2%. The total spectrum for the antenna was TS = 22% with a tunable impedance bandwidth of BW_{TZ} = 14%. A 3.6V battery was used to activate the switch. The measured radiation efficiency was above 72% for the cellular frequency bands of interest. Similar examples of transistor switches integrated with antennas designed for the cellar bands have also been reported, including PIFAs [26, 27], a microstrip patch antenna [28], and a slot antenna [29].

8.3 MEMS SWITCH CONTROL

The main attributes and performance of MEMS RF switches are reviewed in Section 4.3.2. RF MEMS switches have advantages over other solid-state switches including lower insertion loss, higher isolation, low DC power consumption, and

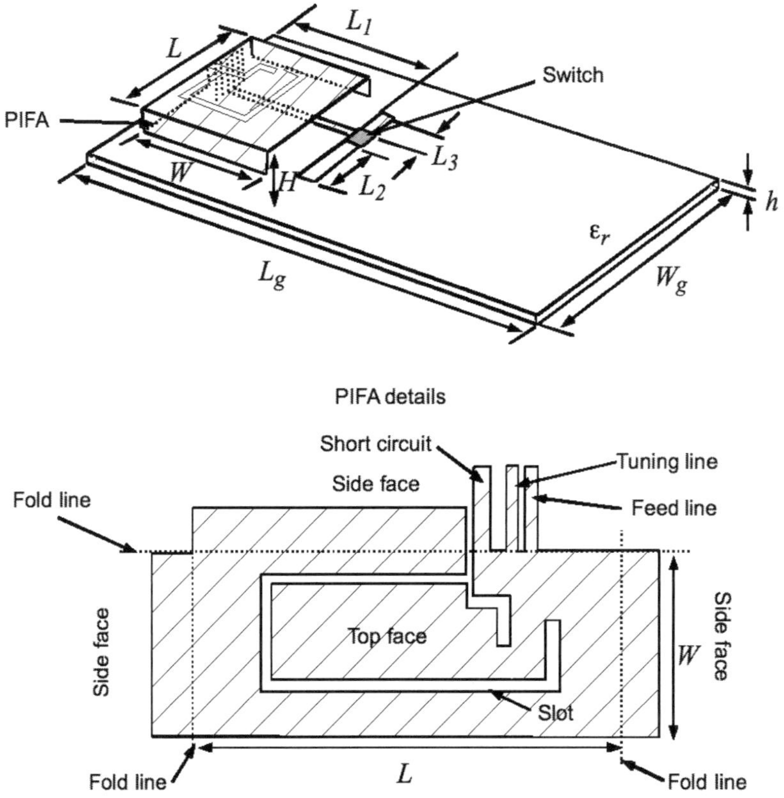

Figure 8.19 PIFA loaded with FET switch (L_g = 110 mm, W_g = 40 mm, h = 0.79 mm, ε_r = 2.33, L = 38 mm, W = 16 mm, H = 8 mm, L_1 = 27.5 mm, L_2 = 13.5 mm, L_3 = 2 mm). (After [25].)

higher power handling ability. Drawbacks for these devices include higher activation voltages, higher cost, lower reliability, and limited commercial availability. This section examines various examples of FAAs using MEMS devices to achieve discrete tuning including microstrip patch antennas, PIFAs, printed monopoles and dipoles, and slot antennas.

8.3.1 Microstrip Patch Antennas

Two examples of microstrip patch antennas loaded with RF MEMS switches are considered in this section. The first is illustrated in Figure 8.20, where two MEMS

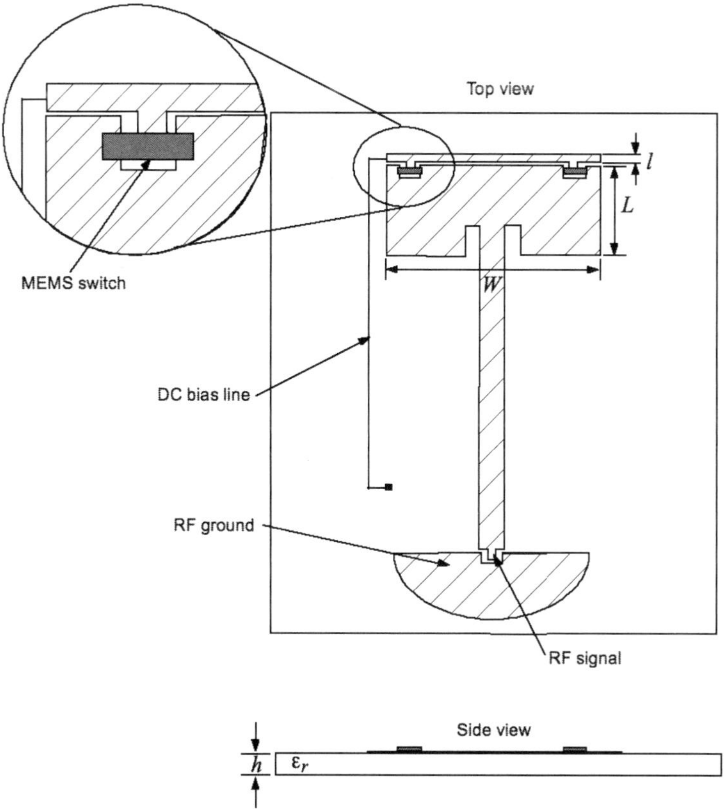

Figure 8.20 Microstrip patch with MEMS switch (L = 1.5 mm, W = 2.6 mm, l = 0.1 mm, h = 0.4 mm, and ε_r = 11.7). (After [30].)

switches were integrated near the two top corners of the patch [30]. The switches were of the fixed-fixed beam design [see Figure 4.20(b)], and when an activation voltage of 55V was applied, the center of the beam deflected down and made a connection with the thin metallic strip located adjacent to the top edge of the patch. When this strip was connected, the effective resonant length of the microstrip patch increased, resulting in a lowering of the resonant frequency from 25.4 to 21.5 GHz, corresponding to a tuning range of TR = 16.6%. Since the impedance was not matched at the lower frequency, the total spectrum and tunable impedance bandwidth for this antenna were only TS = BW_{TZ} = 3%. The gain and radiation efficiency of the antenna were not reported.

The second example, shown in Figure 8.21, consisted of a MEMS switch connected across a rectangular slot in the microstrip patch [31]. When the switch was off, the RF current was forced to flow around the slot, resulting in a longer effective path length and thus a lower resonant frequency. When the MEMS

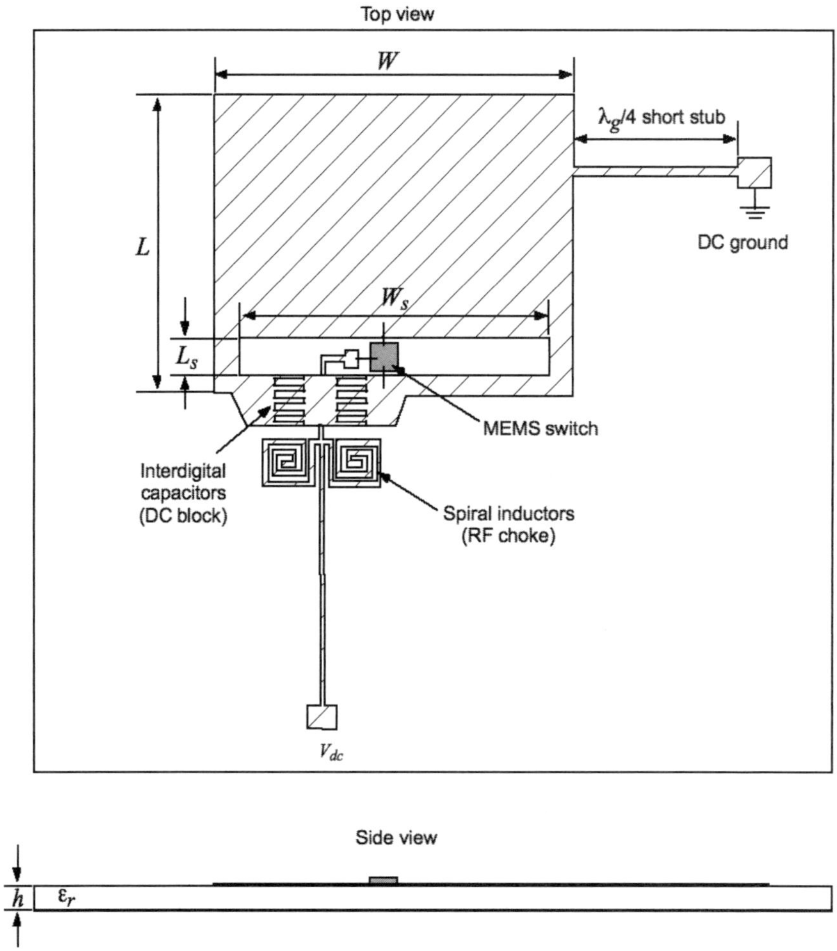

Figure 8.21 Microstrip patch antenna with MEMS switch (L = 11 mm, W = 20 mm, L_s = 3 mm, W_s = 17 mm, h = 3.18 mm, and ε_r = 2.2). (After [31].)

switch was activated by an applied 90-V DC voltage, a connection was made across the slot, shortening the RF current path and raising the resonant frequency. Instead of lumped components, etched distributed elements were used for the biasing circuitry with interdigital capacitors for DC blocks and spiral inductors and a $\lambda_g/4$ short stub for RF choke. When the MEMS switch was biased from the Off state to the On state, the resonant frequency of the antenna shifted from 4.62 to 4.82 GHz, for a tuning range of TR = 4.2%. The antenna exhibited a total spectrum of TS = 7% and a tunable impedance bandwidth of BW_{TZ} = 6%. The measured gain and radiation efficiency for the antenna were not reported.

8.3.2 PIFAs

This section examines three examples of the use of MEMS switches with PIFA antennas for discrete frequency tuning. In the first design, shown in Figure 8.22, the top face of the PIFA consisted of a metal-clad dielectric sheet with a thin L-shaped slot etched on the top surface [32]. The RF MEMS switch was connected across the slot and when switched on with a 20-V DC activation voltage, it short-circuited the slot. The PIFA was designed to resonate at three frequencies. The lowest frequency of f_1 = 800 MHz was unaffected by the state of the switch. The higher resonant frequencies occurred at f_2 = 1.25 GHz and f_3 = 2.7 GHz when the MEMS switch was in the Off state and increased to f_2 = 1.6 GHz and f_3 = 2.9 GHz when the switch was in the On state, corresponding to tuning ranges of TR_2 = 24.5% and TR_3 = 7.1%. Consideration was not given to matching the input impedance of the PIFA; thus the total spectrum and tunable impedance bandwidth values are not applicable. The radiation performance of the antenna was not reported.

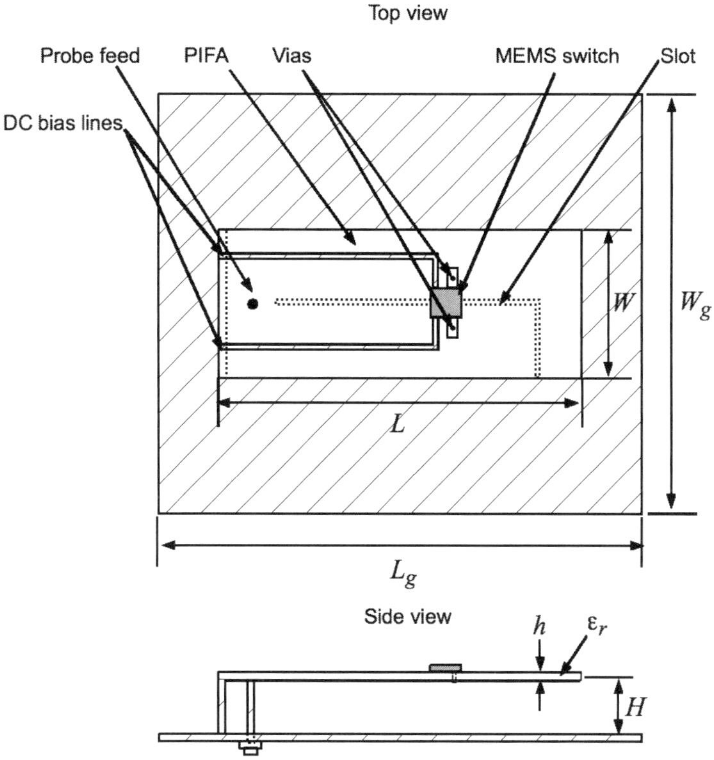

Figure 8.22 PIFA with MEMS switch (L = 75 mm, W = 30 mm, H = 8 mm, L_g = 100 mm, W_g = 85 mm, h = 0.762 mm, and ε_r = 2.2). (After [32].)

In the second example, shown in Figure 8.23, an RF MEMS direct-contact switch was used to alter the effective electrical length of the PIFA's tuning stub, which, in turn, shifted the resonant frequency of the antenna [33]. The switch was connected across a short gap in a microstrip transmission line located on a small circuit board located on the ground plane. The MEMS switch was piezoelectrically activated, with an activation voltage of 2.5V. Switching from the Off to On states, the resonant frequency of the PIFA decreased from 840 to 760 MHz, resulting in a tuning range of TR = 10%, with a total spectrum and tunable impedance bandwidth of TS = BW_{TZ} = 13%. Measured gain and radiation efficiency values were not reported.

The final example of a PIFA antenna loaded with RF MEMS switches is shown in Figure 8.24 [34, 35]. The PIFA has a set of four nested L-shaped slots etched into the top metal patch, with a switch connecting across each slot along the top edge of the patch. The slots allow for multiband operation, and the PIFA was

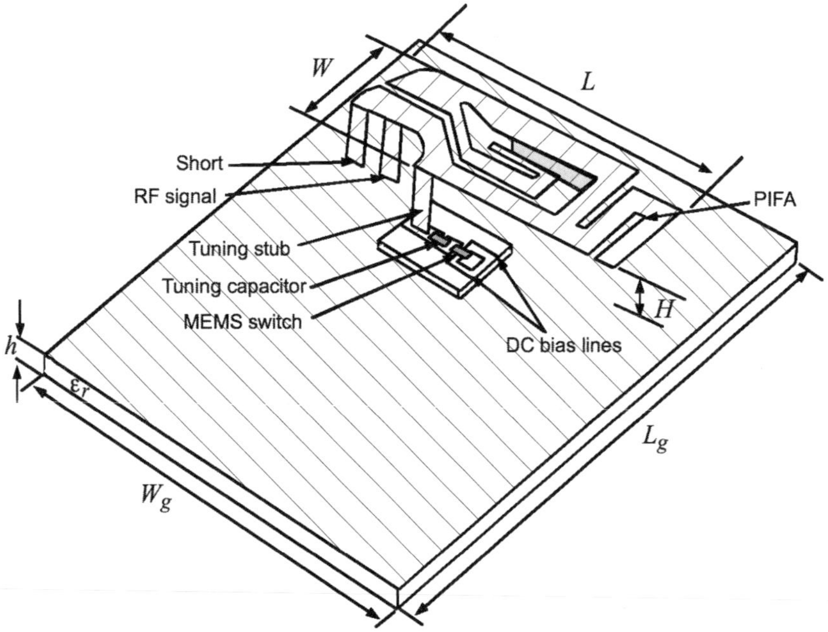

Figure 8.23 PIFA with MEMS switch (L = 35 mm, W = 14 mm, H = 7 mm, L_g = 80 mm, W_g = 44 mm, h = 0.8 mm, and ε_r = 4.4). (After [33].)

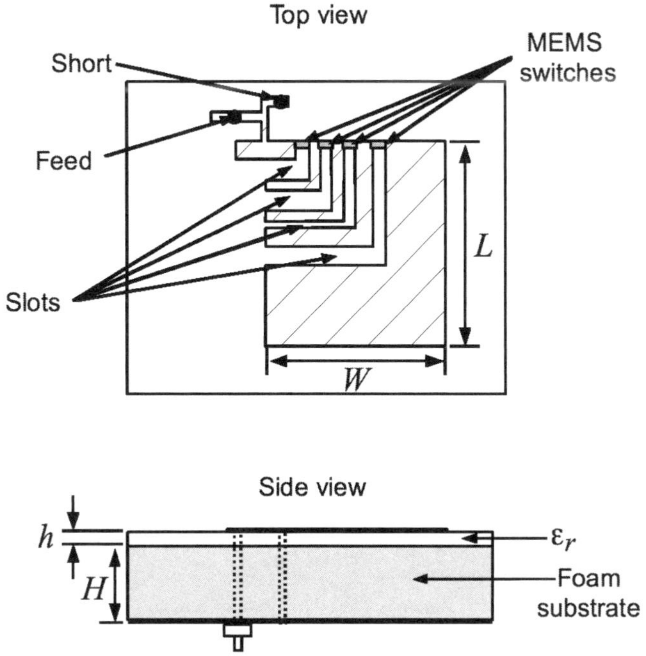

Figure 8.24 PIFA with MEMS switch (L = 21.55 mm, W = 23.05 mm, H = 6.4 mm, h = 1.57 mm, and $\varepsilon_r \sim 4.4$). (After [34, 35].)

designed to cover various services within the 1–5 GHz range including global positioning system (GPS), GSM, DCS, WLAN, and WiMax. Bias control wires (not shown) were connected from the bottom to each of the MEMS switches. Four resonant frequencies were shown for the different switch states (the activation voltage was not reported) ranging from 0.8 to 2.3 GHz, for a tuning range of TR = 96.8%. The total spectrum for the antenna was TS = 112%, with a tunable impedance bandwidth of BW_{TZ} = 48%. The measured peak gain values over the various frequency bands of interest all exceeded 4 dBi.

8.3.3 Printed Wideband Antennas

Two examples of printed wideband antennas with integrated RF MEMS switches are examined in this section. The first is a wideband printed bowtie antenna, shown in Figure 8.25 [36, 37]. Two pairs of RF MEMS switches were connected across small gaps in the bowtie antenna. Series cantilever ohmic contact switches [see Figure 4.20(a)] were used in this design. The DC bias lines required for the switches were etched on the same surface as the bowtie antenna. When an activation voltage of 40V was applied to the RF MEMS switch, the cantilever was pulled down, making a connection across the gap. This increased the electric

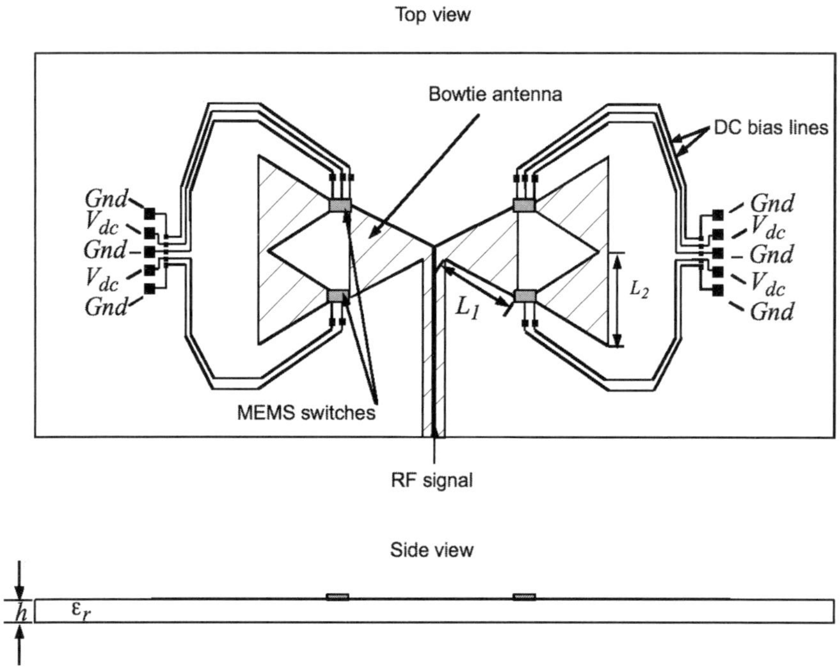

Figure 8.25 Wideband printed bowtie antenna loaded with MEMS switches (L_1 =2 mm, L_2 =2.1 mm, h = 0.4 mm, and ε_r = 11.9). (After [36,37].)

length of the bowtie antenna and resulted in a decrease in the resonant frequency from 15 GHz down to 9.2 GHz, for a tuning range of TR = 47.9%. The total spectrum for the antenna was TS = 58% with a tunable impedance bandwidth of BW_{TZ} = 49%. The measured gain and radiation efficiency of the antenna were not reported.

The second example is a wideband printed monopole, shown in Figure 8.26, that is based on a Sierpinski triangle pattern [38]. As in the previous example, cantilever-type ohmic contact RF MEMS switches were connected across small gaps in the antenna. The interesting difference with this second example was the elimination of the bias lines for activating the switches. This was achieved by designing each consecutive pair of switches (labeled S1, S2, and S3 in Figure 8.26) with increasing activation voltage thresholds. As the applied DC

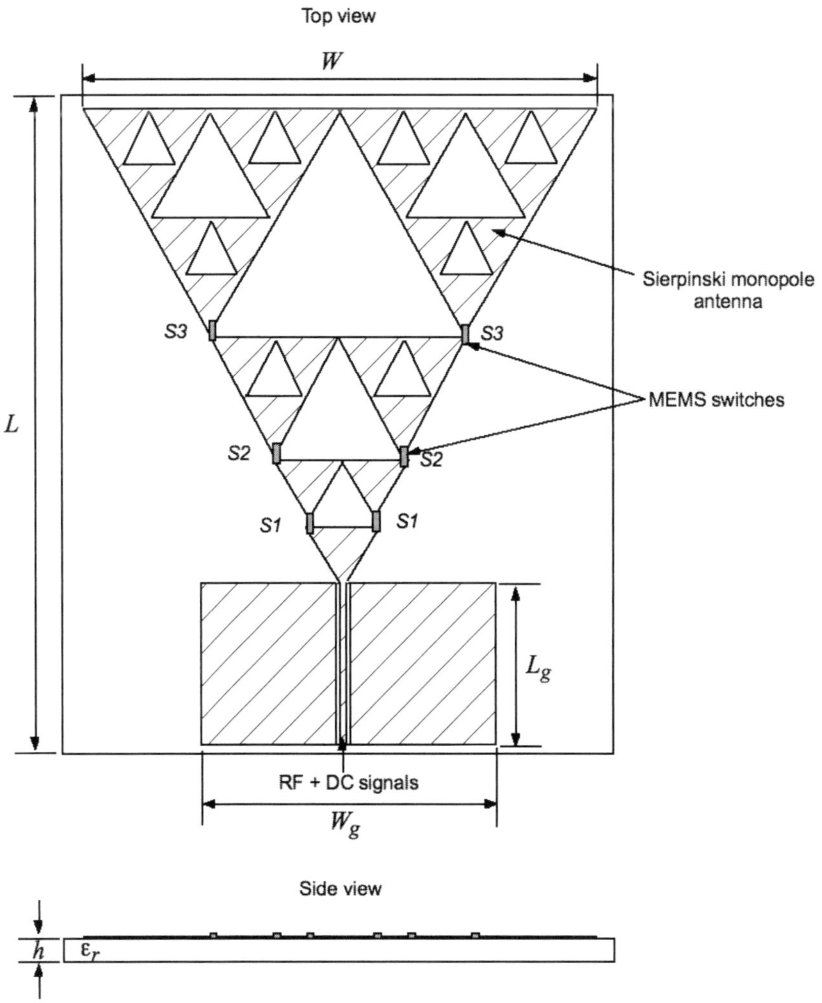

Figure 8.26 Sierpinski monopole antenna loaded with MEMS switches (L =25 mm, W =20 mm, L_g =6 mm, W_g =11 mm, h = 0.1 mm, and ε_r = 3.0). (After [38].)

voltage (which was injected on the RF signal path through a bias-T) was increased from 0V (all switches off) to the first activation threshold (18V), the S1 switches turned on, while S2 and S3 remained off. When the applied voltage was increased to the second activation threshold (28V), the S2 switches also turned on while the S3 switches remained off. Increasing the applied voltage to 38V caused the S3 switches to also turn on. As each pair of switches turned on, the effective electrical length of the monopole increased, resulting in a lowering of the resonant frequency. The resonant frequencies and 10-dB return loss bandwidths for each state are listed in Table 8.10. The antenna achieved a tuning range of TR = 153%, a total spectrum of TS = 166%, and a tunable impedance bandwidth of BW_{TZ} = 97%. Values for measured gain or radiation efficiency were not reported.

8.3.4 Slot Antennas

This section looks at two examples of slot antennas with integrated RF MEMS switches to achieve discrete frequency tuning. The first consists of two concentric annular slots, of radii (r_1, r_2) for the inner slot and (r_3, r_4) for the outer slot as shown in Figure 8.27 [39]. The antenna contained three RF MEMS switches. One switch (S1) was connected across a small gap in the microstrip feed line, located below the slot and was a single-arm cantilever. When no bias was applied, the switch was open and the microstrip line excited the outer ring. When the activation voltage (20V) was applied, the switch closed, extending the microstrip line to excite the inner slot. The two other MEMS switches (S2) were located across the outer slot and were dual-arm cantilevers. These switches were activated at the same time as S1 and caused the outer ring to be short-circuited, preventing it from radiating. The resonant frequency of the antenna was 2.4 GHz when the switches were off and 5.2 GHz when the switches were on, corresponding to a tuning range of TR = 73.7%. The antenna achieved a total spectrum of TS = 81% and a tunable impedance bandwidth of BW_{TZ} = 21%. The measured gain for the antenna was 1.85 dBi at 2.4 GHz and 2.75 dBi at 5.2 GHz, with a radiation efficiency of greater than 90%.

Table 8.10

Resonant Frequency and Bandwidth for the RF MEMS Switch Loaded Monopole Antenna

Voltage (V)	Switch States	Resonant Frequency (GHz)	Bandwidth (GHz)
0	S1: Off S2: Off S3: Off	18	15.5–21.6
18	S1: On S2: Off S3: Off	9.4	8.2–11.6
28	S1: On S2: On S3: Off	5.7	5.2–6.6
38	S1: On S2: On S3: On	2.4	2.0–2.6

(From [38])

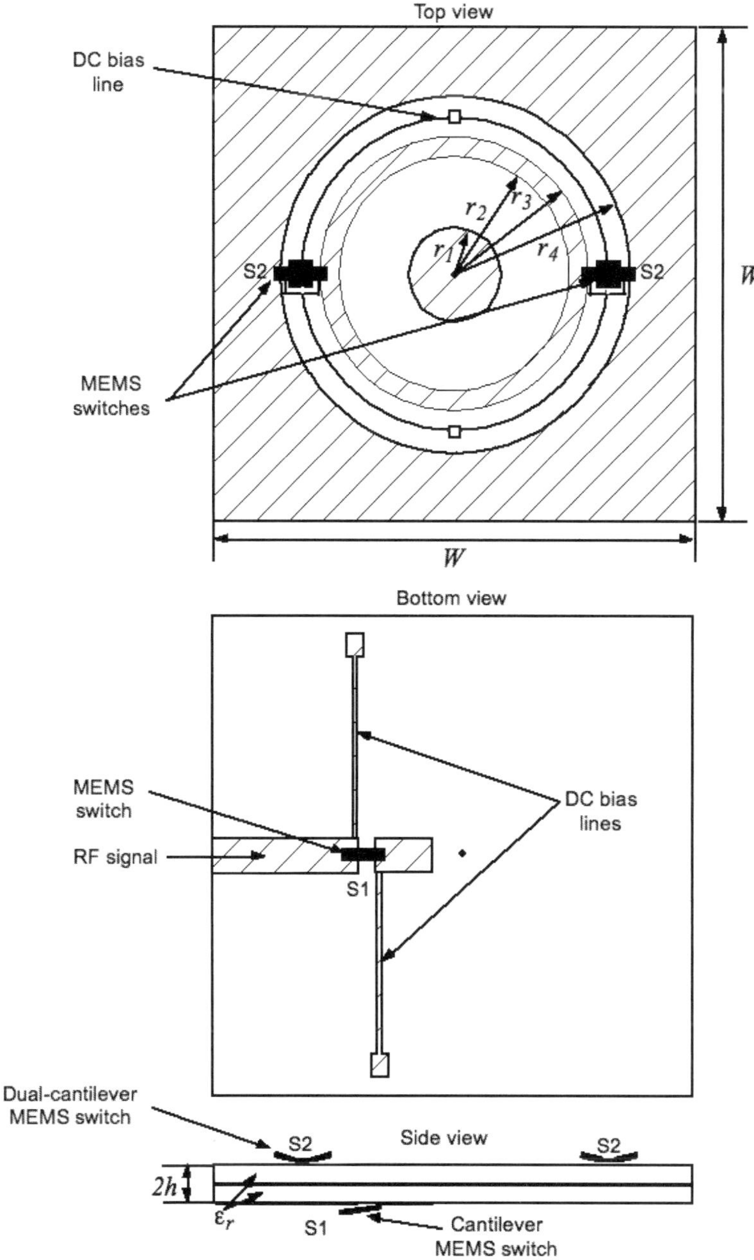

Figure 8.27 Ring slot with MEMS switch (W = 40 mm, r_1 = 3.2 mm, r_2 = 7.8 mm, r_3 = 9 mm, r_4 = 11 mm, h = 0.635 mm, and ε_r = 9.8). (After [39].)

The second example is shown in Figure 8.28 and consists of a circular arc slot, fed by a CPW line [40]. Two dual-arm cantilever MEMS switches were located across the meandering slot segments feeding the arc slot. An activation voltage of between 18 and 21V was required to pull down the cantilever arms, causing a short circuit across the meander line sections. When both switches were in the Off state, the slot resonated at 34.9 GHz. When both switches were in the On state, the resonant frequency dropped to 27.7 GHz, resulting in a tuning range of TR = 23%. The antenna achieved a total spectrum of TS = 27% and a tunable impedance bandwidth of BW_{TZ} = 9%. The measured gain and radiation efficiency of the antenna were not reported.

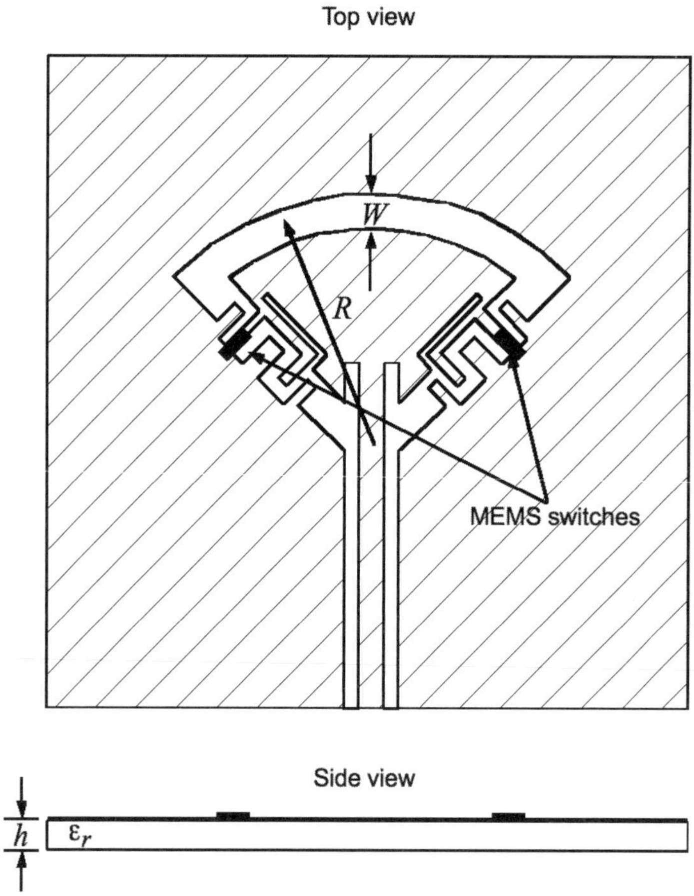

Figure 8.28 Arc slot with MEMS switch (W = 0.135 mm, R = 0.82 mm, h = 0.5 mm, and ε_r = 11.7). (After [40].)

8.4 OPTICAL SWITCH CONTROL

This section examines the use of optically controlled switches to achieve discrete frequency agility. Optoelectronic switches were briefly described in Section 4.3.4. These semiconductor devices can be controlled either by directly illuminating the device or indirectly by illuminating a photovoltaic cell, which, in turn, generates a DC voltage that activates the switch. The examples that are illustrated in this section are all of the direct-illumination type and include a monopole, dipole, and microstrip patch antenna.

Figure 8.29 illustrates a monopole antenna where an optical switch was connected across a small gap in the monopole arm [41]. An optical fiber was used to carry light from a laser up to the switch. When the laser was off, the switch was open and only the lower half (of length L) of the monopole radiated, at a resonant frequency of 375 MHz. When the laser was on, it delivered up to 10 mW of optical power to the switch, which became conductive, allowing the full length ($2L$) of the monopole to resonate at a frequency of 210 MHz. The resultant tuning range was TR = 56.4%, with a total spectrum of TS = 74% and a tunable impedance bandwidth of BW_{TZ} = 37%. Due to the poor power handling capabilities of the particular optical switch used (with a peak signal limited to less than ~ 1V), the antenna could only be considered for receiving applications. Gain and radiation efficiency values of the antenna were not reported.

The second example is a dipole antenna, shown in Figure 8.30, where an optical switch was connected across a small gap in each of the dipole arms [42]. The principle of operation is the same as for the monopole example. When no optical power was applied, the switches were open, and the radiating length of the dipole was (L – $2s$). When the switches were biased with optical power, they formed a short circuit across the gap, and the entire length L of the dipole radiated, shifting the resonant frequency down. In this example, an optical power of 200 mW, supplied by a 980-nm laser, was used to activate the switches. The resonant frequency of the dipole with the switches in the Off state was 3.15 GHz, which shifted down to 2.26 GHz when the switches were in the On state, resulting

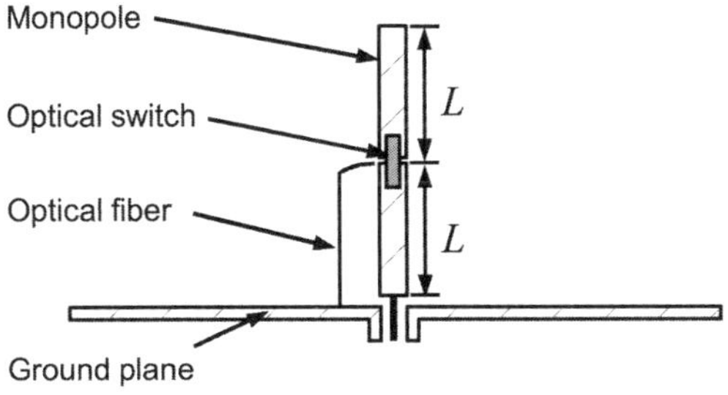

Figure 8.29 Monopole antenna loaded with an optical switch (L = 150 cm). (After [41].)

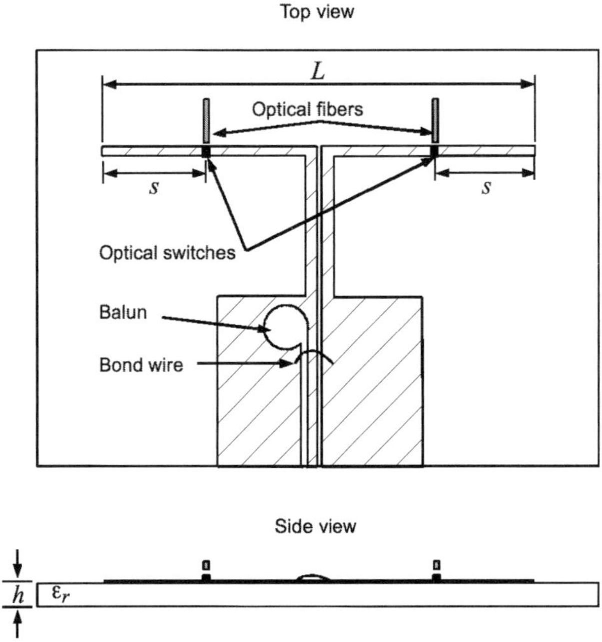

Figure 8.30 Dipole antenna loaded with an optical switch (L = 62.3 mm, s = 14.4 mm, h = 1.17 mm, and ε_r = 2.2). (After [42].)

in a tuning range of TR = 32.9%. The total spectrum for the dipole was TS = 44% with a tunable impedance bandwidth of BW_{TZ} = 23%. The measured gain for the dipole was 4.5 dBi at 3.15 GHz (Off state) and 1 dBi at 2.26 GHz (On state). The drop in gain from the Off to the On state suggested a substantial reduction in radiation efficiency, no doubt due to losses in the switches.

The last example is of a microstrip patch antenna, shown in Figure 8.31 [43]. A small rectangular patch was located adjacent to the main patch, separated by a small gap (g). An optical switch (S1) bridged this gap, connecting the small patch to the main radiator. A rectangular slot was etched into the main radiating patch across which a second optical switch (S2) was connected. In this example, the switches were fabricated from a phosphorus-doped silicon wafer with a 3,000 Ω-cm resistivity and were made conductive using a 980-nm laser with 143.7 mW of optical power. When both switches were off, the antenna resonated at 4.78 GHz. With S1 on and S2 off, the resonant frequency increased to 5.1 GHz, while with S1 off and S2 on, the resonant frequency dropped to 4.55 GHz. The tuning range for the antenna was TR = 11.4%. Since the antenna was only matched at the center frequency, the total spectrum and tunable impedance bandwidth were low, with TS = BW_{TZ} = 7%. The radiation efficiency of the antenna was not reported.

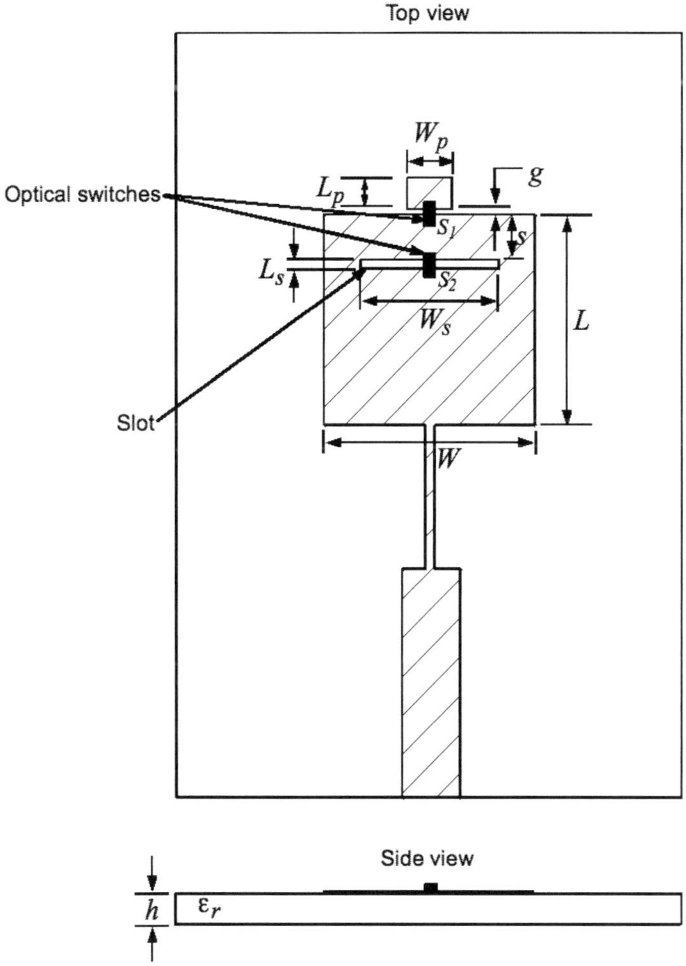

Figure 8.31 Microstrip patch antenna loaded with an optical switch ($L = W = 17.59$ mm, $L_s = 0.75$ mm, $W_s = 11.43$ mm, $L_p = 2.54$ mm, $W_p = 3.81$ mm, $g = 0.5$ mm, $s = 3.81$ mm, $h = 1.57$ mm, and $\varepsilon_r = 2.2$). (After [43].)

8.5 HYBRID SWITCH CONTROL

This final section considers antennas that incorporate both PIN-diode switches and varactor diodes to allow for a combination of discrete and continual frequency tuning. Two examples were selected to illustrate these designs. The first, shown in Figure 8.32, consists of a printed meander-line monopole antenna where both a PIN diode and varactor were connected across small gaps in the line [44].

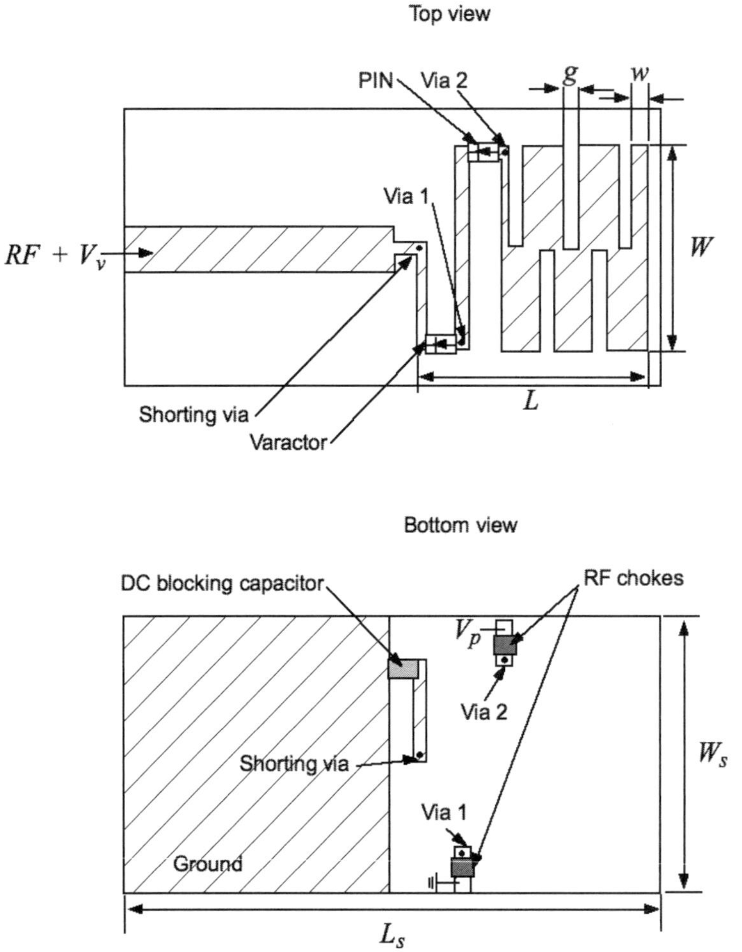

Figure 8.32 Meanderline antenna loaded with a PIN diode and varactor (L = 10 mm, W = 8 mm, L_s = 24 mm, W_s = 12 mm, w = 0.5 mm, g = 0.7 mm, h = 1 mm, and ε_r = 4.4). (After [44].)

Switching between the PIN-diode states switches between two frequency bands (2.3 and 5.15 GHz), while varying the voltage on the varactor allows for fine-tuning across these two bands. A DC bias of V_p = 1V turned the PIN diode on, connecting the meandering line section to the folded monopole, causing the antenna to resonate at 2.3 GHz. When V_p = 0V, the PIN diode was off, the meandering line section was not connected, and the antenna resonated at 5.15 GHz, for a tuning range of TR = 76.5%. When the applied DC voltage to the varactor increased from V_v = 0–3V (applied through a bias-T along with the RF signal), its capacitance decreased from 57 to 10 pF, shifting the resonant frequency upward. The antenna achieved a total spectrum of TS = 85% and a tunable impedance bandwidth of BW_{TZ} = 25%. The measured gain of the antenna remained fairly constant at about 2.6 dBi in the lower frequency band when the varactor voltage was varied between 0.6 and 2.1V and at 2.5 dBi in the upper band for varactor voltage variation of 1.2–2.3V.

The second example is of a PIFA antenna, shown in Figure 8.33, where a PIN diode was connected across a slot gap on the top surface, while a varactor was connected across a small gap in the shorting arm [45]. The PIFA was designed to resonate at two frequencies where the lower frequency (1.9 GHz) was not affected by the state of the PIN diode, while the upper frequency resonated at 5.5 GHz when the PIN diode was off (V_p = 0V) and at 3.5 GHz when the PIN diode was on (V_p = 1V). The tuning range for the upper frequencies was TR = 44%. Varying the voltage (V_v) on the varactor between 0 and 4V caused a shift in frequency in both

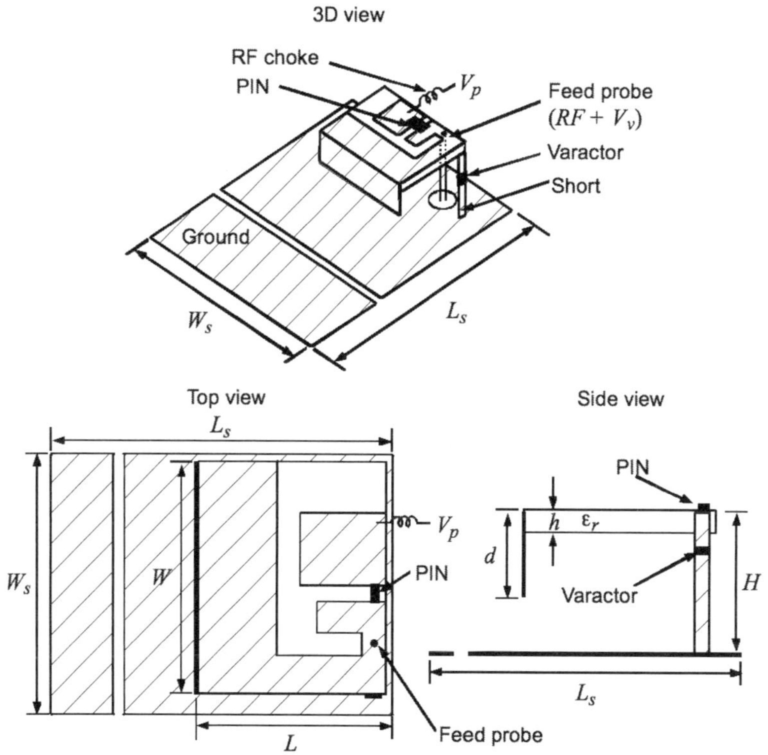

Figure 8.33 PIFA loaded with a PIN diode and varactor (L = 12 mm, W = 15 mm, L_s = 70 mm, W_s = 30 mm, H = 9 mm, d = 5.5 mm, h = 0.8 mm, and ε_r = 4.4). (After [45].)

the lower and upper bands. Only the 6-dB return loss bandwidths for this antenna, which varied from about 7 to 16%, were reported. The radiation efficiency ranged from 63% for operation at 3.5 GHz, to over 90% for the 1.9-GHz and 5.5-GHz bands.

8.6 SUMMARY

The use of integrated electronic components for enabling FAAs with discrete tuning is reviewed in this chapter. The selected examples demonstrate the wide variety of devices and antenna elements that could be combined to exhibit frequency agility, with tuning ranges from a few percent to nearly one hundred percent. A summary of the performance of the selected examples is provided in Table 8.11.

Table 8.11

Performance Summary of FAAs With Discrete Tuning

Tuning Technique	Antenna	Bias	Frequency Range (GHz)	TR (%)	TS (%)	BW_{TZ} (%)	Reference
PIN	Microstrip patch	N/A	3.51–4.28	19.8	23	6	[1]
		60 mA/PIN	1.87–3.19	52.2	56	24	[4]
		2V	4.38–4.82	9.6	16	14	[5]
		N/A	10.2–21.3	70.5	72	16	[9]
		N/A	5.0–5.6	11.3	24	24	[10]
			5.7–6.2	8.4	38	38	
	IFA	±1V	0.858–0.956	10.8	15	9	[11]
			1.82–1.86	2.2	8	8	
		±1V	0.879–2.50	96	108	39	[12]
			1.915–3.50	56	70	23	
		100 mA/PIN	See Table 8.5	-	142	43	[13]
		±1V	1.80–2.39	28.2	40	23	[14]
		10 mA, 20V	0.898–0.941	4.7	9	7	[15]
	PIFA	10 mA, ±5V	0.730–0.806	10	6	6	[16]
			1.023–1.250	20	6	4	
		10 mA, ±1V	0.845–0.931	9.7	12	9	[17]
			1.770–2.025	13.4	20	13	
		N/A	0.9–2.0	75.9	90	41	[18]
			2.4–3.5	37.3	44	14	
	Printed dipole	N/A	5.6–5.8	5.2	N/A	N/A	[19]
	Slot	N/A	2.96–7.96	84.3	102	60	[20]
		N/A	0.537–0.887	49.2	52	12	[21]
		>±0.7V	3.45–4.42	24.7	44	29	[22]
	DRA	43 mA	3.0–8.0	91	95	32	[23]
FET	PIA	3.6V	0.85–1.0	16.2	22	14	[25]
MEMS	Microstrip patch	55V	21.5–25.4	16.6	3	3	[30]
		90V	4.62–4.82	4.2	7	7	[31]
	PIFA	20V	1.25–1.6	24.5	N/A	N/A	[32]
			2.7–2.9	7.1			
		2.5V	0.76–0.84	10	13	13	[33]
		N/A	0.8–2.3	96.8	112	48	[34, 35]
	Printed bowtie	40V	9.2–15	47.9	58	49	[36, 37]
	Printed monopole	38V	2.4–18	153	166	97	[38]
	Slot	20V	2.4–5.2	73.7	81	21	[39]
		21V	27.7–34.9	23	27	9	[40]
Optical	Monopole	10 mW	0.210–0.375	56.4	74	37	[41]
	Dipole	200 mW	2.26–3.15	32.9	44	23	[42]
	Microstrip patch	143.7 mW	4.55–5.1	11.4	7	7	[43]
PIN + Varactor	Meanderline antenna	2.3V	2.3–5.15	76.5	85	25	[44]
	PIFA	4V	3.5–5.5	44	N/A	N/A	[45]

References

[1] Luxey, C., et al., "Dual-Frequency Operation of CPW-Fed Antenna Controlled by PIN Diodes," *Electronics Letters,* Vol. 36, No. 1, January 2000, pp. 2-3.

[2] Laheurte, J. M., H. Tosi, and J. L. Dubard, "Microstrip Antenna Controlled by P-I-N Diodes: Influence of the Bias Current on the Antenna Efficiency," *Microwave and Optical Technology Letters,* Vol. 33, No. 1, April 2002, pp. 44-47.

[3] Menzel, W., and W. Grabuer, "A Microstrip Patch Antenna with Coplanar Feed Line," *IEEE Microwave and Guided Wave Letters,* Vol. 1, No. 11, November 1991, pp. 340-342.

[4] Majid, H. A., et al., "Frequency Agile Microstrip Patch to Slot Antenna," *Proc. Asia Pacific Microwave Conference,* Kaohsiung, Taiwan, December 4-7, 2012, pp. 613-615.

[5] Yang, F., and Y. Rahmat-Samii, "Patch Antenna With Switchable Slot (PASS): Dual-Frequency Operation," *Microwave and Optical Technology Letters,* Vol. 31, No. 3, November 2001, pp. 165-168.

[6] Yang, F., and Y. Rahmat-Samii, "Patch Antennas With Switchable Slots (PASS) in Wireless Communications: Concepts, Designs, and Applications," *IEEE Antennas and Propagation Magazine,* Vol. 47, No. 2, April 2005, pp. 13-29.

[7] Jin, N., F. Yang, and Y. Rahmat-Samii, "A Novel Reconfigurable Patch Antenna With Both Frequency and Polarization Diversities for Wireless Communications," *Proc. IEEE International Symposium on Antennas and Propagation,* Vol. 2, Monterey, CA, June 20-25, 2004, pp. 1796-1799.

[8] De Luis, J. R., and F. De Flaviis, "Frequency Agile Switched Beam Antenna Array System," *IEEE Transactions on Antennas and Propagation,* Vol. 58, No. 10, October 2010, pp. 3196-3204.

[9] Le Garrec, L., R. Sauleau, and M. Himdi, "A 2:1 Band Frequency-Agile Active Microstrip Patch Antenna," *Proc. European Conference on Antennas and Propagation EuCAP,* Edinburgh, Scotland, November 11-15, 2007.

[10] AbuTarboush, H. F., et al., "Reconfigurable Wideband Patch Antenna for Cognitive Radio," *Proc. Loughborough Antenna and Propagation Conference LAPC,* Loughborough, United Kingdom, November 16-17, 2009, pp. 141-144.

[11] Komulainen, M., et al., "Frequency-Reconfigurable Dual-Band Monopole Antenna for Mobile Handset," *Proc. IEEE International Symposium on Antennas and Propagation,* Honolulu, Hawaii, June 9-15, 2007, pp. 3289-3292.

[12] Yang, S., A. E. Fathy, and V. K. Nair, "Reconfigurable Multi-Band Twin-PIFA Antennas for Wi-Fi/WiMax and WWAN Services on a Multi-Radio Laptop Platform," *Proc. IEEE International Symposium on Antennas and Propagation,* San Diego, CA, July 5-11, 2008.

[13] Zhang, C., et al., "A Low-Profile Branched Monopole Laptop Reconfigurable Multiband Antenna for Wireless Applications," *IEEE Antennas and Wireless Propagation Letters,* Vol. 8, 2009, pp. 216-219.

[14] Palukuru, V. K., et al., "Frequency-Tunable Planar Monopole Antenna for Mobile Terminals," *Proc. European Conference on Antennas and Propagation EuCAP,* Edinburgh, Scotland, November 11-15, 2007.

[15] Ollikainen, J., O. Kivekas, and P. Vainikainen, "Low-loss Tuning Circuits for Frequency-Tunable Small Resonant Antennas," *Proc. 13th IEEE International Symposium on Personal Indoor Mobile Radio Communications,* Lisbon, Portugal, September 15-18, 2002, Vol. 4, pp. 1882-1887.

[16] Karmakar, N. C., "Shorting Strap Tunable Stacked Patch PIFA," *IEEE Transactions on Antennas and Propagation,* Vol. 52, No. 11, November 2004, pp. 2877-2884.

[17] Komulainen, M., et al., "A Frequency Tuning Method for a Planar Inverted-F Antenna," *IEEE Transactions on Antennas and Propagation,* Vol. 56, No. 4, April, 2008, pp. 944-950.

[18] Jung, C. W., et al., "Reconfigurable Antenna for Concurrent Operation Over Cellular and Connectivity Bands," *Proc. IEEE International Symposium on Antennas and Propagation,* San Diego, CA, July 5-11, 2008.

[19] Roscoe, D. J., et al., "Tunable Dipole Antennas," *Proc. IEEE International Symposium on Antennas and Propagation,* Vol. 2, Ann Arbor, MI, June 28-July 2, 1993, pp. 672-675.

[20] Laheurte, J. M., "Switchable CPW-Fed Slot Antenna for Multifrequency Operation," *Electronics Letters,* Vol. 37, No. 25, December 2001, pp. 1498-1499.

[21] Peroulis, D., K. Sarabandi, and L. P. B. Katehi, "Design of Reconfigurable Slot Antennas," *IEEE Transactions on Antennas and Propagation,* Vol. 53, No. 2, February 2005, pp. 645-654.

[22] Lee, W., H. Kim, and Y. J. Yoon, "Reconfigurable Slot Antenna With Wide Bandwidth," *Proc. IEEE International Symposium on Antennas and Propagation,* Albuquerque, NM, July 9-14, 2006, pp. 3063-3066.

[23] Desjardins, J., et al., "Electronically Frequency-Reconfigurable Rectangular Dielectric Resonator Antenna," *IEEE Transactions on Antennas and Propagation,* Vol. 60, No. 6, June 2012, pp. 2997-3002.

[24] Petosa, A., and S. Thirakoune, "Frequency-Agile Dielectric Resonator Antenna," *Proc. IEEE International Symposium on Antennas and Propagation,* Chicago, IL, July 8-14, 2012.

[25] Kivekas, O., J. Ollikainen, and P. Vainikainen, "Tunable Internal Antenna for Mobile Phones," *Proc. 12th Int. Symposium Antennas JINA,* Nice, France, November 12-14, 2002, pp. 53-56.

[26] Valkonen, R., et al., "Broadband Tuning of Mobile Terminal Antennas," *Proc. European Conference on Antennas and Propagation EuCAP,* Edinburgh, Scotland, November 11-15, 2007.

[27] Feldner, L. C., et al., "Electrically Small Frequency-Agile PIFA-as-a-Package for Portable Wireless Devices," *IEEE Transactions on Antennas and Propagation,* Vol. 55, No. 11, November, 2007, pp. 3310-3319.

[28] Aberle, J. T., et al., "Reconfigurable Antennas for Portable Wireless Devices," *IEEE Antennas and Propagation Magazine,* Vol. 45, No. 6, December 2003, pp. 148-154.

[29] Mak, A. C. K., et al., "Reconfigurable Multiband Antenna Designs for Wireless Communication Devices," *IEEE Transactions on Antennas and Propagation,* Vol. 55, No. 7, July 2007, pp. 1919-1928.

[30] Simons, R. N., D. Chun, and L. P. B. Katehi, "Microelectromechanical Systems (MEMS) Actuators for Antenna Reconfigurability," *Proc. IEEE Microwave Theory and Techniques Symposium,* Phoeniz, AZ, May 20-24, 2001, pp. 215-218.

[31] Kim, I., and Y. Rahmat-Samii, "Integrated DC Bias Line RF MEMS Switch for Reconfiguring the Patch-Slot Antenna: Simulations and Measurements," *Proc. IEEE International Symposium on Antennas and Propagation,* Toronto, Canada, July 11-17, 2010.

[32] Panaia, P., et al., "Multistandard Reconfigurable PIFA Antenna," *Microwave and Optical Technology Letters,* Vol. 48, No. 10, October 2006, pp. 1975-1977.

[33] Park, J. H., et al., "Tunable Planar Inverted-F Antenna Using RF MEMS Switch for the Reduction of Human Hand Effect," *Proc. IEEE 20th International Conference on Microelectromechanical Systsms, ICM MEMS*, Kobe, Japan, January 21-25, 2007, pp. 163-166.

[34] Zhang, C., et al., "Development of Reconfigurable Mini-Nested Patches Antenna for Universal Wireless Receiver Using MEMS," *Proc. IEEE International Symposium on Antennas and Propagation,* Albuquerque, NM, July 9-14, 2006, pp. 205-208.

[35] Zhang, C., et al., "Reconfigurable Antenna for Simultaneous Multi-Service Wireless Applications," *IEEE Radio and Wireless Symposium,* Long Beach, CA, January 9-11, 2007, pp. 543-546.

[36] Anagnostou, D. E., et al., "Silicon-Etched Re-configurable Self-Similar Antenna with RF-MEMS Switches," *Proc. IEEE International Symposium on Antennas and Propagation,* Monterey, CA, June 20-25, 2004, pp. 1804-1807.

[37] Anagnostou, D. E., et al., "Design, Fabrication, and Measurements of an RF-MEMS-Based Self-Similar Reconfigurable Antenna," *IEEE Transactions on Antennas and Propagation,* Vol. 54, No. 2, February 2006, pp. 422-432.

[38] Kingsley, N., et al., "RF MEMS Sequentially Reconfigurable Sierpinski Antenna on a Flexible Organic Substrate With Novel DC-Biasing Technique," *Journal of Micromechanical Systems,* Vol. 16, No. 5, October 2007, pp. 1185-1192.

[39] Cetiner, B. A., et al., "RF MEMS Integrated Frequency Reconfigurable Annular Slot Antenna," *IEEE Transactions on Antennas and Propagation,* Vol. 58, No. 3, March, 2010, pp. 626-632.

[40] Guo, K., et al., "Fitting SOC Miniature MEMS Re-configurable Antenna," *Proc. Int. Conference on ASIC ASIACON,* Shanghai, China, October 24-27, 2005, pp. 1066-1069.

[41] Freeman, J. L., B. J. Lamberty, and G. S. Andrews, "Optoelectronically Reconfigurable Monopole Antenna," *Electronics Letters,* Vol. 28, No. 16, July 1992, pp. 1502-1503.

[42] Panagamuwa, C. J., A. Chauraya, and J.C. Vardaxoglou, "Frequency and Beam Reconfigurable Antenna Using Photoconducting Switches," *IEEE Transactions on Antennas and Propagation,* Vol. 54, No. 2, February 2006, pp. 449-454.

[43] Lavallee, R. N., and B.A. Lail, "Optically-Controlled Reconfigurable Microstrip Patch Antenna," *Proc. IEEE International Symposium on Antennas and Propagation,* San Diego, CA, July 5-11, 2008.

[44] Jung, C. W., et al., "Macro-Micro Frequency Reconfigurable Antenna," *Proc. IEEE International Symposium on Antennas and Propagation,* Honolulu, HI, June 9-15, 2007, pp. 5853-5856.

[45] Lim, J. H., et al., "A Reconfigurable PIFA Using a Switchable PIN-Diode and a Fine-Tuning Varactor for USPCS/WCDMA/m-WiMAX/WLAN," *IEEE Transactions on Antennas and Propagation,* Vol. 58, No. 7, July, 2010, pp. 2404-2411.

List of Acronyms

AC	alternating current
AIA	active integrated antenna
AM	amplitude-modulated
AR	axial ratio
AWS	advanced wireless service
BW	bandwidth
CDMA	code division multiple access
CPW	coplanar waveguide
CRC	Communications Research Centre Canada
CTS	continuous stub tuning
DARPA	Defense Advanced Research Projects Agency
DC	direct current
DCS	digital communications system
DECT	digital enhanced cordless telecommunications
DMB	digital multimedia broadcasting
DRA	dielectric resonator antenna
DVB-H	digital video broadcasting-handheld
FAA	frequency-agile antenna
FET	field effect transistor
FM	frequency-modulated
GPS	global positioning system
GSM	Global System for Mobile Communications
HEM	hybrid electric magnetic
IFA	inverted-F antenna
IL	insertion loss
IS	isolation
ISM	industrial scientific and medical band
LHCP	left-hand circular polarization
LO	local oscillator
LTE	long-term evolution

MEMS	microelectromechanical system
MIMO	multiple-input multiple-output
PCB	printed circuit board
PCS	personal communication services
PIFA	planar inverted-F antenna
PIN	positive intrinsic negative
Q-factor	quality factor
RECAP	Reconfigurable Aperture Program
RF	radio frequency
RHCP	right-hand circular polarization
RL	return loss
TE	transverse electric
TM	transverse magnetic
TR	tuning range
TS	total spectrum
UMTS	Universal Mobile Telecommunications System
WCDMA	wideband code division multiple access
WiBro	wireless broadband internet
WiMax	worldwide interoperability for microwave access
WLAN	wireless local area network

List of Symbols

β	guided wavenumber (rad/m)
γ	Euler's constant (0.577216)
Γ	reflection coefficient
η_T	tuning efficiency
ε_o	permittivity of free space (8.854 x 10^{-12} F/m)
ε_{eff}	effective relative permittivity (or effective dielectric constant)
ε_r	relative permittivity (or dielectric constant)
λ	wavelength (m)
λ_g	guided wavelength (m)
λ_o	free-space wavelength (m)
μ_o	permeability of free space (4π x 10^{-7} H/m)
μ_{eff}	effective relative permeability
μ_r	relative permeability
AR	axial ratio
BW	bandwidth
BW_T	tunable bandwidth
BW_{TG}	tunable gain bandwidth
BW_{TZ}	tunable impedance bandwidth
D	directivity
e_{rad}	antenna radiation efficiency
E	electric field (V/m)
f	frequency (Hz)
f_o	resonant frequency (Hz)
G	gain
H	magnetic field intensity (A/m)
R	minimum far-field distance (m)

R_{in}	input resistance (Ω)
R_s	surface resistance (Ω)
P_{in}	average power at the antenna input terminals (W/m^2)
P_{rad}	average radiated power (W/m^2)
Q	quality factor
U	radiation intensity (W/solid angle)
U_o	radiation intensity of an isotropic source (W/solid angle)
X_{in}	input reactance (Ω)
Z_{in}	input impedance (Ω)
Z_o	characteristic impedance (Ω)

About the Author

Aldo Petosa was born in Montreal, Canada, in 1966. He attended Carleton University in Ottawa from 1985 to 1995, obtaining bachelor's, master's, and Ph.D. degrees in electrical engineering. From 1990 to 1994, he carried out research at CAL Corporation (now Honeywell) in Ottawa on microstrip antennas for cellular and mobile satellite communication applications. He then worked for one year on the development of planar base station antennas for personal communication systems at GANDEC Systems Inc., Ottawa. Since then, Dr. Petosa has been with the Communications Research Centre Canada, where he is currently the team leader for RF subsystems in the RF Technologies unit. Over the years he has worked on various projects involving a wide range of antenna technology including: multilayer microstrip arrays, microwave lenses, DRAs, holographic-based antennas, spatial power combining arrays, and reconfigurable antennas. Since 1997, Dr. Petosa has also been an adjunct professor with the Department of Electronics at Carleton University, where he teaches a graduate course in antenna engineering. Dr. Petosa is a senior member of the IEEE, and served from 2005 to 2012 as the chair for the CNC URSI Commission B (Fields and Waves). Dr. Petosa has authored or coauthored more than 180 papers published in international journals or conferences, is a coinventor of five patents, and is the author of the *Dielectric Resonator Antenna Handbook* (Artech, 2007).

Index

Recent Titles in the Artech House Antennas and Propagation Series

Jennifer T. Bernhard, Series Editor

Adaptive Array Measurements in Communications, M. A. Halim

Advances in Computational Electrodynamics: The Finite-Difference Time-Domain Method, Allen Taflove, editor

Advances in FDTD Computational Electrodynamics: Photonics and Nanotechnology, Allen Taflove, editor; Ardavan Oskooi and Steven G. Johnson, coeditors

Analysis Methods for Electromagnetic Wave Problems, Volume 2, Eikichi Yamashita, editor

Antenna Design with Fiber Optics, A. Kumar

Antenna Engineering Using Physical Optics: Practical CAD Techniques and Software, Leo Diaz and Thomas Milligan

Antennas and Propagation for Body-Centric Wireless Communications, Second Edition, Peter S. Hall and Yang Hao, editors

Antennas and Site Engineering for Mobile Radio Networks, Bruno Delorme

Analysis of Radome-Enclosed Antennas, Second Edition, Dennis J. Kozakoff

Applications of Neural Networks in Electromagnetics, Christos Christodoulou and Michael Georgiopoulos

AWAS for Windows Version 2.0: Analysis of Wire Antennas and Scatterers, Antonije R. Djordjević, et al.

Broadband Microstrip Antennas, Girsh Kumar and K. P. Ray

Broadband Patch Antennas, Jean-François Zürcher and Fred E. Gardiol

CAD of Microstrip Antennas for Wireless Applications, Robert A. Sainati

The CG-FFT Method: Application of Signal Processing Techniques to Electromagnetics, Manuel F. Cátedra, et al.

Computational Electrodynamics: The Finite-Difference Time-Domain Method, Third Edition, Allen Taflove and Susan C. Hagness

Electromagnetic Modeling of Composite Metallic and Dielectric Structures, Branko M. Kolundzija and Antonije R. Djordjević

Electromagnetic Waves in Chiral and Bi-Isotropic Media, I. V. Lindell, et al.

Electromagnetics, Microwave Circuit and Antenna Design for Communications Engineering, Peter Russer

Engineering Applications of the Modulated Scatterer Technique, Jean-Charles Bolomey and Fred E. Gardiol

Fast and Efficient Algorithms in Computational Electromagnetics, Weng Cho Chew, et al., editors

Frequency-Agile Antennas for Wireless Communications, Aldo Petosa

Fresnel Zones in Wireless Links, Zone Plate Lenses and Antennas, Hristo D. Hristov

Handbook of Antennas for EMC, Thereza MacNamara

Handbook of Reflector Antennas and Feed Systems, Volume I: Theory and Design of Reflectors, Satish Sharma, Sudhakar Rao, and, Lotfollah Shafai, editors

Handbook of Reflector Antennas and Feed Systems, Volume II: Feed Systems, Lotfollah Shafai, Satish Sharma, and Sudhakar Rao, editors

Handbook of Reflector Antennas and Feed Systems, Volume III: Applications of Reflectors, Sudhakar Rao, Lotfollah Shafai, and Satish Sharma, editors

Introduction to Antenna Analysis Using EM Simulators, Hiroaki Kogure, Yoshie Kogure, and James C. Rautio

Iterative and Self-Adaptive Finite-Elements in Electromagnetic Modeling, Magdalena Salazar-Palma, et al.

LONRS: Low-Noise Receiving Systems Performance and Analysis Toolkit, Charles T. Stelzried, Macgregor S. Reid, and Arthur J. Freiley

Measurement of Mobile Antenna Systems, Second Edition, Hiroyuki Arai

Microstrip Antenna Design Handbook, Ramesh Garg, et al.

Microwave and Millimeter-Wave Remote Sensing for Security Applications, Jeffrey A. Nanzer

Mobile Antenna Systems Handbook, Third Edition, Kyohei Fujimoto, editor

Multiband Integrated Antennas for 4G Terminals, David A. Sánchez-Hernández, editor

Noise Temperature Theory and Applications for Deep Space Communications Antenna Systems, Tom Y. Otoshi

Phased Array Antenna Handbook, Second Edition, Robert J. Mailloux

Phased Array Antennas with Optimized Element Patterns, Sergei P. Skobelev

Plasma Antennas, Theodore Anderson

Quick Finite Elements for Electromagnetic Waves, Giuseppe Pelosi, Roberto Coccioli, and Stefano Selleri

Radiowave Propagation and Antennas for Personal Communications, Second Edition, Kazimierz Siwiak

Reflectarray Antennas: Analysis, Design, Fabrication and Measurement, Jafar Shaker, Mohammad Reza Chaharmir, and Jonathan Ethier

Solid Dielectric Horn Antennas, Carlos Salema, Carlos Fernandes, and Rama Kant Jha

Switched Parasitic Antennas for Cellular Communications, David V. Thiel and Stephanie Smith

Understanding Electromagnetic Scattering Using the Moment Method: A Practical Approach, Randy Bancroft

Wavelet Applications in Engineering Electromagnetics, Tapan Sarkar, Magdalena Salazar Palma, and Michael C. Wicks

For further information on these and other Artech House titles, including previously considered out-of-print books now available through our In-Print-Forever® (IPF®) program, contact:

Artech House
685 Canton Street
Norwood, MA 02062
Phone: 781-769-9750
Fax: 781-769-6334
e-mail: artech@artechhouse.com

Artech House
16 Sussex Street
London SW1V HRW UK
Phone: +44 (0)20 7596-8750
Fax: +44 (0)20 7630 0166
e-mail: artech-uk@artechhouse.com

Find us on the World Wide Web at: www.artechhouse.com